Cosmology and Particle Physics

NATO ASI Series

Advanced Science Institutes Series

A Series presenting the results of activities sponsored by the NATO Science Committee, which aims at the dissemination of advanced scientific and technological knowledge, with a view to strengthening links between scientific communities.

The Series is published by an international board of publishers in conjunction with the NATO Scientific Affairs Division

A Life Sciences	Plenum Publishing Corporation
B Physics	London and New York
C Mathematical	Kluwer Academic Publishers
and Physical Sciences	Dordrecht, Boston and London
D Behavioural and Social Sciences	
E Applied Sciences	
F Computer and Systems Sciences	Springer-Verlag
G Ecological Sciences	Berlin, Heidelberg, New York, London,
H Cell Biology	Paris and Tokyo
I Global Environmental Change	

NATO-PCO-DATA BASE

The electronic index to the NATO ASI Series provides full bibliographical references (with keywords and/or abstracts) to more than 30000 contributions from international scientists published in all sections of the NATO ASI Series.
Access to the NATO-PCO-DATA BASE is possible in two ways:

– via online FILE 128 (NATO-PCO-DATA BASE) hosted by ESRIN,
Via Galileo Galilei, I-00044 Frascati, Italy.

– via CD-ROM "NATO-PCO-DATA BASE" with user-friendly retrieval software in English, French and German (© WTV GmbH and DATAWARE Technologies Inc. 1989).

The CD-ROM can be ordered through any member of the Board of Publishers or through NATO-PCO, Overijse, Belgium.

Series C: Mathematical and Physical Sciences - Vol. 427

Cosmology and Particle Physics

edited by

Venzo de Sabbata

Department of Physics,
University of Bologna,
Bologna, Italy
and
Department of Physics,
University of Ferrara,
Ferrara, Italy

and

Ho Tso-Hsiu

Institute of Theoretical Physics,
Academia Sinica,
Beijing, People's Republic of China

Kluwer Academic Publishers

Dordrecht / Boston / London

Published in cooperation with NATO Scientific Affairs Division

Proceedings of the NATO Advanced Study Institute on
Cosmology and Particle Physics (13th Course on the International School of
Cosmology and Gravitation of the Ettore Majorana Centre for Scientific Culture)
Erice, Italy
May 3–14, 1993

A C.I.P. Catalogue record for this book is available from the Library of Congress.

ISBN 0-7923-2768-3

Published by Kluwer Academic Publishers,
P.O. Box 17, 3300 AA Dordrecht, The Netherlands.

Kluwer Academic Publishers incorporates the publishing programmes of
D. Reidel, Martinus Nijhoff, Dr W. Junk and MTP Press.

Sold and distributed in the U.S.A. and Canada
by Kluwer Academic Publishers,
101 Philip Drive, Norwell, MA 02061, U.S.A.

In all other countries, sold and distributed
by Kluwer Academic Publishers Group,
P.O. Box 322, 3300 AH Dordrecht, The Netherlands.

Printed on acid-free paper

CONTENTS

PREFACE

This Course has provided an up-to-date understanding of the progress and current problems of the interplay between particle physics and cosmology. In fact, during recent years, the gap between these two fields has considerably narrowed at both the theoretical and experimental levels.

Particle physics has been provided with new experimental data from some of the big accelerators in operation and already the next generation of accelerators is being seriously discussed. Much new data has also been accumulated through space satellite and astronomical observations. Such interplay between these two types of observations fruitfully influences both. Nevertheless, cosmology is still one of the main 'laboratories' for testing particle theory. All this was fully discussed during the Course.

As is usual with this type of course, which brings together individuals from different levels of scientific maturity, ranging from postdoctorate research students to well-established researchers, we were fully introduced to all aspects of cosmology and particle physics, going through inflation, strings, dark matter, neutrinos and gravitational wave physics in the very early Universe, field theory at Planck's scale, and high energy physics.

Moreover, particular emphasis was placed on a new topology for spatial infinity, on the relation between temperature and gravitational potential, on a canonical formulation of General Relativity, on the mass of a neutrino, on the spin in the early Universe, on the gravity measurements in the 10 - 100 m range of distances, on the problem of galaxy-galaxy and cluster-cluster correlation, on black holes, on string theory in cosmology, and string/string duality, always stressing the interplay between cosmology and particle physics.

Excellent reviews of high-energy elementary particle physics were presented, entreating the meaning, status, and perspectives of the unification of standard model gauge couplings, and during each lecture given, emphasis was placed on both the experimental and theoretical aspects of the subject.

The dark matter problem was studied both from the particle physics and astrophysics points of view.

So, in this the XIIIth NATO ASI Course, it was stressed that the new concepts of cosmology generally have their origins in both particle physics and field theory.

The editors wish to conclude by thanking the NATO Scientific Affairs Division which provided the basic funding of the School, and the Ettore Majorana Center for Scientific Culture which made a considerable financial contribution. We should like to express our great appreciation to Dr Gabriele and his staff at the Center in Erice who provided excellent administrative services and continuous assistance.

Finally, we wish to thank the lecturers and seminar speakers who did so much to make this School successful, and all the participants for contributing to the very stimulating scientific and human atmosphere.

Venzo de Sabbata
Bologna, Italy

Ho Tso-hsiu
Beijing, People's Republic of China

November 1993

A NEW TOPOLOGY FOR SPATIAL INFINITY ?

Peter G. Bergmann
Departments of Physics,
Syracuse University and New York University,
and
Gerrit J. Smith
Department of Philosophy, Fordham University

For spatial infinity we introduce the topology of a projective Lorentz sphere. This topology is indicated by the reduced variety of physically admissible solutions of both the electromagnetic and the gravitational field equations. In this topology past and future are fused, so that the notions of cause and effect lose their intuitive meanings.

1.   INTRODUCTION.   In an asymptotically flat space-time both gravitational and electromagnetic fields at spatial infinity can be obtained as scalar solutions of linear wave equations on three-dimensional Lorentz spheres, with one time-like and two space-like dimensions.[1,2] The appropriate wave equation for electromagnetic fields is

$$\nabla^2 u = o, \qquad (1.1)$$

the one and for gravitational fields

$$\nabla^2 u - 3u = o. \qquad (1.2)$$

Both equations are manifestly invariant with respect to the six-parametric group of rotations and boosts, 0(3,1). However, further examination reveals that in both cases one-half of the solutions do not lend themselves to a continuation from spatial infinity to finite regions of space-time, or to light-like infinity.

When the wave equations (1.1) and (1.2) are solved by expansions into spherical harmonics, with time-dependent coefficients, these coefficients satisfy ordinary second-order differential equations, which in each case admit pairs of

1

V. de Sabbata and H. Tso-Hsiu (eds.), Cosmology and Particle Physics, 1–8.
© 1994 Kluwer Academic Publishers. Printed in the Netherlands.

solutions, one even in the time coordinate, the other odd. In the electromagnetic case the physically admissible solutions are even for odd values of ℓ, the order of the spherical harmonic, and odd for even values of ℓ. The reverse holds for the solutions of the gravitational wave equation (1.2).

The Lorentz-invariant character of these conditions is made evident if they are formulated in terms of <u>antipodal</u> points. Pairs of points are called antipodal to each other if in the embedding pseudo-Euclidian space-time they lie on a straight line passing through the center of symmetry of the Lorentz sphere. In terms of pseudo-Cartesian coordinates of the imbedding Minkowski manifold antipodal points have coordinates that pairwise add up to zero.

Admissible solutions of the respective equations (1.1) and (1.2) can be characterized in terms of their properties at pairs of antipodal points: In electrodynamics the values at antipodal points must differ in sign only (u'=-u); such solutions will be called <u>odd</u>. Gravitational solutions must have the same values at antipodal pairs of points (u'=u); they will be called <u>even</u>. the pairing of antipodal points is invariant with respect to rotations and boosts. Hence the evenness or oddness of scalar fields is an invariant property.

2. ATTEMPT AT QUANTIZATION. In terms of Cauchy data on a space-like hyperplane through the center of symmetry of a Lorentz sphere the restriction to even or to odd fields is equivalent to the imposition of constraints on the canonical (i.e. Hamiltonian) variables. In odd fields the configuration variables vanish on the chosen hyperplane, whereas in even fields the canonical momentum variables are zero.

If one proceeds strictly in formal analogy between Poisson brackets and quantum operators, the surviving dynamical variables all commute with each other, hence there are no Heisenberg uncertainty requirements. If instead one uses time-dependent annihilation and creation operators to construct Heisenberg analogs of the physically acceptable solutions, one finds that the products of the minimal uncertainties of **E** and **B** of Maxwell-Lorentz theory, and of the corresponding variables in Einstein gravitational theory, differ from the values at light-like infinity ("scri"). In particular, as one approaches the light cone boundary of space-like infinity, these uncertainties tend to zero, whereas on scri they are constant, and non-zero.

These negative outcomes have convinced us that before attempting to quantize fields at spatial infinity one should understand the non-quantum theory. In the sections that follow we shall confine ourselves to this less ambitious task.

3. HARMONIC FUNCTIONS ON THE TWO-DIMENSIONAL LORENTZ SPHERE. Functions on Lorentz spheres which assume either identical or opposite values on pairs of antipodal points can be defined, without loss of information, as single-valued or as double-valued functions on projective Lorentz spheres. A projective Lorentz sphere is defined as a manifold whose points represent pairs of antipodal points of (ordinary) Lorentz spheres. As they are multiply connected, functions can have more than a single value at the same point without necessitating discontinuities. Projective Lorentz spheres cannot be oriented, in some respects they resemble Möbius bands.

Though primary physical interest rests with three-dimensional projective Lorentz spheres, we shall deal with the two-dimensional projective Lorentz sphere in some detail, as it lends itself readily to a complete treatment. A space-like Lorentz sphere imbedded in a three-dimensional Minkowski manifold can be identified by its equation in standard Lorentz coordinates as

$$x^2 + y^2 - t^2 = 1. \qquad (3.1)$$

It is a single hyperboloid of rotation, ruled by two sets of straight null lines, each of which covers the whole Lorentz sphere once. That is to say, through each point of the Lorentz sphere pass exactly two of these null lines. If we identify a particular null line by its intersection with the circle t=0, $x^2 + y^2 = 1$, denoting the point of intersection by α for one set, β for the other, then these two angles, each ranging from 0 to 2π will serve as a coordinate system on the (1,1) sphere. The line element is

$$d\delta^2 = \cos^{-2}\gamma \, d\alpha \, d\beta,$$
$$\gamma = \tfrac{1}{2}(\alpha - \beta). \qquad (3.2)$$

Next we shall examine d'Alembert's equation,

$$\nabla^2 u = 0 \qquad (3.3)$$

on the (1,1) sphere. In terms of a, β that equation has the form

$$\frac{\partial^2 u}{\partial\alpha\,\partial\beta} = 0, \qquad (3.4)$$

with the general solution

$$u(\alpha,\beta) = f(\alpha) + g(\beta). \qquad (3.5)$$

f and g are arbitrary (continuous, differentiable) functions of their respective arguments, periodic with the period 2π if their arguments are to be extended beyond their original range.

Next we shall restrict ourselves to even, and to odd solutions. These are:

$$u(\alpha,\beta) = f(\alpha) + f(\beta + \pi) \qquad (3.6)$$

and

$$u(\alpha,\beta) = f(\alpha) - f(\beta + \pi), \qquad (3.7)$$

respectively. Thus, by requiring either evenness or oddness we have reduced the two arbitrary functions to one. Moreover, though f is a function of but one argument, it determines the solutions in a two-dimensional domain.

Can these relationships be converted into the dependence of $u(\alpha,\beta)$ on Cauchy data on a space-like one-dimensional domain, such as half the circle t = 0, corresponding in our coordinate system to $\alpha=\beta$? (Obviously, only the data on half the circle will do, as the other half of the circle is antipodal to the first.) This is indeed the case.

Denote the value of $u(\theta, \theta)$ on that half-circle by $v(\theta)$, and its normal derivative by $\dot{v}(\theta)$,

$$\dot{v}(\theta) = \frac{\partial u}{\partial \alpha} - \frac{\partial u}{\partial \beta},$$

$$\theta = \alpha = \beta. \qquad (3.8)$$

As antipodal points are related to each other by the equalities

$$\alpha' = \beta + \pi, \quad \beta' = \alpha + \pi, \qquad (3.9)$$

the values of the Cauchy data on the entire circle must satisfy the conditions:

$$v(\theta + \pi) = v(\theta), \qquad \dot{v}(\theta + \pi) = -\dot{v}(\theta) \qquad (3.10)$$

for even fields, and

$$v(\theta + \pi) = -v(\theta), \qquad \dot{v}(\theta + \pi) = \dot{v}(\theta) \qquad (3.11)$$

for odd fields.

The derivative of the function $v(\theta)$ along the circle, $v' \equiv \frac{dv}{d\theta}$, is related to those of $u(\alpha, \beta)$ at the same locations by the formula

$$v'(\theta) = \frac{\partial u}{\partial \alpha} + \frac{\partial u}{\partial \beta}. \qquad (3.12)$$

Combining this expression with Eq. (3.8), as well as with (3.6) or with (3.7), we obtain for $df/d\theta$ an expression that depends only on Cauchy data:

$$\frac{df}{d\theta} \equiv f'(\theta) = \frac{1}{2}(\dot{v} + v'). \qquad (3.13)$$

f itself is thereby determined up to a constant of integration.

For odd fields that constant of integration is irrelevant. Substituting into Eq. (3.7) one obtains immediately:

$$u(\alpha, \beta) = \frac{1}{2}[v(\alpha) - v(\beta + \pi) + \int_{\theta = \beta + \pi}^{\alpha} \dot{v}\,d\theta]$$

$$= \frac{1}{2}[v(\alpha) + v(\beta) + \int_{\theta = \beta + \pi}^{\alpha} \dot{v}(\theta)\,d\theta]. \qquad (3.14)$$

For even fields we can make use of Eqs. (3.10) in conjunction with (3.13). The result is:

$$f(\theta) = \tfrac{1}{2} v(\theta) + \tfrac{1}{2}\int_{\theta'=0}^{\theta} \dot{v}d\theta' - \tfrac{1}{4}\int_{\theta'=0}^{\pi} \dot{v}d\theta' \qquad (3.15)$$

and

$$u(a,\beta) = \tfrac{1}{2}[v(\alpha) + v(\beta) + \int_{\theta=\beta}^{\alpha} \dot{v}(\theta)\,d\theta] . \qquad (3.16)$$

A few examples will illustrate the role of the Cauchy data.

(i) Even field: $v(\theta) = \cos 2n\theta$, $\dot{v}(\theta) = o$.

$\qquad\qquad u(\alpha,\beta) = \cos 2n\gamma \cos 2n\theta$, $\gamma = \tfrac{1}{2}(\alpha-\beta)$, $\theta = \tfrac{1}{2}(\alpha+\beta)$.

(ii) Even field: $v(\theta) = o$, $\dot{v}(\theta) = \cos(2n+1)\theta$.

$\qquad u(\alpha,\beta) = (2n+1)^{-1}\sin(2n+1)\gamma \cos(2n+1)\theta$.

(iii) Odd field: $v(\theta) = \cos(2n+1)\theta$, $\dot{v}(\theta) = o$.

$\qquad u(\alpha,\beta) = (2n)^{-1}\sin 2n\gamma \cos 2n\theta$.

(iv) Odd field: $v(\theta) = o$, $\dot{v}(\theta) = \cos 2n\theta$.

$\qquad u(\alpha,\beta) = (2n)^{-1}\sin 2n\gamma \cos 2n\theta$.

In all these examples n is to be a non-zero integer. As the field $u(\alpha, \beta)$ depends on the Cauchy data linearly, these examples also serve as guides to the Fourier decomposition of the data and of the resulting field.

4.   THE PROJECTIVE LORENTZ SPHERE. When pairs of antipodal points are mapped into a single point of a target manifold, the result is a projective Lorentz sphere. Conversely, the (ordinary) Lorentz sphere is a covering (though not the universal covering) manifold of the projective Lorentz sphere.   The contraction preserves the two sets of ruling null lines as separate sets, though it cuts each set in half. Projective Lorentz spheres are not orientable, nor are they time-orientable [3]: One cannot tell future and past apart. Though the Cauchy data on the half-circle determine the field $u(\alpha, \beta)$ everywhere, that half-circle does not cut the projective Lorentz sphere in two. Thus the intuitive distinction between cause and effect loses all meaning.

On the projective Lorentz sphere _even_ fields are simply those fields which at each point of the manifold have but a single value. _Odd_ fields have two values at every point, which differ only by their signs. This change of sign occurs if the point in question is connected with itself by a closed curve that passes an odd number of times through the Cauchy surface. The Cauchy surface itself in the case of the (1,1) projective Lorentz sphere is a closed curve, it has no end points.

5. THE THREE-DIMENSIONAL PROJECTIVE LORENTZ SPHERE. The notion of antipodal points permit the construction of (2,1) projective Lorentz spheres in perfect analogy to the (1,1) construction. Three-dimensional projective Lorentz spheres also are not orientable, nor are they time-orientable. Their Cauchy surfaces are (1,1) projective Lorentz spheres. Even and odd fields are defined the same as in the two-dimensional case, and the Cauchy data are again the values of the fields, and of their normal derivatives, on the Cauchy surface.

In this paper we have primarily dealt with the two-dimensional structure because of its relative ease of treatment in terms of the arbitrary functions $f(\theta)$. We have been able to study the properties of solutions in three dimensions only by expanding them with respect to spherical harmonics on the Cauchy surface. The expansion coefficients, which are functions of the third dimension and which obey ordinary differential equations, with initial values on the Cauchy surface itself, play the same role in three dimensions as the arbitrary function $f(\theta)$ that we were able to introduce explicitly in the two-dimensional problem.

The proposed topology of spatial infinity differs significantly from the topology of light-like infinity ("scri"), and that difference is related somehow to the _prima facie_ conflict between the irreducible quantum uncertainties on spi and on scri. conceivably, this conflict may be mitigated by an appropriate quantum theory addressing itself to the altered causality on an unoriented manifold. It remains to be seen whether this hope can be satisfied.

8

REFERENCES

1. N. Alexander and P.G. Bergmann, Found. Phys. $\underline{16}$, 445 (1986).
2. P.G. Bergmann, Gen. Rel. Grav. $\underline{19}$, 371 (1987), $\underline{21}$, 271 (1989).
3. S.W. Hawking and G.F.R. Ellis, <u>The Large-Scale Structure of Space-Time</u>. Cambridge University Press, Cambridge 1973. p. 181f.

TEMPERATURE AND GRAVITATIONAL POTENTIAL
Peter G. Bergmann
Syracuse University and New York University

It is argued that thermal equilibrium of spatially
extended systems is characterized by the presence of a
"global" temperature, but that there also exists a "local"
temperature, which is sensitive to the value of the
gravitational potential. It is further asserted that
thermometers read out the local temperature.

1.   INTRODUCTION. With   any   reasonable   definition   of
"thermodynamic equilibrium" it would appear that both matter and
radiation fields in an extended spatial domain can be in
equilibrium only if the temperature (and other relevant
thermodynamic potentials) is well defined and has the same
numerical value throughout that domain. But if the gravitational
potential has different values at different locations, then general
relativity predicts a "gravitational red shift" even if there is
thermodynamic equilibrium.
     Suppose, for instance, that chunks of Fe_{57} interact with each
other by emitting and absorbing Mössbauer x-rays.   Then theory
predicts,   and   experiment   confirms,   that   resonance   can   be
established only by offsetting the gravitational red shift by a
compensating Doppler shift.   If the sharp Mössbauer line is
replaced by black-body thermal radiation, we must extrapolate the
behavior of the single sharp line to each small frequency segment,
resulting in a shift of the broadband spectrum as a whole, which
will be perceived as a change in temperature associated with the
change in gravitational potential.
     Seemingly, we are confronted by a paradox.   On the one hand,
the very notion of temperature as a thermodynamic potential appears
to call for a single temperature in an extended spatial domain as
a condition for thermodynamic equilibrium, regardless of a
spatially variable gravitational potential.   On the other, the
phenomenon of gravitational red shift inexorably involves a
dependence of the temperature on the local value of the
gravitational potential if equilibrium is to be achieved.
     This problem is resolved by the introduction of two different
quantities, both exhibiting characteristics ordinarily associated
with the notion of temperature.   One of these quantities will need
to be constant throughout the spatial domain in question as a
precondition for thermal equilibrium; it will be called the "global
temperature".   The other will exhibit the dependence on the
gravitational potential that accounts for the gravitational red
shift; it will be referred to as the "local" temperature.
     In what follows, these two temperature concepts will be
connected with each other in examples whose properties are well

9

V. de Sabbata and H. Tso-Hsiu (eds.), Cosmology and Particle Physics, 9–12.
© 1994 Kluwer Academic Publishers. Printed in the Netherlands.

known from statistical mechanics. A final section will be devoted to the question which temperature will be recorded by a thermometer.

2. THE SPACE-TIME MODEL. Consider a three-dimensional extended region, with time forming the fourth dimension, which is endowed with properties that enable thermodynamic equilibrium to be formed, and which otherwise is kept as simple as possible. This four-dimensional manifold will serve as a stage for the construction of microdynamic models of macroscopic systems.

Thermodynamic equilibrium requires that the background space-time be stationary, which is another way of saying that our manifold must possess a time-like Killing field. If the Killing field is surface-forming, then the metric tensor can be given the form

$$g_{oo} = 1 + 2V(\vec{x}), \; g_{ok} = o, \; k = 1, 2, 3,$$
$$g_{ik} = -h_{ik}(\vec{x}), \; i, k = 1, 2, 3. \qquad (2.1)$$

All components are functions of the three spatial coordinates only; they are independent of x^0. For the present investigation the spatial structure is of no interest. Accordingly we shall specialize the metric further by requiring

$$h_{ik} = \delta_{ik}. \qquad (2.2)$$

As a final simplification of our model of space-time we shall disregard any effect of the dynamic system back onto the metric. Thus the gravitational potential $V(\vec{x})$ will not be regarded as a dynamical variable but is to be considered given ab initio. This, then, is to be our model for a background space-time, on which equilibirum physical systems are to be examined.

3. PERFECT RELATIVISTIC NON-QUANTUM GAS. Let N mass points of equal masses m travel on time-like geodesics in the space-time defined in the preceding section. along each curve the line element will be given by the expression

$$d\tau^2 = (1 + 2V) \, dt^2 - d\vec{x}^2, \qquad (3.1)$$

hence the action for one "molecule" will be:

$$S = \int L \, dt,$$
$$L = -m(1 + 2V(\vec{x}) - \dot{\vec{x}}^2)^{1/2}. \qquad (3.2)$$

The Hamiltonian of a molecule is obtained by the usual Laplace transform. It turns out to have the form:

$$H = [(1+2V(\vec{x})(m^2+\vec{p}^2)]^{\frac{1}{2}}, \qquad (3.3)$$
$$\vec{p} = (1+2V-\dot{\vec{x}}^2)^{-\frac{1}{2}} m\dot{\vec{x}}.$$

Incidentally, the canonical momentum is independent of the choice of the variable of integration in the action integral. The choice of $t=x^0$ does not destroy the essentially relativistic character of the result, which establishes H as the constant of the motion, the one that generates Killing displacements along the time axis. The Hamiltonian of the whole gas consists of the N-fold repetition of Eq. (3.3).

Consider now the perfect gas in a Gibbs canonical distribution. In the non-relativistic gas the Hamiltonian is the sum of two terms, one of which involves only the configuration coordinates $\vec{x}$, the other only the momentum coordinates $\vec{p}$. As a result, the momentum distribution at a fixed location does not depend on the gravitational potential; only the density varies. For the relativistic gas the Hamiltonian does not decompose that simply. At a fixed location the momentum spectrum is

$$\sigma' = \frac{1}{Z'} \exp(-\sqrt{m^2+\vec{p}^2}/kT'),$$
$$T' = \frac{T}{\sqrt{1+2V(\vec{x})}}. \qquad (3.4)$$

Z' is a normalizing coefficient, T the global temperature, which by assumption is independent of location, and T' the local temperature, which is sensitive to the local value of the gravitational potential $V(\vec{x})$.

At a location with large negative potential T' will be greater than T. That is to say, if an observer stationed at a location where V=0 monitors black-body radiation that has its origin near a large mass, he will find that to maintain radiative equilibrium he must permit the incoming radiation to interact with a body at the temperature T<T'.

4. RADIATION FIELDS. Consider a radiation field of arbitrary integral spin that propagates at the speed of light, in an arbitrary but finite region, as in the preceding sections, but with some definite boundary conditions. Such a field can be decomposed into normal modes, each with a sinusoidal dependence on the time coordinate t. For the n-th mode the characteristic frequency will be designated as ω_n.

If the radiation is bosonic, the probability of there being ℓ_n quanta in the n-th normal mode will be

$$p_{\ell_n} = (1 - e^{-\hbar\omega_n/kT}) \, e^{-\ell_n \frac{\hbar\omega}{kT}}. \qquad (4.1)$$

This number cannot depend on the location of the observer, as the outcomes of any measurements of the numbers ℓ_n are discrete. But an observer who uses his own clock to determine the characteristic frequency of the n-th normal mode will obtain the result

$$\omega_n' = (1 + 2V)^{-\frac{1}{2}} \omega_n. \qquad (4.2)$$

The probabilities (4.1), associated with the locally determined frequencies (4.2), are consistent with a temperature T',

$$T' = (1 + 2V)^{-\frac{1}{2}} T. \qquad (4.3)$$

the same outcome as in the preceding section.

5.   THERMOMETERS. A thermometer, having a small heat capacity, upon made to interact with a system or subsystem that is in thermodynamic equilibrium with its surroundings, will itself tend to attain equilibrium conditions without unduly changing the other system's conditions, and will provide a reading of the temperature. But which temperature, global or local?

Once equilibrium has been achieved, the thermometer must be at the same temperature as its surroundings, which is T. But there remains a distinction between the thermometer being at the temperature T and its exhibiting (on its read-out) that, or any other temperature. The location of the thermometer remains characterized by the value of $g_{00}=1+2V$. According to the principle of equivalence one must introduce a locally Minkowski coordinate system if ordinary laws of physics are to apply.

The examples in the preceding two sections demonstrate that for the laws of statistical mechanics to take their "ordinary" form, that is to say the form prevailing in the absence of gravitation, one needs to introduce the temperature T' of Eqs. (3.4) and (4.3), and a Minkowski coordinate system. The conclusion is that the thermometer will exhibit local temperature, T'.

I wish to thank Art Komar for many and sometimes agitated discussions, which have helped me to clarify and to improve my understanding of the topics of this paper.

How can General Relativity be formulated most canonically?

Horst-Heino v. Borzeszkowski
WIP Gravitation Project
An der Sternwarte 16
D-14482 Potsdam

This paper is devided into three sections: The first introductorily rediscusses some steps toward the standard (holonomic) canonical formulation of Einstein's 1915 theory of general relativity in order to recall the origin of the problems arising for the canonical quantization of this theory. The second section discusses anholonomic representations of general relativity theory (this section is a review of a paper by H.-J. Treder and the author [1]), and the third reconsiders the Einstein-Schrödinger purely affine theory as a general-relativistic model of gravity that, to some extend, realizes a canonical formulation of the non-Riemannian part of gravity (this section reviews some points of a paper by H.-J. Treder, V. de Sabbata, C. Sivaram, and the author [2]).

The latter two sections demonstrate, first, that the problems arising in the canonical quantization of the Einstein-Riemann part of gravity can hardly be avoided and, second, that this fact is neither a defect nor an inconsistency , but a physically understandable feature of Einstein-Riemann gravity and thus of Einstein's 1915 theory.

1. Some steps toward the standard (holonomic) formulation of Einstein's 1915 theory of general relativity

During the last decades it became more and more clear that all efforts made to arrive at a theory of quantum gravity by using field-theoretic methods encounter obstacles seemingly unsurmountable. This failure seems to be due to the fact that such methods use perturbation expansions linearizing Einstein's essentially nonlinear theory. In 1976, Penrose [3] referring to P. G. Bergmann reminded of that if one removes "the life from Einstein's beautiful theory by steam-rolling it first to flatness and linearity, then we shall learn nothing from attemting to wave the magic wand of quantum theory over the resulting corpse". Therefore, the search for a non-perturbative framework is on the agenda.

In the last years, in particular, Ashtekar and his group [4,5] has done a lot of work in order to establish methods of quantization non-perturbatively. In this context, Ashtekar also stres-

13

V. de Sabbata and H. Tso-Hsiu (eds.), Cosmology and Particle Physics, 13–24.
© 1994 *Kluwer Academic Publishers. Printed in the Netherlands.*

sed the point that, once one accepts the need of non-perturbative approach, general relativity should be reconsidered as the point of departure for quantization. For, from the point of view that the failure of quantization of general relativity is due the perturbative methods the initial arguments for abandoning general relativity in favour of other theories lose their force.

This line of arguments leads (back) to general relativity and its canonical quantization. The latter point, namely the preference of the canonical conception, is motivated by that the other main procedure, the background quantization, has to introduce an exterior metric from the outset. This assumption is, however, criticizable because, even in the case that one considers *covariant* background methods instead of the weak-field perturbation theory developing the metric field around a given background,[1] such a background assumption is not compatible with the principles of equivalence and general relativity (cf., Ref. 6). Maybe, there are mathematical or physical reasons requiring to introduce, by hand or by more intrinsic methods, a background structure of space-time. But the point to be emphasised is that this would modify general relativity; and if one wishes to ask to what results a non-perturbative quantization scheme of non-modified general relativity leads then canonical quantization seems to be preferable.

The preference of canonical schemes is, however, only then justified when one is able to show that these schemes really can avoid a background structure and, furthermore, that the transition to a canonical formulation does not change or destroy fundamentals of Einstein's theory of general relativity. As to the first point, canonical formulations of general relativity and the formal transition to quantum theory do not require any background structure and, the less so, a development around a background metric. It remains, however, some doubt. Maybe, at later stages, when one wants to draw physical conclusions, a background somehow has to be introduced for reasons of measurement or as an interpretative framework.

As to the second point, it seems very questionable whether there is a canonical formulation that is equivalent to original Einstein's 1915 general relativity. In considering attempts to find such formulations one gets the impression that one only arrives at a physically meaningful canonical formulation if one accepts a modification of this theory. Altogether, the statement that canonical have an advantage over background schemes should be considered with caution.

In section 2, we shall give some arguments in favour of the conjecture that the Einstein-Riemann part of gravity and thus Einstein's 1915 theory of general relativity cannot be given a canonical formulation without changing its fundamentals. But before doing that, some words on the standard approach to canonical general relativity to show the origin of these problems.

The canonical analysis of general relativity proceeds as follows (see, e. g., Refs. 4, 7, 8). First one fixes a foliation $X: \Sigma \times R = M$ of the four-dimensional globally hyperbolic mani-

---

[1] As was emphasized by Isham [7], the covariant background approach to quantum gravity uses methods other than weak-field perturbation theory which often have a fairly obvious translation into the canonical framework.

fold M with the metric g_{ab} (whose signature be $(-+++)$), where Σ is a three-dimensional spacelike manifold assumed to be compact or diffeomorphic to R outside a compact region. If one confines oneself, for simplicity, to Einstein's vacuum equations $G_{ab} = 0$ then one has to find canonical variables and first-order differential equations that determine how the canonical variables change with t , where t denotes a parameter counting the leafs of the foliation.

As was shown by Arnowitt, Deser, and Misner [9], one can choose the metric g_{ab} on Σ defined by

$$q_{ab} = g_{ab} + n_a n_b \qquad (a, b, = 0, ..., 3) \tag{1}$$

(n_a is normal to Σ) as canonical configuration variables.[2] Using q_{ab} and the extrinsic curvature of Σ defined as $K_{ab} := q_a^m q_b^n n_{n;m}$ one can derive from the Einstein-Hilbert Lagrangian $L_{EH} = {}^4R$ the first-order expression

$$L(t) = \frac{1}{\kappa} \int N \, (det \, q_{ab})^{1/2} \, [K_{ab} K^{ab} - (K_a^a)^2 + {}^3R(q)] \, d^3x \tag{2}$$

($\kappa := 16\pi G/c^4$, ${}^4R$ is the four-dimensional Ricci scalar formed from g_{ab}, ${}^3R$ is the three-dimensional Ricci scalar formed from q_{ab} , and N is defined as $t^a = Nn^a + N^a$ via the decomposition of the time-like, future-directed vector field t^a with the affine parameter t). This Lagrangian L enables us to determine the variables conjugate to q_{ab} :

$$p^{ab} = \frac{\partial L}{\partial \dot{q}_{ab}} = \frac{(det \, q_{ab})^{1/2}}{\kappa} \, (K^{ab} - Kq^{ab}). \tag{3}$$

With the variables (1) and (3), one can show that the whole set of Einstein's equations, i. e., the arising constraints (in the coordinates representation, the equations $G_{0a} = 0$) and the true equation of motion (corresponding to $G_{\mu\nu} = 0$) can be derived from the first-order action

$$S[q, p, N, \vec{N}] = \int dt \int d^3x \, [p^{ab} \dot{q}_{ab} - NC(q, p) - N^a C_a(q, p)], \tag{4}$$

where

$$C_a(q, p) := -2q_{ab} D'_e p^{bc} = 0, \qquad (D_a q_{bc} = 0) \tag{5a}$$

---

[2] The original ADM approach was formulated in terms of coordinates, where q_{ab} is given by $g_{\mu\nu}$ ($\mu, \nu = 1, 2, 3$)

$$C(q, p) := (det \ q_{ab})^{1/2} \ 3R + (det \ q_{ab})^{-1/2} (p^{ab} p_{ab} - \frac{1}{2} p^2) = 0 \qquad (5b)$$

are the primary constraint equations. This shows that N and N^a are no real variables but Lagrange multipliers $(p_n := \frac{\delta L}{\delta \dot{N}} = 0$, $p_{\vec{N}} := \frac{\delta L}{\dot{\vec{N}}^a} = 0)$.

Hitherto, the procedure strongly resembles the usual transition from the Lagrangian to the Hamiltonian formulatism describable by a Legendre transformation. At this stage, one however encounters the problem signaled by eq. (4). According to (4), the canonical Hamiltoian of general relativity reads

$$H[N, \vec{N}] = \int_{\Sigma} [N(x) \ C(x) + N^a(x) \ C_a(x)] \ d^3x. \qquad (6)$$

Discussing this point, Isham [7] calls this a fact that has a remarkable implication for the putative "Schrödinger equation"

$$i \frac{d}{dt} \ \psi_t = \hat{H} [N, \vec{N}] \psi_t \qquad (7)$$

since if the state ψ_t satisfies the constraint equations,

$$C_a (\hat{q}, \hat{p}) = 0 \ ,$$
$$C (\hat{q}, \hat{p}) = 0 \ , \qquad (8)$$

we see that it has no time dependence at all. And he adds that it is no longer confusing but understood to reflect the absence of any intrinsic definition of time in general relativity. (For approaches to resolving the problem of time in quantum gravity, see Refs. 8 and 10.)

It is precisely the problem of (missing) time fostering the doubt about the physical meaning of the transition to a canonical formulation of general relativity. This becomes fairly obvious on the classical level where H given by (6) is equal to zero when the constraints (5) are satisfied. Therefore, one can formally pass from the Einstein-Hilbert Lagrangian L_{EH} to some Hamiltonian-type function H satisfying the condition

$$\dot{p}^{ab} (x, t) = - \frac{\delta H [N, \vec{N}]}{\delta q_{ab}(x, t)} , \qquad (9)$$

but this H is "watt-less". By adding an appropriate surface integral, any desirable value can be given to it. There is no notion that can be rigorously identified with energy. To arrive, at least in some situations, at an energy notion one defines therefore energy as a surface integral "at infinity" (cf., Refs. 4, 5, 11). For this purpose, one considers a function T on Σ which is constant outside a compact region so that, in the space-time picture, Tn^a represents an asymptotic time-translation. Its generator

$$H_r(q, p) = \oint_{\partial\Sigma} dS^b T (\partial_a q_{bc} - \partial_b q_{ac}) e^{ac} \tag{10}$$

($\partial\Sigma$ is a surface enclosing the compact region of Σ and e^{ac} the Euclidean metric) provides, for $T = 1$, ADM energy. This is a way out of the dilemma arising in a theory having no background geometry. but it does not remedy the defect of this canonical formalism that there is not (and can not be) a Hamiltonian that may be generally, i. e., also locally, identified with energy. This lack can already be seen in the four-dimensional formulation of general relativity where one cannot define an energy-momentum tensor; and one cannot, of course, expect to improve this situation by a (3+1) decomposition.

The problems arising in the holonomic version of general relativity become even more obvious if one considers anholonomic representations and corresponding extended phase spaces.

2. Anholonomic formulations of Einstein's 1915 theory of general relativity

Anholonomic canonical formulations are mainly in the center of current interest because of the progress made in this field by Ashtekar and his colleagues [4, 5]. The basic idea is to in-troduce, instead of of the space-time metric g_{ab} and the affine connection Γ^a_{bc} , the basic space-time field consisting of a pair, $(e^a{}_A , {}^4A_a{}^{AB})$, of a (complex) tetrad $e^a{}_A$ and a connection 1-form ${}^4A_a{}^{AB}$ which is self-dual in the internal (Lorentz) indices A , B . Then the Einstein-Hilbert actin can be rewritten as

$$S(e, {}^4A) = \int d^4x \, (det \, e^c{}_B) e^a{}_A \, e^b{}_B \, {}^4F_{ab}{}^{AB} , \tag{11}$$

where ${}^4F_{ab}{}^{AB}$ is the curvature of ${}^4A_a{}^{AB}$ and $(det \, e^c{}_D)$ the determinant of $e^c{}_D$ (cf. Refs. 12, 13). The independent variation of (11) with respect to ${}^4A_a{}^{AB}$ and $e^a{}_A$ leads to the connection ${}^4A_a{}^{AB}$ compatible with $e^a{}_A$ and the Einstein vacuum equations.

Now, one can repeat all steps done in the holonomic approach (for this procedure, cf. Refs. 4, 5). The first one is to carry out a (3+1) decomposition. Omitting again a divergence term in the Lagrangian, (11) can be rewritten in a form which shows that the configuration variable turns out to be a SO(3) connection $A^i{}_a$ and its canonical momentum turns out to be a triad $\widetilde{E}^a{}_i$, with density weight one (i is here the SO(3) internal index running from 1 to 3). This formalism has an equivalent SU(2) spinor version.

So far, so good. But, by this procedure, one changed the Lagrangian to obtain, instead of the phase space $\widetilde{\Gamma}$, whose points are given by the pairs (p^{ab} , q_{ab}) , the enlarged phase space Γ with points given as pairs $(A^i{}_a , \widetilde{E}^a{}_i)$ or $(A_a{}^{AB} , \widetilde{\sigma}^a{}_{AB})$. Therefore, the two questions closely related are (i) whether the procedure necessary to gain Γ really did not change the physical content of the theory and (ii) whether one can indeed reduce Γ to $\tilde{\Gamma}$.

In discussing in Ref. 1 some aspects of these questions, we argued as follows. To arrive at first-order canonical general relativity one has to substract divergence terms from L_{EH} . By this procedure, at least, one of the invariance properties of L_{EH} in lost. Indeed, in the holonomic representation the Lagrange density L_{EH} ,

$$L_{EH} = (det\ g_{ab})^{1/2}\ g^{ac}g^{bd}\ 4R_{abcd}\ , \tag{12}$$

is covariant under coordinate transformations and invariant under local Lorentz transformations; by substracting a divergence term at least coordinate covariance is lost. And in the anholonomic formulation using the tetrads $e^a{}_A$ (with $\eta_{AB} = g_{ab}e^a{}_A e^b{}_B$) one obtains,

$$L_{EH} = (det\ e^c{}_D)\ \eta^{LK}\eta^{PQ}e^a{}_L e^b{}_P e^c{}_Q e^d{}_R\ 4R_{abcd}\ , \tag{13}$$

(or the corresponding SL(C ,2) spinor representation). L_{EH} is invariant with respect to coordinate transformations and covariant with respect to local Lorentz transformations; by subtracting now a divergence term from (13) the proparty of local Lorentz covariance is lost. For instance, if one subtracts the term

$$D_1 = \partial_a[(-det\ g_{cb})^{1/2}\ (-g^{il}\Gamma^a{}_{il} + g^{ia}\Gamma^m{}_{im})] \tag{14}$$

from (12) one obtains the Einstein Lagrange density [14],

$$L_E = (-det\ g_{cd})^{1/2}\ g^{ab}(\Gamma^l{}_{ab}\Gamma^m{}_{lm} - \Gamma^l{}_{am}\Gamma^m{}_{bl}) \tag{15}$$

which is not coordinate covariant. And if one omits the divergence term

$$D_2 = \partial_a[det\ e^c{}_D\ (\gamma^l{}_{PQ}\eta^{PQ}e^a{}_L - \gamma^{AP}{}_A e^a{}_P)] \tag{16}$$

in (13) one is led to Møller's Lagrange density [15]

$$L_M = (det\ e^c{}_D)\ (\gamma^L{}_{PM}\gamma^{MP}{}_L - \gamma^A{}_{LA}\gamma^{BL}{}_B) \tag{17}$$

($\gamma^l{}_{PM} = -e^b{}_M e^a{}_{P;b}e^L{}_a$ are the Ricci rotation coeffients) which is not covariant under local Lorentz transformations. The latter is also true when one starts with the self-dual form of L_{EH} given by (11) and subtracts a divergence to arrive at a canonical formulation.

Thus, one obtains in all cases Lagrangians that can be used to derive Einstein's vacuum equations from them. (For this purpose, one has now, of course, to assume that the connections are compatible with g_{ab}, e^a_A, and $^4\sigma_a{}^{AB'}$, respectively; therefore, the field equations cannot be derived from the new Lagrangians via Palatini variation.) The new first-order Lagrangians are, however, of no further use. In particular, they are not available for constructing a physically meaningful canonical energy-momentum complex or a related Hamiltonian.[3] The above-discussed holonomic and the anholonomic approaches considered here enable us to construct *formally* Hamliltonians satisfying relations like (9) but the physical meaning of such expressions remains questionable. And it has to be added that, when one uses the first-order Lagrangians for more than deriving the field equations, one either changes the content of the theory or establishes a mathematically underdetermined framework which is physically hardly interpretable. In other words, Γ can only be reduced to $\tilde{\Gamma}$ by the framework of general relativity.

To demonstrate the latter point let us return to the self-dual form (11) of the Einstein-Hilbert action and the corresponding Lagrangian. In this version, L_{EH} is represented in an extend space of variables. Instead of the ten g_{ab} one uses now the sixteen e^a_A functions. Since L_{EH} is locally Lorentz covariant there exist corresponding constraints. (They are identically satisfied by Einstein's vacuum equations.) If it is of practical use one can choose gauge conditions: six for fixing six of the e^a_A and, in the (3+1) decomposition, three for fixing three of the $\tilde{E}^i{}_a$. This situaton changes, however, drastically when one passes to a Lagrangian that is no longer Lorentz covariant. Now one has again an extended space of variables but one cannot reduce them by requiring gauge conditions.In the (3+1) decomposition this means: one has to consider an extended phase space which results by the transition from the cotangent bundle T^*C of the configuration space to a phase space Γ by introducing the freedom to perform internal SO(3) or SU(2) rotations. But now one can only reduce Γ by exploiting the constraints corresponding to the internal rotations when the demolition of the local Lorentz covariance did not destroy the SO(3) or SU(2) invariance of the formalism.

Thus, the transition to anholonomic variables leads to the following dilemma: Either one tries to work with the SO(3) or SU(2) variables without constraining them by other than the conditions (3); then the arising framework generally is underdetermined. Or one requires SO(3) and, respectively, SU(2) constraints; then one goes over to a theory that in general differs from Einstein's 1915 theory of general relativity. It is a theory with a Lagragian that is not locally Lorentz covariant and, as a possible consequence, even not SO(3) invariant. But even in the case that the expressions derived by a (3+1) decomposition are SO(3) or SU(2) ivariant one cannot be satisfied. Expressions resulting from a Lagrangian that does not satisfy Einstein's principle of general relativity can hardly be considered as fundamentals of a theory of general relativiy.

---

[3] It should, however, be stressed here that L_{EH}, L_E and L_M all imply one physically meaningful statement on gravitational energy-momentum, namely Einstein's vacuum equations saying that the metrical energy-momentum tensor of gravity is equal to zero.

To discuss some consequences of a quantum theory of gravity that bases on such an anholonomic "canonical" formalism, in Ref. 1 we considered flat space-time, where in pseudocartesian coordinates generally the relations

$$g_{ab} = \eta_{ab}, \quad \Gamma_{bc}^a = 0, \quad h_a^A(x^i) = \omega_B^A(x^i)\delta_a^A \quad (\neq \delta_a^A) \tag{18}$$

are satisfied. Since this choice of coordinates provides a natural foliation the (00)-component of Møller's energy-momentum complex (which is coordinate but not Lorentz covariant) leads to the Hamiltonian

$$\kappa H = 2(\omega^D{}_{A,M}\delta^0{}_D - \omega^D{}_{A,D}\delta^0{}_M)\omega^{MA}{}_{,0} - L_M , \tag{19}$$

where the configuration variables ω^{lA},

$$\omega^{lA} := \omega^{BA}\delta^l{}_B \tag{20}$$

and the momentum variables p_{Al},

$$\frac{1}{2}c\kappa p_{Al} := \omega^{00}{}_{A,l} - \delta^0_l \, \omega^C{}_{A,C} \tag{21}$$

satisfy the Poisson bracket relations

$$\begin{aligned} \{\omega^{lA}(x), \, p_{Bk}(x')\} &= \delta^l_\kappa \delta^A_B \delta(x, x'), \\ \{\omega^{lA}(x), \, \omega^{kB}(x')\} &= 0, \\ \{p_{Al}(x), \, p_{Bk}(x')\} &= 0 \end{aligned} \tag{22}$$

On the quantum level, the corresponding relations state that there exist quanta attributed to the Lorentz rotations $\omega^A_B(\omega^A_C\omega^C_B = \delta^A_B$ which, due to the missing local Lorentz covariance cannot be gauged away. This example shows especially clear that the quantization of Einstein's 1915 theory of general relativity leads to results that are not physically interprtable.

It should be stressed here that the problems encountered by canonical quantization do not mean a lack of self-consistency of general relativity. They are nothing but an implication of the fact that Einstein's 1915 theory is no genuine field theory. (For a more detailed discussion of this aspect and, in particular, of its relation to Rosenfeld's uncertainty relations, cf. Ref. 1.)

3. The Einstein-Schrödinger purely affine model

The obstacles that one encounters in attempting to quantize Einstein's 1915 theory of general relativity are due to several reasons. Two of them should be recalled here, namely

1. the basic invariance group, i. e., the group $Diff(M)$ of diffeomorphisms of the space-time manifold M, and

2. the structure of the Einstein-Hilbert Lagrangian.

As to the group $Diff(M)$, on the one hand, it is analogous to that of the gauge group in Yang-Mills theory. Otherwise, it is quite different for it does not act on gravitational fields at a fixed point. It is the group of active point transformations moving the points of the space-time around. This brings forth a variety of problems for defining observables, two-point functions, time-ordered products etc. (cf. Refs. 7 and 10). This situation is however unavoidable as long as one takes seriously the priciple of general relativity. Modifications of this framework (caused, for instance, by the introduction of a background structure) contradict this priciple.

Furthermore, if one intends to realize the principle of general relativity by means of a metric theory one has to start with the Einstein-Hilbert Lagrangian because it is the only invariant providing second-order differential equations for the g_{ab} field. This Lagrangian is however of no canonical form, and thus one is led to the supplementary problems described in section 2.

In the light of the above, it is a suggestive idea to consider others than metric realizations of general relativity. Such modified theories cannot of course overcome the difficulties caused by the diffeomorphism group invariance, but they can test whether such approaches allow for a canonical formulation of general-relativistic theory of gravity.[4]

In order to unify gravitational and electromagnetic interactions generalizations of the 1915 theory were proposed Weyl, Eddington, Einstein himself, Schrödinger, and others. In Ref. 2, we discussed the Einstein-Schrödinger purely affine theory [16, 17], but now not as a gravoelectromagnetic theory but as a toy model describing Riemannian and non-Riemannian parts of gravity. This model is mostly interesting, first, because, in cotrast to "mixed" theories based on the metric g_{ab} and the affine connection Γ^i_{kl}, it contains only one set of field variables. Second, because it implies torsion and thus is a promising candidate for realizing an idea of V. de Sabbata and C. Sivaram [18] according to which torsion and curvature could be brought in the mutual relation of canonically conjugate variables. - The following, remarks present a summary of some points discussed in Ref 2 and concerning this topic.

The Einstein-Schrödinger Lagrangian $\Im$ reads

$$\Im = \frac{2}{\lambda}(-det\ R_{ik})^{1/2}, \tag{23}$$

---

[4] This and earlier formulations emphasizing the point that Einstein's 1915 theory is only one of the possible realizations of the relativity principle regard alredy some of the comments made by Professor P. Bergmann with respect to our section 2.

where

$$R_{ik} = \delta_n^m R^n{}_{ikm},$$

$$det\ R_{ik} = \frac{1}{24}\, \eta^{iklm} \eta^{abcd} R_{ia} R_{kb} R_{lc} R_{md}\,, \tag{24}$$

$$\lambda = const.$$

The field equations resulting via

$$\frac{\delta \Im}{\delta R_{mn}} \frac{\delta R_{nm}}{\delta \Gamma^i_{kl}} = 0 \tag{25}$$

from (23) have the form

$$R_{ik,l} - R_{rk}\Gamma^r_{il} - R_{ir}\Gamma^r_{lk} = \frac{2}{3}(R_{ik}\Gamma_l + R_{il}\Gamma_k)\,, \tag{26}$$

with the torsion vector

$$\Gamma_l = \delta_n^m \Gamma^n_{[ml]} = \frac{1}{2}\delta_n^m(\Gamma^n_{lm} - \Gamma^n_{ml})\,. \tag{27}$$

Using Schrödinger's relation

$$\frac{2}{\lambda}\, \frac{\partial \Im}{\partial R_{ik}} = g^{ik} \tag{28}$$

defining a non-symmetric fundamental tensor g^{ab} (with g_{ab} given by $g_{ac}g^{bc} = \delta_a^b$) as a reduced minor, the equations (26) can also be written as

$$g_{ikl} - g_{rk}\Gamma^r_{kl} - g_{ir}\Gamma^r_{lk} = \frac{2}{3}(g_{ik}\Gamma_i + g_{il}\Gamma_i)\,. \tag{29}$$

These relations represent 60 conditions on the 64 components of $\Gamma^i_{k_l}$ so that the 4 components of the torsion vector remain undetermined. This fact is a consequence of that this theory is gauge invariant with respect to transformations

$$\Gamma'^i{}_{kl} = \Gamma^i_{kl} + \delta_k^i \phi_l\,. \tag{30}$$

As an expression of this additional invariance there occur, beside the Bianchi identities resulting from the $Diff(\,M\,)$ invarance, the relations

$$(\sqrt{-g}\,R^{[ik]})_{,k} = \lambda(\sqrt{-g}\;g^{[ik]})_{,k} = 0 \,. \tag{31}$$

Using now Schrödinger's gauge condition $\phi_i = \frac{2}{3}\Gamma_i$, the affine connection is given by the star connection

$$\Gamma^{*i}{}_{kl} = \Gamma^i_{kl} + \frac{2}{3}\delta^i_k\Gamma_l \,. \tag{32}$$

Defining furthermore the tensor F_{ik} $(= -F_{ik})$,

$$F_{ik} = R^*{}_{ik} - R_{ik} = \frac{2}{3}(\Gamma_{i,k} - \Gamma_{k,i}) \,, \tag{33}$$

one is led to the equations

$$F_{ik,l} + F_{kl,i} + F_{li,k} = 0 \,,$$
$$(\sqrt{-g}\;F^{ik})_{,k} = \sqrt{-g}\;I^i \quad \text{with} \quad \sqrt{-g}\;I^i = -(\sqrt{-g}\;R^{*[ki]})_{,k} \,. \tag{34}$$

The first group of equations results from (33) and the second is identical with (31).

In the Schrödinger gauge one arrives at Maxwell-like equations in a polarized and charged medium. Now the torsion vector $\Gamma_i = \Gamma^k_{[ik]}$ can be considered as the vector potential $\Gamma_i = \Gamma^k_{[ik]}$,and the tensor $F_{ik} = R^*{}_{ik} - R_{ik}$ plays the role of the field strength. Thus, Γ_i and F_{ik} can be regarded as canonically conjugate variables, in the same sense as the vector potential and the field strength are such variables in electrodynamics. This means that this realization of the principle of general relativity leads to a "more canonical" formalism. It must however be mentioned that this statement concerns only the non-Riemannian part of the geometric field. For the Riemannian part one meets the same situation as in Einstein's 1915 theory of gravity.

References

1. H.-H. v. Borzeszkowski and H.-J. Treder, *Gen. Rel. Grav.* 25, 291 (1993).
2. H.-H. v. Borzeszkowski, V. de Sabbata, C. Sivaram, and H.-J. Treder, Torsion and Curvature in Quantum Gravity (to be published).
3. R. Penrose, *Gen. Rel. Grav.* 1, 31 (1976).
4. A. Ashtekar (with invited contributions), *New Perspectives in Canonical Gravity* (Bibliopolis, Naples 1988).
5. A. Ashtekar, *Lectures on Non-Perturbative Canonical Gravity* (World Scientific, Singapore et al., 1991).
6. H.-H. v. Borzeszkowski and H.-J. Treder, *Found. Phys.* (in print).

24

7. C. J. Isham,Conceptual and Geometrical Problems in Quantum Gravity, *Imperial College Preprint* TP/90-91/14 (1991).

8. C. J. Isham, Canonical Quantum Gravity and the Problem of Time, *Imperial College Preprint* TP/91-92/25 (1992).

9. R. Arnowitt, S. Deser, and C. W. Misner, in *Gravitation: An Introduction to Current Research*, L. Witten, ed. (Wiley, New York, (1962).

10. A Ashtekar and J. Stachel, eds. *Conceptual Problems of Quantum Gravity*, (Birkhäuser, Boston et al., 1991).

11. R. Arnowitt, S. Deser, and C. W. Misner, in *Recent Developments in General Relativity* (Pergamon, Oxford et al., (1962).

12. J. Samuel, *Pramana J. Phys.* **28**, 429 (1987).

13. T. Jacobsen and L Smolin, *Phys. Lett.* **196B**, 39 (1987).

14. A. Einstein, *Sitzungsber. Preuss. Akad. Wiss.* , p. 1111 (1916).

15. C. Møller, *Nat. Fys. Skr. Dan Vid, Selskab.* 1, 10 (1961).

16. F. Schrödinger, *Space-Time Structure*, (Cambridge U. P., 1950).

17. A. Einstein, *The Meaning of Relativity* (Princeton U. P., 1950, 1951).

18. V. de Sabbata and C. Sivaram, *Found. Phys. Lett.* **5**, 579 (1992).

PHASE TRANSITIONS OUT OF EQUILIBRIUM

D. Boyanovsky[a] and H. J. de Vega[b]

(a) Department of Physics and Astronomy
University of Pittsburgh, Pittsburgh, P. A. 15260, U.S.A.
(b) Laboratoire de Physique Théorique et Hautes Energies[*]
Université Paris VI
Tour 16, 1er. étage, 4, Place Jussieu
75252 Paris cedex 05, FRANCE.

Abstract

The dynamics of typical phase transitions is studied out of equilibrium in weakly coupled inflaton-type scalar field theories in Minkowski space. The shorcomings of the effective potential and equilibrium descriptions are pointed out. A case of a rapid supercooling from $T_i > T_c$ to $T_f \ll T_c$ is considered. The equations of motion up to one-loop for the order parameter are obtained and integrated for the case of "slow rollover initial conditions". It is shown that the instabilities responsible for the process of phase separation introduce dramatic corrections to the evolution. Domain formation and growth (spinodal decomposition) is studied in a non-perturbative self-consistent approximation. The size of domains is shown to grow as $\xi_D(t) \approx \sqrt{t\xi(0)}$. The amplitude of the fluctuations grow up to a maximum time t_s which in weakly coupled theories is estimated to be

$$t_s \approx -\xi(0)\ln\left[\left(\frac{3\lambda}{4\pi^3}\right)^{\frac{1}{2}}\left(\frac{(\frac{T_i}{2T_c})^3}{[\frac{T_i^2}{T_c^2}-1]}\right)\right]$$

with $\xi(0)$ the zero temperature correlation length. For very weakly coupled theories, their final size is several times the zero temperature correlation length. For strongly coupled theories the final size of the domains is comparable to the zero temperature correlation length and the transition proceeds faster. We also obtain the evolution equations to one-loop order and in a non-perturbative Hartree approximation in spatially flat FRW cosmologies.

1 Introduction and Motivation

Inflationary cosmological models provide very appealing scenarios to describe the early evolution of our universe[1]. Since the original model proposed by Guth[2], several alternative scenarios have been proposed to overcome some of the difficulties with the original proposal.

Among them, the new inflationary model[3, 4, 5, 6] is perhaps one of the most attractive. The essential ingredient in the new inflationary model is a scalar field (the "inflaton") that undergoes a second order phase transition from a high temperature symmetric phase to a low temperature broken symmetry phase. The expectation value (or thermal average) of the

V. de Sabbata and H. Tso-Hsiu (eds.), Cosmology and Particle Physics, 25–48.

undergoes a second order phase transition from a high temperature symmetric phase to a low temperature broken symmetry phase. The expectation value (or thermal average) of the scalar field ϕ serves as the order parameter. Initially at high temperatures, the scalar field is assumed to be in thermal equilibrium and $\phi \approx 0$. The usual field- theoretic tool to study the phase transition is the effective potential[7, 8, 9].

At high temperatures, the global minimum of the effective potential is at $\phi = 0$, whereas at low temperatures there are two degenerate minima.

The conjectured behavior of the phase transition in the new inflationary model is the following: as the universe cools down the expectation value of the scalar field remains close to zero until the temperature becomes smaller than the critical temperature, when the effective potential develops degenerate minima away from the origin. When this happens, the scalar field begins to "roll down the potential hill". In the new inflationary scenario, the effective potential below the critical temperature is extremely flat near the maximum, and the scalar field remains near the origin, i.e. the false vacuum, for a very long time and eventually rolls down the hill very slowly. This scenario receives the name of "slow rollover". During the slow rollover stage, the energy density of the universe is dominated by the constant vacuum energy density $V_{eff}(\phi = 0)$, and the universe evolves rapidly into a de Sitter space (see for example the reviews by Kolb and Turner[10], Linde[11] and Brandenberger[12]). Perhaps the most remarkable consequence of the new inflationary scenario and the slow rollover transition is that they provide a calculational framework for the prediction of density fluctuations[13]. The coupling constant in the typical zero temperature potentials must be fine tuned to a very small value to reproduce the observed limits on density fluctuations[10, 11].

This picture of the slow rollover scenario, is based on the *static* effective potential, its usefulness in a time dependent situation has been questioned by Mazenko, Unruh and Wald[14]. These authors argued that the *dynamics* of the cooling down process is very similar to the process of phase separation in statistical mechanics, and argued that the system will form domains within which the scalar field will relax to the values at the minima of the potential very quickly. It is now accepted that this picture may be correct for very strongly coupled theories but not for weakly coupled scenarios as required in inflation.

Guth and Pi[15], studied the effects of quantum fluctuations on the time evolution below the critical temperature by treating the potential near the origin as an *inverted harmonic oscillator*. They recognized that the instabilities associated with these upside-down oscillators lead to an exponential growth of the quantum fluctuations at long times and to a classical description of the probability distribution function. Guth and Pi also recognized that the *static* effective potential is not appropriate to describe the dynamics, that must be treated as a time dependent process.

Subsequently, Weinberg and Wu[16], have studied the effective potential, particularly in the situation when the tree level potential allows for broken symmetry ground states. In this case, the effective potential becomes complex. These authors identified the imaginary part of the effective potential with the decay rate of a particular initial state.

In statistical mechanics, the imaginary part is a consequence of a succession of thermodynamically unstable states.

2 Necessity for a Non-Equilibrium description:

The effective potential offers no information on the *dynamics* of the process of the phase transition. As mentioned above it becomes complex in a region in field space corresponding to thermodynamically unstable states.

Furthermore, in an expanding gravitational background the notion of a static effective potential is at best an approximate one. To understand whether one may treat the problem in the approximation of local thermodynamic equilibrium, the time scales involved must be understood carefully. In a typical FRW cosmology the important time scale is determined by the Hubble expansion factor $H(t) = \dot{a}(t)/a(t)$ whereas the equilibration processes are determined by the typical collisional relaxation rates $\Gamma(T) \approx \lambda^2 T$ in typical scalar theories. Local thermodynamic equilibrium will prevail if $\Gamma(T) \gg H(t)$. In de Sitter evolution, after a typical phase transition at $T \approx T_c \approx 10^{14}$ Gev, $H \approx 10^{-5} T_c$[11] and local thermodynamic equilibrium *will not* prevail for weakly coupled inflaton theories in which $\lambda \approx 10^{-12} - 10^{-14}$[11, 13, 20].

Thus in these scenarios, weakly coupled scalar field theories *cannot be assumed* to be in local thermal equilibrium. Typical expansion rates are much larger than typical equilibration rates and the phase transition will occur very rapidly. The long-wavelength fluctuations will be strongly out of equilibrium as they typically have very slow dynamics (see[17] and references therein). The phase transition *must* be studied away from equilibrium.

A complete discussion of these issues and the complex effective potential may be found in the articles by Boyanovsky and de Vega[18] and Boyanovsky et. al.[17, 19].

3 Non-Equilibrium time evolution

The dynamics will be determined once the evolution Hamiltonian and *the initial state* are prescribed.

Let us consider the situation in which for time $t < 0$ the system is in *equilibrium* at an initial temperature $T_i > T_c$ where T_c is the critical temperature. At time $t = 0$ the system is rapidly "quenched" to a final temperature below the critical temperature $T_f < T_c$ and evolves thereafter out of equilibrium.

What we have in mind in this situation, is a cosmological scenario with a period of rapid inflation in which the temperature drops very fast compared to typical relaxation times of the scalar field.

Precisely because of the weak couplings and critical slowing down of long-wavelength fluctuations, we conjecture that an inflationary period at temperatures near the critical temperature, may be described in this "quenched" approximation. Another situation that may be described by this approximation is that of a scalar field again at $T_i > T_c$ suddenly coupled to a "heat bath" at a much lower temperature (below the transition temperature) and evolving out of equilibrium. The "heat bath" may be other fields at a different temperature.

Certainly this is only a plausibility argument, a deeper understanding of the initial conditions must be pursued to obtain a more precise knowledge of the cooling down process.

Although we are currently studying the case of inflationary cosmologies, we will here concentrate on the dynamics of a supercooled phase transition in Minkowsky space.

This situation is modelled by introducing a Hamiltonian with a *time dependent mass term*

$$H(t) = \int_\Omega d^3x \left\{ \frac{1}{2}\Pi^2(x) + \frac{1}{2}(\vec{\nabla}\Phi(x))^2 + \frac{1}{2}m^2(t)\Phi^2(x) + \frac{\lambda}{4!}\Phi^4(x) \right\} \tag{1}$$

$$m^2(t) = m^2\Theta(-t) + (-\mu^2)\Theta(t) \tag{2}$$

where both m^2 and μ^2 are positive. We assume that for all times $t < 0$ there is thermal equilibrium at temperature T_i, and the system is described by the density matrix

$$\hat{\rho}_i = e^{-\beta_i H_i} \tag{3}$$

$$H_i = H(t < 0) \tag{4}$$

In the Schrödinger picture, the density matrix evolves in time as

$$\rho(\hat{t}) = U(t)\hat{\rho}_i U^{-1}(t) \tag{5}$$

with $U(t)$ the time evolution operator.

An alternative and equally valid interpretation (and the one that we like best) is that the initial condition being considered here is that of a system in equilibrium in the symmetric phase, and evolved in time with a Hamiltonian that allows for broken symmetry ground states, i.e. the Hamiltonian (1, 2) for $t > 0$.

The expectation value of any operator is thus

$$< \mathcal{O} > (t) = Tre^{-\beta_i H_i}U^{-1}(t)\mathcal{O}U(t)/Tre^{-\beta_i H_i} \tag{6}$$

This expression may be written in a more illuminating form by choosing an arbitrary time $T < 0$ for which $U(T) = \exp[-iTH_i]$ then we may write $\exp[-\beta_i H_i] = \exp[-iH_i(T - i\beta_i - T)] = U(T - i\beta_i, T)$. Inserting in the trace $U^{-1}(T)U(T) = 1$, commuting $U^{-1}(T)$ with $\hat{\rho}_i$ and using the composition property of the evolution operator, we may write (6) as

$$< \mathcal{O} > (t) = TrU(T - i\beta_i, t)\mathcal{O}U(t, T)/TrU(T - i\beta_i, T) \tag{7}$$

The numerator of the above expression has a simple meaning: start at time $T < 0$, evolve to time t, insert the operator $\mathcal{O}$ and evolve backwards in time from t to $T < 0$, and along the negative imaginary axis from T to $T - i\beta_i$. The denominator, just evolves along the negative imaginary axis from T to $T - i\beta_i$. The contour in the numerator may be extended to an arbitrary large positive time T' by inserting $U(t, T')U(T', t) = 1$ to the left of $\mathcal{O}$ in (7), thus becoming

$$< \mathcal{O} > (t) = TrU(T - i\beta_i, T)U(T, T')U(T', t)\mathcal{O}U(t, T)/TrU(T - i\beta_i, T) \tag{8}$$

The numerator now represents the process of evolving from $T < 0$ to t, inserting the operator $\mathcal{O}$, evolving further to T', and backwards from T' to T and down the negative imaginary axis to $T - i\beta_i$. Eventually we take $T \to -\infty$; $T' \to \infty$. It is straightforward to generalize to real time correlation functions of Heisenberg picture operators.

As usual, the insertion of an operator may be achieved by inserting sources in the time evolution operators, defining the generating functionals and eventually taking functional derivatives with respect to these sources. Notice that we have three evolution operators, from T to T', from T', back to T (inverse operator) and from T to $T - i\beta_i$, since each of these operators has interactions and we want to use perturbation theory and generate the diagrammatics from the generating functionals, we use *three different sources*. A source J^+ for the evolution $T \to T'$, J^- for the branch $T' \to T$ and finally J^β for $T \to T - i\beta_i$. The denominator may be obtained from the numerator by setting $J^+ = J^- = 0$. Finally the generating functional $Z[J^+, J^-, J^\beta] = TrU(T - i\beta_i, T; J^\beta)U(T, T'; J^-)U(T', T; J^+)$, may be written in term of path integrals as (here we neglect the spatial arguments to avoid cluttering of notation)

$$Z[J^+, J^-, J^\beta] = \int D\Phi D\Phi_1 D\Phi_2 \int \mathcal{D}\Phi^+ \mathcal{D}\Phi^- \mathcal{D}\Phi^\beta e^{i\int_T^{T'} \{\mathcal{L}[\Phi^+, J^+] - \mathcal{L}[\Phi^-, J^-]\}} \times$$
$$e^{i\int_T^{T-i\beta_i} \mathcal{L}[\Phi^\beta, J^\beta]} \tag{9}$$

with the boundary conditions $\Phi^+(T) = \Phi^\beta(T - i\beta_i) = \Phi$; $\Phi^+(T') = \Phi^-(T') = \Phi_2$; $\Phi^-(T) = \Phi^\beta(T) = \Phi_1$. As usual the path integrals over the quadratic forms may be done and one obtains the final result for the partition function

$$Z[J^+, J^-, J^\beta] = e^{\left\{i\int_T^{T'} dt[\mathcal{L}_{int}(-i\delta/\delta J^+) - \mathcal{L}_{int}(i\delta/\delta J^-)]\right\}} e^{\left\{i\int_T^{T-i\beta_i} dt\mathcal{L}_{int}(-i\delta/\delta J^\beta)\right\}} \times$$
$$e^{\left\{\frac{i}{2}\int_c dt_1 \int_c dt_2 J_c(t_1)J_c(t_2)G_c(t_1,t_2)\right\}} \tag{10}$$

Where J_c are the currents defined on the segments of the contour[18] $J^\pm$, J^β[23] and G_c is the Green's function on the contour (see below), and again the spatial arguments have been suppressed.

In the two contour integrals (on t_1; t_2) in (10) there are altogether nine terms, corresponding to the combination of currents in each of the three branches. However, in the limit $T \to -\infty$, the contributions arising from the terms in which one current is on the $(+)$ or $(-)$ branch and another on the imaginary time segment (from T to $T - i\beta_i$), go to zero when computing correlation functions in which the external legs are at finite *real time*. For this *real time correlation functions* there is no contribution from the J^β terms, that cancel between numerator and denominator, and the information on finite temperature is encoded in the boundary conditions on the Green's functions (see below). Then for the calculation of finite *real time* correlation functions the generating functional simplifies to[24, 25]

$$Z[J^+, J^-] = e^{\left\{i\int_T^{T'} dt[\mathcal{L}_{int}(-i\delta/\delta J^+) - \mathcal{L}_{int}(i\delta/\delta J^-)]\right\}} \times$$
$$e^{\left\{\frac{i}{2}\int_T^{T'} dt_1 \int_T^{T'} dt_2 J_a(t_1)J_b(t_2)G_{ab}(t_1,t_2)\right\}} \tag{11}$$

with $a, b = +, -$.

This formulation in terms of time evolution along a contour in complex time has been used many times in non-equilibrium statistical mechanics. To our knowledge the first to use this formulation were Schwinger[26] and Keldysh[27] (see also Mills[28]). There are many articles in the literature using these techniques to study time dependent problems, some of the more clear articles are by Jordan[31], Niemi and Semenoff[23], Landsman and van Weert[29],Semenoff and Weiss[30], Kobes and Kowalski[32], Calzetta and Hu[25],Paz[33] and references therein (for more details see[18, 17, 19].

The Green's functions that enter in the integrals along the contours in (10, 11) are given by (see above references)

$$G^{++}(t_1, t_2) = G^>(t_1, t_2)\Theta(t_1 - t_2) + G^<(t_1, t_2)\Theta(t_2 - t_1) \tag{12}$$

$$G^{--}(t_1, t_2) = G^>(t_1, t_2)\Theta(t_2 - t_1) + G^<(t_1, t_2)\Theta(t_1 - t_2) \tag{13}$$

$$G^{+-}(t_1, t_2) = -G^<(t_1, t_2) \tag{14}$$

$$G^{-+}(t_1, t_2) = -G^>(t_1, t_2) = -G^<(t_2, t_1) \tag{15}$$

$$G^<(T, t_2) = G^>(T - i\beta_i, t_2) \tag{16}$$

As usual $G^<, G^>$ are homogeneous solutions of the quadratic form with appropriate boundary conditions. We will construct them explicitly later. The condition (16) is recognized as the periodicity condition in imaginary time (KMS condition)[34].

To obtain the evolution equations we use the tadpole method[9], and write

$$\Phi^\pm(\vec{x}, t) = \phi(t) + \Psi^\pm(\vec{x}, t) \tag{17}$$

Where, again, the $\pm$ refer to the branches for forward and backward time propagation. The reason for shifting both ($\pm$) fields by the *same* classical configuration, is that ϕ enters in the time evolution operator as a background c-number variable, and time evolution forward and backwards are now considered in this background.

The evolution equations are obtained with the tadpole method by expanding the Lagrangian around $\phi(t)$ and considering the *linear*, cubic, quartic, and higher order terms in $\Psi^\pm$ as perturbations and requiring that

$$< \Psi^\pm(\vec{x}, t) >= 0.$$

It is a straightforward exercise to see that this is equivalent to extremizing the one-loop effective action in which the determinant (in the logdet) incorporates the boundary condition of equilibrium at time $t < 0$ at the initial temperature.

To one loop we find the equation of motion

$$\frac{d^2\phi(t)}{dt^2} + m^2(t)\phi(t) + \frac{\lambda}{6}\phi^3(t) + \frac{\lambda}{2}\phi(t) \int \frac{d^3k}{(2\pi)^3}(-i)G_k(t, t) = 0 \tag{18}$$

with $G_k(t, t) = G_k^<(t, t) = G_k^>(t, t)$ is the spatial Fourier transform of the equal-time Green's function.

Notice that

$$(-iG_k(t,t)) = < \Psi_k^+(t)\Psi_{-k}^+(t) >$$

is a *positive definite quantity* (because the field Ψ is real) and as we argued before (and will be seen explicitly shortly) this Green's function grows in time because of the instabilities associated with the phase transition and domain growth[15, 16].

These Green's functions are constructed out of the homogeneous solutions to the operator of quadratic fluctuations

$$\left[\frac{d^2}{dt^2} + \vec{k}^2 + M^2(t)\right]\mathcal{U}_k^\pm = 0 \tag{19}$$

$$M^2(t) = (m^2 + \frac{\lambda}{2}\phi_i^2)\Theta(-t) + (-\mu^2 + \frac{\lambda}{2}\phi^2(t))\Theta(t) \tag{20}$$

The boundary conditions on the homogeneous solutions are

$$\mathcal{U}_k^\pm(t < 0) = e^{\mp i\omega_<(k)t} \tag{21}$$

$$\omega_<(k) = \left[\vec{k}^2 + m^2 + \frac{\lambda}{2}\phi_i^2\right]^{\frac{1}{2}} \tag{22}$$

where ϕ_i is the value of the classical field at time $t < 0$ and is the initial boundary condition on the equation of motion. Truly speaking, starting in a fully symmetric phase will force $\phi_i = 0$, and the time evolution will maintain this value, therefore we admit a small explicit symmetry breaking field in the initial density matrix to allow for a small ϕ_i. The introduction of this initial condition seems artificial since we are studying the situation of cooling down from the symmetric phase.

However, we recognize that the phase transition from the symmetric phase occurs via formation of domains (in the case of a discrete symmetry) inside which the order parameter acquires non-zero values. The domains will have the same probability for either value of the field and the volume average of the field will remain zero. These domains will grow in time, this is the phenomenon of phase separation and spinodal decomposition familiar in condensed matter physics. Our evolution equations presumably will apply to the coarse grained average of the scalar field inside each of these domains. This average will only depend on time. Thus, we interpret φ_i as corresponding to the coarse grained average of the field in each of these domains. The question of initial conditions on the scalar field is also present (but usually overlooked) in the slow-rollover scenarios but as we will see later, it plays a fundamental role in the description of the evolution.

The identification of the initial value φ_i with the average of the field in each domain is certainly a plausibility argument to justify an initially small asymmetry in the scalar field which is necesary for the further evolution of the field, and is consistent with the usual assumption within the slow rollover scenario.

The boundary conditions on the mode functions $\mathcal{U}_k^\pm(t)$ correspond to "vacuum" boundary conditions of positive and negative frequency modes (particles and antiparticles) for $t < 0$.

Finite temperature enters through the periodicity conditions (16) and the Green's functions are

$$G_k^>(t, t') = \frac{i}{2\omega_<(k)} \frac{1}{1 - e^{-\beta_i \omega_<(k)}} \left[\mathcal{U}_k^+(t) \mathcal{U}_k^-(t') + e^{-\beta_i \omega_<(k)} \mathcal{U}_k^-(t) \mathcal{U}_k^+(t') \right] \quad (23)$$

$$G_k^<(t, t') = G^>(t', t) \quad (24)$$

The effective equations of motion to one loop that determine the time evolution of the scalar field are

$$\frac{d^2 \phi(t)}{dt^2} + m^2(t) \phi(t) + \frac{\lambda}{6} \phi^3(t) +$$
$$\frac{\lambda}{2} \phi(t) \int \frac{d^3 k}{(2\pi)^3} \frac{\mathcal{U}_k^+(t) \mathcal{U}_k^-(t)}{2\omega_<(k)} \coth \left[\frac{\beta_i \omega_<(k)}{2} \right] = 0 \quad (25)$$

$$\left[\frac{d^2}{dt^2} + \vec{k}^2 + M^2(t) \right] \mathcal{U}_k^\pm = 0 \quad (26)$$

with (20) , (21).

This set of equations is too complicated to attempt an analytic solution, we will study this system numerically shortly.

However, before studying numerically these equations, one recognizes that there are several features of this set of equations that reveal the basic physical aspects of the dynamics of the scalar field.

i): The effective evolution equations are **real**. The mode functions $\mathcal{U}_k^\pm(t)$ are complex conjugate of each other as may be seen from the time reversal symmetry of the equations, and the boundary conditions (21). This situation must be contrasted with the expression for the effective potential for the *analytically continued modes*.

ii): Consider the situation in which the initial configuration of the classical field is near the origin $\phi_i \approx 0$, for $t > 0$, the modes for which $\vec{k}^2 < (k_{max})^2$; $(k_{max})^2 = \mu^2 - \frac{\lambda}{2} \phi_i^2$ are *unstable*.

In particular, for early times $(t > 0)$, when $\phi_i \approx 0$, these unstable modes behave approximately as

$$\mathcal{U}_k^+(t) = A_k e^{W_k t} + B_k e^{-W_k t} \quad (27)$$

$$\mathcal{U}_k^-(t) = (\mathcal{U}_k^+(t))^* \quad (28)$$

$$A_k = \frac{1}{2} \left[1 - i \frac{\omega_<(k)}{W_k} \right] \ ; \ B_k = A_k^* \quad (29)$$

$$W_k = \left[\mu^2 - \frac{\lambda}{2} \phi_i^2 - \vec{k}^2 \right]^{\frac{1}{2}} \quad (30)$$

Then the early time behavior of $(-iG_k(t, t))$ is given by

$$(-iG_k(t, t)) \approx \frac{1}{2\omega_<(k)} \left[1 + \frac{\mu^2 + m^2}{\mu^2 - \frac{\lambda}{2} \phi_i^2 - k^2} [\cosh(2W_k t) - 1] \right] \coth[\beta_i \omega_<(k)/2] \quad (31)$$

This early time behavior coincides with the Green's function of Guth and Pi[15] and Weinberg and Wu[16] for the inverted harmonic oscillators when our initial state (density matrix) is taken into account.

Our evolution equations, however, permit us to go beyond the early time behavior and to incorporate the non-linearities that will eventually shut off the instabilities.

These early-stage instabilities and subsequent growth of fluctuations and correlations, are the hallmark of the process of phase separation, and precisely the instabilities that trigger the phase transition.

It is clear from the above equations of evolution, that the description in terms of inverted oscillators will only be valid at very early times.

At early times, the *stable* modes for which $\vec{k}^2 > (k_{max})^2$ are obtained from (27) , (28) , (29) by the analytic continuation $W_k \rightarrow -i\omega_>(k) = \left[\vec{k}^2 - \mu^2 + \frac{\lambda}{2}\phi_i^2\right]^{\frac{1}{2}}$.

For $t < 0$, $\mathcal{U}_k^+(t)\mathcal{U}_k^-(t) = 1$ and one obtains the usual result for the evolution equation

$$\frac{d^2\phi(t)}{dt^2} + \frac{dV_{eff}(\phi)}{d\phi} = 0$$

with $V_{eff}(\phi)$ the finite temperature effective potential but for $t < 0$ there are no unstable modes.

It becomes clear, however, that for $t > 0$ there are no *static* solutions to the evolution equations for $\phi(t) \neq 0$.

iii)Coarsening: as the classical expectation value $\phi(t)$ "rolls down" the potential hill, $\phi^2(t)$ increases and $(k_{max}(t))^2 = \mu^2 - \frac{\lambda}{2}\phi^2(t)$ *decreases*, and only the very long-wavelength modes remain unstable, until for a particular time t_s ; $(k_{max}(t_s))^2 = 0$. This occurs when $\phi^2(t_s) = 2\mu^2/\lambda$, this is the inflexion point of the tree level potential. In statistical mechanics this point is known as the "classical spinodal point" and t_s as the "spinodal time"[21, 22]. When the classical field reaches the spinodal point, all instabilities shut- off. From this point on, the dynamics is oscillatory and this period is identified with the "reheating" stage in cosmological scenarios[11, 12].

It is clear from the above equations of evolution, that the description in terms of inverted oscillators will only be valid at small positive times, as eventually the unstable growth will shut-off.

The value of the spinodal time depends on the initial conditions of $\phi(t)$. If the initial value ϕ_i is very near the classical spinodal point, t_s will be relatively small and there will not be enough time for the unstable modes to grow too much. In this case, the one-loop corrections for small coupling constant will remain perturbatively small. On the other hand, however, if $\phi_i \approx 0$, and the initial velocity is small, it will take a very long time to reach the classical spinodal point. In this case the unstable modes **may grow dramatically making the one-loop corrections non-negligible even for small coupling.** These initial conditions of small initial field and velocity are precisely the "slow rollover" conditions that are of interest in cosmological scenarios of "new inflation".

The renormalization aspects have been studied in reference[18] and we refer the reader to that article for details.

4   Analysis of the Evolution

As mentioned previously within the context of coarsening, when the initial value of the scalar field $\phi_i \approx 0$, and the initial temporal derivative is small, the scalar field slowly rolls down the potential hill. But during the time while the scalar field remains smaller than the "spinodal" value, the unstable modes grow and the one-loop contribution grows consequently. For a "slow rollover" condition, the field remains very small ($\phi^2(t) \ll 2\mu^2/\lambda$) for a long time, and during this time the unstable modes grow exponentially. After renormalization, the stable modes give an oscillatory contribution which is bound in time, and for weak coupling remains perturbatively small at all times.

Then for a "slow rollover" situation and for weak coupling, only the unstable modes will yield to an important contribution to the one-loop correction. Thus, in the evolution equation for the scalar field, we will keep only the integral over the *unstable modes* in the one loop correction.

Phenomenologically the coupling constant in these models is bound by the spectrum of density fluctuations to be within the range $\lambda_R \approx 10^{-12} - 10^{-14}$[11]. The stable modes will always give a perturbative contribution, whereas the unstable modes grow exponentially in time thus raising the possibility of giving a non- negligible contribution.

With the purpose of numerical analysis of the effective equations of motion, it proves convenient to introduce the following dimensionless variables

$$\tau = \mu_R t \; ; \; q = k/\mu_R \tag{32}$$

$$\eta^2(t) = \frac{\lambda_R}{6\mu_R^2}\phi^2(t) \; ; \; L^2 = \frac{m_R^2 + \frac{1}{2}\lambda_R\phi_i^2}{\mu_R^2} \tag{33}$$

and to account for the change from the initial temperature to the final temperature ($T_i > T_c$; $T_f < T_c$) we parametrize[35]

$$m^2 = \mu_R(0)\left[\frac{T_i^2}{T_c^2} - 1\right] \tag{34}$$

$$\mu_R = \mu_R(0)\left[1 - \frac{T_f^2}{T_c^2}\right] \tag{35}$$

where the subscripts (R) stand for renormalized quantities, and $-\mu_R(0)$ is the renormalized zero temperature "negative mass squared" and $T_c^2 = 24\mu_R^2(0)/\lambda_R$. Furthermore, because $(k_{max}(t))^2 \leq \mu_R^2$ and $T_i > T_c$, for the unstable modes $T_i \gg (k_{max}(t))$ and we can take the high temperature limit $\coth[\beta_i\omega_<(k)/2] \approx 2T_i/\omega_<(k)$. Finally the effective equations of evolution for $t > 0$, keeping in the one-loop contribution only the unstable modes as explained above ($q^2 < (q_{max}(\tau))^2$) become, after using $\omega_<^2 = \mu_R^2(q^2 + L^2)$

$$\frac{d^2}{d\tau^2}\eta(\tau) - \eta(\tau) + \eta^3(\tau) +$$
$$g\eta(\tau)\int_0^{(q_{max}(\tau))} q^2\frac{\mathcal{U}_q^+(\tau)\mathcal{U}_q^-(\tau)}{q^2 + L^2}dq = 0 \tag{36}$$

$$\left[\frac{d^2}{d\tau^2} + q^2 - (q_{max}(t))^2\right]\mathcal{U}_q^{\pm}(\tau) = 0 \qquad (37)$$

$$(q_{max}(\tau))^2 := 1 - 3\eta^2(\tau) \qquad (38)$$

$$g = \frac{\sqrt{6\lambda_R}}{2\pi^2}\frac{T_i}{T_c\left[1 - \frac{T_f^2}{T_c^2}\right]} \qquad (39)$$

For $T_i \geq T_c$ and $T_f \ll T_c$ the coupling (39) is bound within the range $g \approx 10^{-7}-10^{-8}$. The dependence of the coupling with the temperature reflects the fact that at higher temperatures the fluctuations are enhanced.

From (36) we see that the quantum corrections act as a *positive dynamical renormalization* of the "negative mass" term that drives the rolling down dynamics. It is then clear, that the quantum corrections tend to *slow down the evolution.*

In particular, if the initial value $\eta(0)$ is very small, the unstable modes grow for a long time before $\eta(\tau)$ reaches the spinodal point $\eta(\tau_s) = 1/\sqrt{3}$ at which point the instabilities shut off. If this is the case, the quantum corrections introduce substantial modifications to the classical equations of motion, thus becoming non- perturbative. If $\eta(0)$ is closer to the classical spinodal point, the unstable modes do not have time to grow dramatically and the quantum corrections are perturbatively small.

Thus we conclude that the initial conditions on the field determine whether or not the quantum corrections are perturbatively small.

Figures (1,2) depict the solutions for the classical (solid lines) and quantum (dashed lines) evolution.

For the numerical integration we have chosen $L^2 = 1$, the results are only weakly dependent on L, and taken $g = 10^{-7}$, we have varied the initial condition $\eta(0)$ but used $\frac{d\eta(\tau)}{d\tau}|_{\tau=0} = 0$.

We recall, from a previous discussion that $\eta(\tau)$ should be identified with the average of the field within a domain. We are considering the situation in which this average is very small, according to the usual slow-rollover hypothesis, and for which the instabilities are stronger.

In figure (1) it is shown η vs τ for $g = 10^{-7}$, $\eta(0) = 2.3 \times 10^{-5}$ $\eta'(0) = 0$; $L = 1$, the solid line is the classical evolution, the dashed line is the evolution from the one-loop corrected equation of motion. We begin to see that the quantum corrections become important at $t \approx 10/\mu_R$ and slow down the dynamics. By the time that the *classical* evolution reaches the minimum of the classical potential at $\eta = 1$, the quantum evolution has just reached the classical spinodal $\eta = 1/\sqrt{3}$. The quantum correction becomes large enough to change the sign of the "mass term"[18], the field continues its evolution towards larger values, however, because the velocity is different from zero. As η gets closer to the classical spinodal point, the instability shuts off and the quantum correction arising from the unstable modes become small. From the spinodal point onwards, the field evolves towards the minimum and begins to oscillate around it, the quantum correction will be perturbatively small, as all the instabilities had shut-off. Higher order corrections, will introduce a damping term as quanta may decay in elementary excitations of the true vacuum.

Figure (2) shows a dramatic behavior for $\eta(0) = 2.258 \times 10^{-5}$; $\frac{d\eta(0)}{d\tau} = 0$ for the same values of the parameters as for figure (1). The unstable modes have enough time to grow so dramatically that the quantum correction becomes extremely large overwhelming the "negative mass" term near the origin. The dynamical time dependent potential, now becomes a *minimum* at the origin and the quantum evolution begins to *oscillate* near $\eta = 0$. The contribution of the unstable modes has become *non-perturbative*, and certainly our one- loop approximation breaks down.

As the initial value of the field gets closer to zero, the unstable modes grow for a very long time. At this point, we realize, however, that this picture cannot be complete. To see this more clearly, consider the case in which the initial state or density matrix corresponds exactly to the symmetric case. $\eta = 0$ is necessarily, by symmetry, a fixed point of the equations of motion. Beginning from the symmetric state, the field will *always remain* at the origin and though there will be strong quantum and thermal fluctuations, these are symmetric and will sample field configurations with opposite values of the field with equal probability.

In this situation, and according to the picture presented above, one would then expect that the unstable modes will grow indefinetely because the scalar field does not roll down and will never reach the classical spinodal point thus shutting-off the instabilities. What is missing in this picture and the resulting equations of motion is a self-consistent treatment of the unstable fluctuations, that must necessarily go beyond one- loop.

5   Domain formation and growth:

The instabilities correspond to the growth of long-wavelength fluctuations and the formation of domains within which the field is correlated. These domains will grow as long as the instabilities persist. In order to understand the process of domain growth when the average value of the field remains zero it proves illuminating to understand the tree level correlations.

The relevant quantity of interest is the *equal time* correlation function

$$S(\vec{r}; t) = \langle \Phi(\vec{r}, t) \Phi(\vec{0}, t) \rangle \tag{40}$$

$$S(\vec{r}; t) = \int \frac{d^3k}{(2\pi)^3} e^{i\vec{k}\cdot\vec{r}} S(\vec{k}; t) \tag{41}$$

$$S(\vec{k}; t) = \langle \Phi_{\vec{k}}(t) \Phi_{-\vec{k}}(t) \rangle = (-iG_{\vec{k}}^{++}(t; t)) \tag{42}$$

where we have performed the Fourier transform in the spatial coordinates (there still is spatial translational and rotational invariance). Notice that at equal times, all the Green's functions are equal, and we may compute any of them.

Clearly in an equilibrium situation this equal time correlation function will be time independent, and will only measure the *static correlations*. In the present case, however, there is a non trivial time evolution arising from the departure from equilibrium of the initial state.

The unstable contribution to the Green function at equal times is given by (31), in this case for $\phi_i = 0$.

It is convenient to introduce the following dimensionless quantities

$$\kappa = \frac{k}{m_f} \;\; ; \;\; L^2 = \frac{m_i^2}{m_f^2} = \frac{[T_i^2 - T_c^2]}{[T_c^2 - T_f^2]} \;\; ; \;\; \tau = m_f t \;\; ; \;\; \vec{x} = m_f \vec{r} \tag{43}$$

Furthermore for the unstable modes $\vec{k}^2 < m_f^2$, and for initial temperatures larger than the critical temperature $T_c^2 = 24\mu^2/\lambda$, we can approximate $\coth[\beta_i\omega_<(k)/2] \approx 2T_i/\omega_<(k)$. Then, at tree-level, the contribution of the unstable modes to the subtracted structure factor (42) $S^{(0)}(k,t) - S^{(0)}(k,0) = (1/m_f)S^{(0)}(\kappa,\tau)$ becomes

$$S^{(0)}(\kappa,\tau) = \left(\frac{24}{\lambda[1 - \frac{T_f^2}{T_c^2}]}\right)^{\frac{1}{2}} \left(\frac{T_i}{T_c}\right) \frac{1}{2\omega_\kappa^2}\left(1 + \frac{\omega_\kappa^2}{W_\kappa^2}\right)[\cosh(2W_\kappa\tau) - 1] \tag{44}$$

$$\omega_\kappa^2 = \kappa^2 + L^2 \tag{45}$$

$$W_\kappa = 1 - \kappa^2 \tag{46}$$

To obtain a better idea of the growth of correlations, it is convenient to introduce the scaled correlation function

$$\mathcal{D}(x,\tau) = \frac{\lambda}{6m_f^2}\int_0^{m_f}\frac{k^2 dk}{2\pi^2}\frac{\sin(kr)}{(kr)}[S(k,t) - S(k,0)] \tag{47}$$

The reason for this is that the minimum of the tree level potential occurs at $\lambda\Phi^2/6m_f^2 = 1$, and the inflexion (spinodal) point, at $\lambda\Phi^2/2m_f^2 = 1$, so that $\mathcal{D}(0,\tau)$ measures the excursion of the fluctuations to the spinodal point and beyond as the correlations grow in time.

At large τ (large times), the product $\kappa^2 S(\kappa,\tau)$ in (47) has a very sharp peak at $\kappa_s = 1/\sqrt{\tau}$. In the region $x < \sqrt{\tau}$ the integral may be done by the saddle point approximation and we obtain for $T_f/T_c \approx 0$ the large time behavior

$$\mathcal{D}(x,\tau) \approx \mathcal{D}(0,\tau)\exp[-\frac{x^2}{8\tau}]\frac{\sin(x/\sqrt{\tau})}{(x/\sqrt{\tau})} \tag{48}$$

$$\mathcal{D}(0,\tau) \approx \left(\frac{\lambda}{12\pi^3}\right)^{\frac{1}{2}}\frac{(\frac{T_i}{2T_c})^3}{[\frac{T_i^2}{T_c^2} - 1]}\frac{\exp[2\tau]}{\tau^{\frac{3}{2}}} \tag{49}$$

Restoring dimensions, and recalling that the zero temperature correlation length is $\xi(0) = 1/\sqrt{2}\mu$, we find that for $T_f \approx 0$ the amplitude of the fluctuation inside a "domain" $\langle\Phi^2(t)\rangle$, and the "size" of a domain $\xi_D(t)$ grow as

$$\langle\Phi^2(t)\rangle \approx \frac{\exp[\sqrt{2}t/\xi(0)]}{(\sqrt{2}t/\xi(0))^{\frac{3}{2}}} \tag{50}$$

$$\xi_D(t) \approx (8\sqrt{2})^{\frac{1}{2}}\xi(0)\sqrt{\frac{t}{\xi(0)}} \tag{51}$$

An important time scale corresponds to the time τ_s at which the fluctuations of the field sample beyond the spinodal point. Roughly speaking when this happens, the instabilities should shut-off as the mean square root fluctuation of the field $\sqrt{\langle \Phi^2(t) \rangle}$ is now probing the stable region. This will be seen explicitly below when we study the evolution non-perturbatively in the Hartree approximation and the fluctuations are incorporated self-consistently in the evolution equations. In zeroth order we estimate this time from the condition $3\mathcal{D}(0,t) = 1$, we use $\lambda = 10^{-12}$; $T_i/T_c = 2$, as representative parameters (this value of the initial temperature does not have any particular physical meaning and was chosen only as representative). We find

$$\tau_s \approx 10.15 \tag{52}$$

or in units of the zero temperature correlation length

$$t \approx 14.2\xi(0) \tag{53}$$

for other values of the parameter τ_s is found from the above condition on (49).

Clearly any perturbative expansion will fail because the propagators will contain unstable wavelengths, and the one loop term will grow faster than the zero order, etc.[17]. We now turn to a non-perturbative analysis (for details see[17]).

As the correlations and fluctuations grow, field configurations start sampling the stable region beyond the spinodal point. This will result in a slow down in the growth of correlations, and eventually the unstable growth will shut-off. When this happens, the state may be described by correlated domains with equal probability for both phases inside the domains. The expectation value of the field in this configuration will be zero, but inside each domain, the field will acquire a value very close to the value in equilibrium at the minimum of the *effective potential*. The size of the domain in this picture will depend on the time during which correlations had grown enough so that fluctuations start sampling beyond the spinodal point.

Since this physical picture may not be studied within perturbation theory, we now introduce a *non-perturbative* method based on a self-consistent Hartree approximation, which is implemented as follows: in the initial Lagrangian write

$$\frac{\lambda}{4!}\Phi^4(\vec{r},t) = \frac{\lambda}{4}\langle\Phi^2(\vec{r},t)\rangle\Phi^2(\vec{r},t) + \left(\frac{\lambda}{4!}\Phi^4(\vec{r},t) - \frac{\lambda}{4}\langle\Phi^2(\vec{r},t)\rangle\Phi^2(\vec{r},t)\right) \tag{54}$$

the first term is absorbed in a shift of the mass term

$$m^2(t) \to m^2(t) + \frac{\lambda}{2}\langle\Phi^2(t)\rangle$$

(where we used spatial translational invariance). The second term in (54) is taken as an interaction with the term $\langle\Phi^2(t)\rangle\Phi^2(\vec{r},t)$ as a "mass counterterm". The Hartree approximation consists of requiring that the one loop correction to the two point Green's functions must be cancelled by the "mass counterterm". This leads to the self consistent set of equations

$$\langle\Phi^2(t)\rangle = \int \frac{d^3k}{(2\pi)^3}\left(-iG_k^<(t,t)\right) = \int \frac{d^3k}{(2\pi)^3}\frac{1}{2\omega_<(k)}\mathcal{U}_k^+(t)\mathcal{U}_k^-(t)\coth[\beta_i\omega_<(k)/2] \tag{55}$$

$$\left[\frac{d^2}{dt^2} + \vec{k}^2 + m^2(t) + \frac{\lambda}{2} \langle \Phi^2(t) \rangle \right] \mathcal{U}_k^{\pm} = 0 \tag{56}$$

Before proceeding any further, we must address the fact that the composite operator $\langle \Phi^2(\vec{r}, t) \rangle$ needs one subtraction and multiplicative renormalization. As usual the subtraction is absorbed in a renormalization of the bare mass, and the multiplicative renormalization into a renormalization of the coupling constant.

At this stage our justification for using this approximation is based on the fact that it provides a non-perturbative framework to sum an infinite series of Feynman diagrams of the cactus type[8, 36]. Alternatively, it is equivalent to introduce a N-component scalar field and take the infinite N limit.

It is clear that for $t < 0$ there is a self-consistent solution to the Hartree equations with equation (55) and

$$\begin{aligned}
\langle \Phi^2(t) \rangle &= \langle \Phi^2(0^-) \rangle \\
\mathcal{U}_k^{\pm} &= \exp[\mp i\omega_<(k)] \\
\omega_<^2(k) &= \vec{k}^2 + m_i^2 + \frac{\lambda}{2} + \langle \Phi^2(0^-) \rangle = \vec{k}^2 + m_{i,R}^2
\end{aligned} \tag{57}$$

$$\tag{58}$$

where the composite operator has been absorbed in a renormalization of the initial mass, which is now parametrized as $m_{i,R}^2 = \mu_R^2[(T_i^2/T_c^2) - 1]$. For $t > 0$ we subtract the composite operator at $t = 0$ absorbing the subtraction into a renormalization of m_f^2 which we now parametrize as $m_{f,R}^2 = \mu_R^2[1 - (T_f^2/T_c^2)]$. We should point out that this choice of parametrization only represents a choice of the bare parameters, which can always be chosen to satisfy this condition. The logarithmic multiplicative divergence of the composite operator will be absorbed in a coupling constant renormalization consistent with the Hartree approximation[36, 37], however, for the purpose of understanding the dynamics of growth of instabilities associated with the long-wavelength fluctuations, we will not need to specify this procedure. After this subtraction procedure, the Hartree equations read

$$[\langle \Phi^2(t) \rangle - \langle \Phi^2(0) \rangle] = \int \frac{d^3k}{(2\pi)^3} \frac{1}{2\omega_<(k)} [\mathcal{U}_k^+(t) \mathcal{U}_k^-(t) - 1] \coth[\beta_i \omega_<(k)/2] \tag{59}$$

$$\left[\frac{d^2}{dt^2} + \vec{k}^2 + m_R^2(t) + \frac{\lambda_R}{2} \left(\langle \Phi^2(t) \rangle - \langle \Phi^2(0) \rangle \right) \right] \mathcal{U}_k^{\pm}(t) = 0 \tag{60}$$

$$m_R^2(t) = \mu_R^2 \left[\frac{T_i^2}{T_c^2} - 1 \right] \Theta(-t) - \mu_R^2 \left[1 - \frac{T_f^2}{T_c^2} \right] \Theta(t) \tag{61}$$

with $T_i > T_c$ and $T_f \ll T_c$. With the self-consistent solution and boundary condition for $t < 0$

$$\begin{aligned}
[\langle \Phi^2(t < 0) \rangle - \langle \Phi^2(0) \rangle] &= 0 \tag{62} \\
\mathcal{U}_k^{\pm}(t < 0) &= \exp[\mp i\omega_<(k)t] \tag{63} \\
\omega_<(k) &= \sqrt{\vec{k}^2 + m_{iR}^2} \tag{64}
\end{aligned}$$

This set of Hartree equations is extremely complicated to be solved exactly. However it has the correct physics in it. Consider the equations for $t > 0$, at very early times, when (the renormalized) $\langle \Phi^2(t) \rangle - \langle \Phi^2(0) \rangle \approx 0$ the mode functions are the same as in the zeroth order approximation, and the unstable modes grow exponentially. By computing the expression (59) self-consistently with these zero-order unstable modes, we see that the fluctuation operator begins to grow exponentially.

As $(\langle \Phi^2(t) \rangle - \langle \Phi^2(0) \rangle)$ grows larger, its contribution to the Hartree equation tends to balance the negative mass term, thus weakening the instabilities, so that only longer wavelengths can become unstable. Even for very weak coupling constants, the exponentially growing modes make the Hartree term in the equation of motion for the mode functions become large and compensate for the negative mass term. Thus when

$$\frac{\lambda_R}{2m_{f,R}^2} \left(\langle \Phi^2(t) \rangle - \langle \Phi^2(0) \rangle \right) \approx 1$$

the instabilities shut-off, this equality determines the "spinodal time" t_s. The modes will still continue to grow further after this point because the time derivatives are fairly (exponentially) large, but eventually the growth will slow-down when fluctuations sample deep inside the stable region.

After the subtraction, and multiplicative renormalization (absorbed in a coupling constant renormalization), the composite operator is finite. The stable mode functions will make a *perturbative* contribution to the fluctuation which will be always bounded in time. The most important contribution will be that of the *unstable modes*. These will grow exponentially at early times and their effect will dominate the dynamics of growth and formation of correlated domains. The full set of Hartree equations is extremely difficult to solve, even numerically, so we will restrict ourselves to account *only* for the unstable modes. From the above discussion it should be clear that these are the only relevant modes for the dynamics of formation and growth of domains, whereas the stable modes, will always contribute perturbatively for weak coupling after renormalization.

Introducing the dimensionless ratios (43) in terms of $m_{f,R}$; $m_{i,R}$, (all momenta are now expressed in units of $m_{f,R}$), dividing (60) by $m_{f,R}^2$, using the high temperature approximation $\coth[\beta_i \omega_<(k)/2] \approx 2T_i/\omega_<(k)$ for the unstable modes, and expressing the critical temperature as $T_c^2 = 24\mu_R/\lambda_R$, the set of Hartree equations (59, 60) become the following integro-differential equation for the mode functions for $t > 0$

$$\left[\frac{d^2}{d\tau^2} + q^2 - 1 + g \int_0^1 dp \left\{ \frac{p^2}{p^2 + L_R^2} [\mathcal{U}_p^+(t)\mathcal{U}_p^-(t) - 1] \right\} \right] \mathcal{U}_q^\pm(t) = 0 \qquad (65)$$

with

$$\mathcal{U}_q^\pm(t < 0) = \exp[\mp i\omega_<(q)t] \qquad (66)$$

$$\omega_<(q) = \sqrt{q^2 + L_R^2} \qquad (67)$$

$$L_R^2 = \frac{m_{i,R}^2}{m_{f,R}^2} = \frac{[T_i^2 - T_c^2]}{[T_c^2 - T_f^2]} \qquad (68)$$

$$g = \frac{\sqrt{24\lambda_R}}{4\pi^2} \frac{T_i}{[T_c^2 - T_f^2]^{\frac{1}{2}}} \tag{69}$$

The effective coupling (69) reflects the enhancement of quantum fluctuations by high temperature effects; for $T_f/T_c \approx 0$, and for couplings as weak as $\lambda_R \approx 10^{-12}$, $g \approx 10^{-7}(T_i/T_c)$.

The equations (65) may now be integrated numerically for the mode functions; once we find these, we can then compute the contribution of the unstable modes to the subtracted correlation function equivalent to (47)

$$\mathcal{D}^{(HF)}(x,\tau) = \frac{\lambda_R}{6m_{f,R}^2}\left[\langle\Phi(\vec{r},t)\Phi(\vec{0},t)\rangle - \langle\Phi(\vec{r},0)\Phi(\vec{0},0)\rangle\right] \tag{70}$$

$$3\mathcal{D}^{(HF)}(x,\tau) = g\int_0^1 dp\left(\frac{p^2}{p^2 + L_R^2}\right)\frac{\sin(px)}{(px)}\left[\mathcal{U}_p^+(t)\mathcal{U}_p^-(t) - 1\right] \tag{71}$$

In figure (3) we show

$$\frac{\lambda_R}{2m_{f,R}^2}(\langle\Phi^2(\tau)\rangle - \langle\Phi^2(0)\rangle) = 3(\mathcal{D}^{HF}(0,\tau) - \mathcal{D}^{HF}(0,0))$$

(solid line) and also for comparison, its zeroth-order counterpart $3(\mathcal{D}^{(0)}(0,\tau) - \mathcal{D}^{(0)}(0,0))$ (dashed line) for $\lambda_R = 10^{-12}$, $T_i/T_c = 2$. (this value of the initial temperature does not have any particular physical significance and was chosen as a representative). We clearly see what we expected; whereas the zeroth order correlation grows indefinitely, the Hartree correlation function is bounded in time and oscillatory. At $\tau \approx 10.52$, $3(\mathcal{D}^{(HF)}(0,\tau) - \mathcal{D}^{(HF)}(0,\tau)) = 1$, fluctuations are sampling field configurations near the classical spinodal, fluctuations continue to grow, however, because the derivatives are still fairly large. However, after this time, the modes begin to probe the stable region in which there is no exponential growth. At this point $\frac{\lambda_R}{2m_{f,R}^2}(\langle\Phi^2(\tau)\rangle - \Phi^2(0))$, becomes small again because of the small coupling $g \approx 10^{-7}$, and the correction term becomes small. When it becomes smaller than one, the instabilities set in again, modes begin to grow and the process repeats. This gives rise to an oscillatory behavior around $\frac{\lambda_R}{2m_{f,R}^2}(\langle\Phi^2(\tau)\rangle - \Phi^2(0)) = 1$ as shown in figure (3). In figure (4), we show the structure factors as a function of x for $\tau = 10$, both for zero-order (tree level) $\mathcal{D}^{(0)}$ (dashed lines) and Hartree $\mathcal{D}^{(HF)}$ (solid lines) for the same value of the parameters as for figure (3). These correlation functions clearly show the growth in amplitude and that the size of the region in which the fields are correlated increases with time. Clearly this region may be interpreted as a "domain", inside which the fields have strong correlations, and outside which the fields are uncorrelated.

We see that up to the spinodal time $\tau_s \approx 10.52$ at which $\frac{\lambda_R}{2m_{f,R}^2}(\langle\Phi^2(\tau_s)\rangle - \Phi^2(0)) = 1$, the zeroth order correlation $3\mathcal{D}^{(0)}(0,\tau)$ is very close to the Hartree result. In fact at τ_s, the difference is less than 15%. For these values of the coupling and initial temperature, the zeroth order correlation function leads to $t_s \approx 10.15$, and we may use the zeroth order correlations to provide an analytic estimate for t_s, as well as the form of the correlation functions and the size of the domains. The fact that the zeroth-order correlation remains very close to the Hartree-corrected correlations up to times comparable to the spinodal is a consequence of the very small coupling. The stronger coupling makes the growth of domains much faster and the departure from tree-level correlations more dramatic[17]. For strong couplings domains will form very rapidly and only grow to sizes of the order of the zero temperature correlation length. The phase transition will occur very rapidly, and our initial assumption of a rapid supercooling will be unjustified. This situation for strong couplings, of domains forming very rapidly to sizes of the order of the zero temperature correlation length is the picture presented by Mazenko and collaborators[14]. However, for very weak couplings (consistent with the bounds from density fluctuations), our results indicate that the phase transition will proceed very slowly, domains will grow for a long time and become fairly large, with a typical size several times the zero temperature correlation length. In a sense, this is a self consistent check of our initial assumptions on a rapid supercooling in the case of weak couplings.

As we argued above, for very weak coupling we may use the tree level result to give an approximate bound to the correlation functions up to times close to the spinodal time using the result given by equation (49), for $T_f \approx 0$. Thus, we conclude that for large times, and very weakly coupled theories ($\lambda_R \leq 10^{-12}$) and for initial temperatures of the order of the critical temperature, the size of the domains $\xi_D(t)$ will grow typically in time as

$$\xi_D(t) \approx (8\sqrt{2})^{\frac{1}{2}}\xi(0)\sqrt{\frac{t}{\xi(0)}} \qquad (72)$$

with $\xi(0)$ the zero temperature correlation length. The maximum size of a domain is approximately determined by the time at which fluctuations begin probing the stable region, this is the spinodal time t_s and the maximum size of the domains is approximately $\xi_D(t_s)$.

An estimate for the spinodal time, is obtained from equation (49) by the condition $3\mathcal{D}(\tau_s) = 1$. For weakly coupled theories and $T_f \approx 0$, we obtain

$$\tau_s = \frac{t_s}{\sqrt{2}\xi(0)} \approx -\ln\left[\left(\frac{3\lambda}{4\pi^3}\right)^{\frac{1}{2}}\left(\frac{(\frac{T_i}{2T_c})^3}{[\frac{T_i^2}{T_c^2}-1]}\right)\right] \qquad (73)$$

6 FRW Cosmologies:

Following the same method, we obtained the evolution equations for the order parameter in the case of a spatially flat FRW cosmology. We used as an initial condition that the ensemble at an early time $t_o \to -\infty$ describes a state of local thermodynamic equilibrium at a temperature $T_o = 1/\beta_o$ for the adiabatic modes[38]. We find

$$\ddot{\phi} + 3\frac{\dot{a}}{a}\dot{\phi} + V'(\phi) + \lambda\phi\frac{\hbar}{2}\int\frac{d^3k}{(2\pi)^3}\frac{[\mathcal{U}_{1k}^2(t)+\mathcal{U}_{1k}^2(t)]}{2a^3(t)W_k(t_o)}\coth\left[\beta_o\hbar W_k(t_o)/2\right] = 0 \qquad (74)$$

where the mode functions $\mathcal{U}_{\alpha,k}(t)$; $\alpha = 1, 2$ satisfy

$$\ddot{\mathcal{U}}_{\alpha k} - \frac{3}{2}\left(\frac{\ddot{a}}{a} + \frac{1}{2}\frac{\dot{a}^2}{a^2}\right)\mathcal{U}_{\alpha k} + \left(\frac{\vec{k}^2}{a^2(t)} + V''(\phi_{cl}(t))\right)\mathcal{U}_{\alpha k} = 0 \tag{75}$$

$$\mathcal{U}_{1k}(t_o) = 1 \ ; \ \mathcal{U}_{2k}(t_o) = 0 \tag{76}$$

$$\dot{\mathcal{U}}_{1k}(t_o) = \frac{3}{2}\frac{\dot{a}(t_o)}{a(t_o)} \ ; \ \dot{\mathcal{U}}_{2k}(t_o) = W_k(t_o) = \left[\frac{\vec{k}^2}{a^2(t_o)} + V''(\phi_{cl}(t_o))\right]^{\frac{1}{2}} \tag{77}$$

where $\phi_{cl}(t)$ is the solution to the *classical* equations of motion. Renormalization forces the introduction of a coupling to the Ricci scalar and the tree level potential considered is thus[38]

$$V(\phi) = \frac{1}{2}\left[m^2 + \xi\mathcal{R}\right]\phi^2 + \frac{\lambda}{4!}\phi^4 \tag{78}$$

We can see clearly that the time dependence of the mode functions is very important, at no times a notion of the effective potential will be valid for the time evolution of the order parameter.

Invoking a Hartree factorization as in a previous section, we also derived the equations of motion in a self-consistent non-perturbative framework. Introducing the fluctuation $\eta(\vec{x}, t) = \Phi(\vec{x}, t) - \phi(t)$ and the Hartree "frequencies"

$$\mathcal{V}^{(2)}(\phi) = V''(\phi) + \frac{\lambda}{2}\langle\eta^2\rangle \tag{79}$$

we find

$$\ddot{\phi} + 3\frac{\dot{a}}{a}\dot{\phi} + V'(\phi) + \lambda\phi\frac{\hbar}{2}\int\frac{d^3k}{(2\pi)^3}\hbar\frac{[\mathcal{U}_{1k}^2(t) + \mathcal{U}_{1k}^2(t)]}{2a^3(t)\bar{W}_k(t_o)}\coth\left[\beta_o\hbar\bar{W}_k(t_o)/2\right] = 0 \tag{80}$$

$$\left[\frac{d^2}{dt^2} - \frac{3}{2}\left(\frac{\ddot{a}}{a} + \frac{1}{2}\frac{\dot{a}^2}{a^2}\right) + \frac{\vec{k}^2}{a^2(t)} + V''(\phi(t)) + \right.$$
$$\left.\lambda\frac{\hbar}{2}\int\frac{d^3k}{(2\pi)^3}\frac{[\mathcal{U}_{1k}^2(t) + \mathcal{U}_{1k}^2(t)]}{2a^3(t)\bar{W}_k(t_o)}\coth\left[\beta_o\hbar\bar{W}_k(t_o)/2\right]\right]\mathcal{U}_{\alpha k} = 0 \tag{81}$$

$$\mathcal{U}_{1k}(t_o) = 1 \ ; \ \mathcal{U}_{2k}(t_o) = 0 \tag{82}$$

$$\dot{\mathcal{U}}_{1k}(t_o) = \frac{3}{2}\frac{\dot{a}(t_o)}{a(t_o)} \ ; \ \dot{\mathcal{U}}_{2k}(t_o) = \bar{W}_k(t_o) = \left[\frac{\vec{k}^2}{a^2(t_o)} + \mathcal{V}^{(2)}(\phi(t_o))\right]^{\frac{1}{2}} \tag{83}$$

These equations present new features: first the "friction" term arising from the expansion, and secondly the fact that physical wave-vectors are red-shifted. The red-shift will tend to enhance the instabilities as more wavelengths become unstable, but the presence of the horizon will ultimately constrain the final size of the domains. We are currently studying the numerical evolutions of these equations and expect to report soon on details of the dynamics of the phase transition in these cosmologies[38].

7 Conclusions:

We provided a dynamical picture of the time evolution of a weakly coupled inflaton scalar field theory undergoing a typical second order phase transition in Minkowski space-time. The shortcomings of the usual approach based on the equilibrium effective and the necessity for a description out of equilibrium were pointed out. The non-equilibrium evolution equations for the order parameter were derived to one-loop and integrated for "slow-rollover" initial conditions. We pointed out that the instabilities responsible for the onset of the phase transition and the process of domain formation, that is the growth of long-wavelength fluctuations, introduced dramatic corrections to the "slow-rollover" picture. The net effect of these instabilities is to *slow even further* the evolution of the order parameter (average of the scalar field) and for "slow rollover" initial conditions (the scalar field very near the false vacuum initially) the dynamics becomes *non-perturbative*.

We introduced a self-consistent approach to study the dynamics out of equilibrium in this situation. It is found that for very weakly coupled theories domains growth to large sizes typically several times the zero temperature correlation length and that the growth obeys a scaling law at intermediate times.

In strongly coupled theories the phase transition will occur much faster and domains will only grow to sizes of the order of the zero temperature correlation length.

The one-loop equations of evolution for the scalar order parameter were obtained in spatially flat FRW cosmologies. These equations reveal that at no times is the approximation of an effective potential valid. We have also obtained the evolution equations in a non-perturbative self-consistent Hartree approximation. These equations present new features that illuminate the fact that the process of phase separation in cosmological settings is more subtle.

The physical wave vectors undergo a red-shift and more wavelengths become unstable, thus enhancing the instability. On the other hand the presence of the horizon and the "friction" term in the equations of motion, prevent domains from growing bigger than the horizon size. Eventually there is a competition of time scales that must be understood deeply to obtain any meaningful conclusion about formation and growth of domains in FRW cosmologies. Work on these issues is in progress[38].

Acknowledgments:

D. B. would like to thank R. Holman, D-S. Lee, and A. Singh for fruitful discussions and remarks, LPTHE at the Universite de Pierre et Marie Curie for hospitality during part of this work, and the N.S.F. for support through Grant: PHY-9302534 and for a Binational Collaboration supported by N.S.F. through Grant: INT-9216755.

References

[*] Laboratoire Associé au CNRS UA280.

[1] For an introductory review on the subject, see for example L. Abbott and S.-Y. Pi, Inflationary Cosmology (World Scientific, 1986).

[2] A. H. Guth, Phys. Rev. D15, 347 (1981); A. H. Guth and E. J. Weinberg, Nucl. Phys. B212, 321 (1983).

[3] A. H. Guth, Phys. Rev. D23, 347 (1981).

[4] A. D. Linde, Phys. Lett. B108, 389 (1982).

[5] A. Albrecht and P. J. Steinhardt, Phys. Rev. Lett. 48, 1220 (1982).

[6] For a review see: A. D. Linde, Rep. Prog. Phys. 47, 925 (1984); P. J. Steinhardt, Comments Nucl. Part. Phys. 12, 273 (1984).

[7] D. A. Kirzhnits and A. D. Linde, Phys. Lett. B42, 471, (1972).

[8] L. Dolan and R. Jackiw, Phys. Rev. D9, 3320 (1974).

[9] S. Weinberg, Phys. Rev. D9, 3357 (1974).

[10] E. W. Kolb and M. S. Turner, "The Early Universe", Addison Wesley (Frontiers in Physics) (1990)

[11] See for example A. Linde, Particle Physics and Inflationary Cosmology, Harwood Academic Publishers (1990), and references therein.

[12] R. H. Brandenberger, Rev. of Mod. Phys. 57, 1 (1985); Int. J. Mod. Phys. 2A, 77 (1987).

[13] A. A. Starobinsky, Phys. Lett. 117B, 175 (1982);, 1110 (1982); J. M. Bardeen, P. J. Steinhardt and M. S. Turner, Phys. Rev. D28, 679 (1983); R. Brandenberger, R. Kahn and W. H. Press, Phys. Rev. D28, 1809 (1983).

[14] G. F. Mazenko, W. G. Unruh and R. M. Wald, Phys. Rev. D31, 273 (1985); G. F. Mazenko, Phys. Rev. Lett. 54, 2163 (1985).

[15] A. Guth and S-Y. Pi, Phys. Rev. D32, 1899 (1985).

[16] E. J. Weinberg and A. Wu, Phys. Rev. D36, 2474 (1987).

[17] D. Boyanovsky, D-S Lee and A. Singh, (to appear in Phys. Rev. D. 1993).

[18] D. Boyanovsky and H. J. de Vega, Phys. Rev. D47, 2343 (1993).

[19] D. Boyanovsky, "Quantum Spinodal Decomposition" (to appear in Phys. Rev. E).

[20] S. W. Hawking, Phys. Lett. B 115, 295 (1982); A. A. Starobinsky, Phys. Lett. B 117, 175, (1982); A. H. Guth and S. Y. Pi, Phys. Rev. Lett. 49, 1110 (1982).

[21] J. S. Langer in "Fluctuations, Instabilities and Phase Transitions) (T. Riste Ed.) page 19, Plenum N.Y. 1975.

[22] J. D. Gunton, M. San Miguel and P.S. Sahni in "Phase Transitions and Critical Phenomena" (C. Domb and J. J. Lebowitz, Eds.) Vol 8, Academic Press, (1983).

[23] A. Niemi and G. Semenoff, Ann. of Phys. (N.Y.) 152, 105 (1984); Nucl. Phys. B [FS10], 181, (1984).

[24] E. Calzetta, Ann. of Phys. (N.Y.) 190, 32 (1989)

[25] E. Calzetta and B. L. Hu, Phys. Rev. D35, 495 (1987); Phys. Rev. D37, 2878 (1988).

[26] J. Schwinger, J. Math. Phys. 2, 407 (1961).

[27] L. V. Keldysh, Sov. Phys. JETP 20, 1018 (1965).

[28] R. Mills, "Propagators for Many Particle Systems" (Gordon and Breach, N. Y. 1969).

[29] N. P. Landsman and C. G. van Weert, Phys. Rep. 145, 141 (1987).

[30] G. Semenoff and N. Weiss, Phys. Rev. D31, 689; 699 (1985).

[31] R. D. Jordan, Phys. Rev. D33, 444 (1986).

[32] R. L. Kobes and K. L. Kowalski, Phys. Rev. D34, 513 (1986); R. L. Kobes, G. W. Semenoff and N. Weiss, Z. Phys. C 29, 371 (1985).

[33] J. P. Paz, Phys. Rev. D41, 1054 (1990); Phys. Rev. D42, 529 (1990).

[34] L. P. Kadanoff and G. Baym, Quantum Statistical Mechanics (Benjamin, N.Y. 1962).

[35] This may be done formally by generalizing the method proposed by S. Weinberg[9] and introduce a time dependent counterterm in the original Hamiltonian of the form $\frac{\lambda}{24}\left(T_i^2\theta(-t) + T_f^2\theta(t)\right)$ and cancelling this counterterm against the one loop contribution for $t < 0$.

[36] For a treatment of a scalar field theory in the Hartree approximation see also S.J. Chang, Phys. Rev. D 12, 1071 (1975).

[37] An alternative renormalization scheme for the effective action is provided by J. Avan and H. J. de Vega, Phys. Rev. D 29, 2891, 2904 (1984).

[38] D. Boyanovsky, H. J. de Vega and R. Holman, "Phase Transitions out of Equilibrium in FRW Cosmologies" (in preparation).

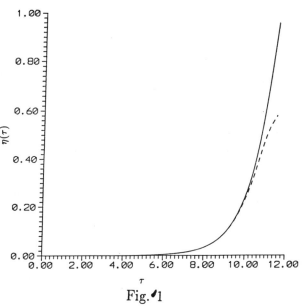

Fig. 1

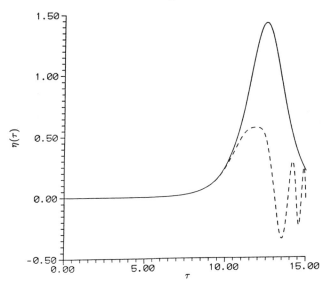

Fig. 2

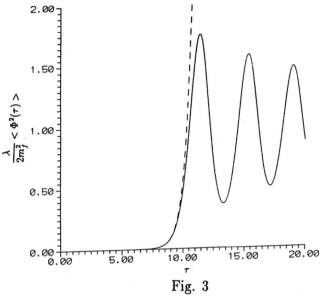

Fig. 3

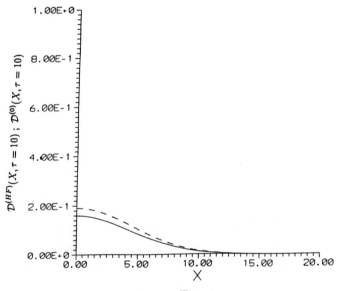

Fig. 4

Some Physical Analyses in Relativistic Heavy Ion Collisions[*]

Chao(Zhao) Wei-Qin

Center of Theoret. Physics, CCAST(World Lab.) Beijing, China

Institute of High Energy Physics, Academia Sinica

P.O.Box 918(4), Beijing 100039, China[†]

Abstract

Some physical analyses in relativistic heavy ion collisions are summarized. Special attention is paid to the energy density analysis and some phenomena which may be related to QGP formation, such as Bose-Einstein correlation, J/ψ suppression and strangeness enhancement.

Invited lecture presented at "International School for Cosmology and Gravitation"
13th course: Cosmology and Particle Physics
Ettore Majorana Center for Scientific Culture, Erice, Italy, 3-14 May, 1993.

[*]Work partly supported by National Natural Science Foundation of China

[†]Mailing address

V. de Sabbata and H. Tso-Hsiu (eds.), Cosmology and Particle Physics, 49–65.

I. Introduction

There is a well known puzzle in the nature: many experiments have shown the evidences of the existence of quarks and gluons. However, no free quarks or gluons have been observed experimentally up to now. They are confined in hadrons. Is it possible to break the hadron bags and let quarks and gluons move freely in a larger space region? It has been predicted based on Lattice calculation that a phase transition from hadronic matter to quark-gluon phase would happen at critical temperature or density.

Two simple estimations of these critical conditions are given in the following. The obtained values are fairly close to the results based on more sophisticated Lattice calculations.

Imagining that hadrons are like bags. If one compresses them to a high enough density the hadron bags would be broken, and quarks and gluons would move freely in the space region containing these broken bags. The radii of a nucleus and a nucleon are, respectively, $R_A = r_0 A^{1/3}$, $r_0 \sim 1.2 \ fm$ and $r_N \sim 0.8 \ fm$. The critical density can be estimated by taking $\rho_c = \rho_N$, which gives

$$\rho_c = (r_0/r_N)^3 \cdot \rho_A \sim 4 \ \rho_A \ . \tag{1}$$

When nuclear matter is compressed to a density $\rho \geq \rho_c$ the nucleon bags start overlapping and the system reaches the critical point to allow quarks and gluons to move freely in it.

On the other hand, if one heats the dilute hadronic matter up many pions will be produced, like in boiled water. When pions are so crowded that the pion bags overlap, quarks and gluons would move freely in the whole region. For ideal pion gas the density can be expressed as

$$\rho_\pi = 3/2\pi^2 \ \cdot \ T^3 \cdot I \ , \tag{2}$$

where $I \sim 2.4$ is an integral. When the system is heated up and reaches the point $\rho_\pi = \rho_c = 1/(\frac{4}{3}\pi\, r_\pi^3)$ which gives the critical temperature T_c the pion bags start overlapping and break up. Taking $r_\pi \sim 0.7fm$ one obtains $T_c \sim 250MeV$.

Although the above estimations are very simple the results are not far from the predictions based on Lattice calculation. With zero baryon number Lattice calculation gives the critical temperature of $T_c \sim 150 - 200\ MeV$ or the critical energy density of $\epsilon_c \sim 1 - 3\ GeV/fm^3$ for the phase transition from hadronic matter to quark-gluon plasma (QGP) phase. To have reliable predictions of the critical conditions huge computers are designed to carry on Lattice calculations with larger Lattice size, to include different number of flavors and to include non-zero baryon number in it.

How to reach the critical condition and really form QGP? It there the possibility for QGP to exist in the nature? These questions are closely related to the topic of this school: cosmology.

First, in the early universe shortly after the big bang its temperature is very high, all elementary particles (e.g., leptons, quarks, gauge bosons) are in equilibrium. No hadrons can exist. The universe expands continuously and its temperature goes down crossing various critical points. The one we are specially interested in is at $T_c \sim 200\ MeV$ when the phase transition from QGP to hadronic matter happens. At this very early period when γ-radiation dominates, the energy density drops as $1/R^3 \cdot 1/R$, where R corresponds to the size of the system. Here $1/R^3$ is due to the volume expansion, while the additional $1/R$ comes from the red shift caused by the expansion. Based on the energy-momentum conservation law and the Hubbe's law one obtains the temperature

$$T \sim 10^{10}/\sqrt{t}\ ^0K\ ,\tag{3}$$

where t is the time interval after the big bang. Based on eq.(3) one obtains a time scale of 10^{-5} sec after the big bang when the universe cools down to $T_c \sim 200\ MeV$ and experiences the phase transition from QGP to hadronic matter. Unfortunately, it happened so long ago that not much information is kept for our analysis now.

Another possibility is the existence of high density objects in our universe. One candidate is the neutron star with density at its core of the order of $10^{16}\ g/cm^3$ which is considerably higher than the density in a nucleon and enough for breaking the nuclear bags. However, neutron stars are too for way for detailed analysis.

People are looking for the possibility to realize the QGP - phase transition in Lab. Various analyses show that relativistic heavy-ion collisions may reach the required critical conditions: a high enough energy density (or temperature) in a relatively large space region, lasting for a long enough time period for the formation of QGP.

Stimulated by the theoretical prediction people summarized their experiences on heavy ion experiments at energies around several GeV/N and started to build new accelerators with energies one or two order of magnitude higher. Two relativistic heavy ion accelerators started operating in 1986. At CERN ^{16}O and ^{32}S are accelerated up to $200 GeV/N$ by SPS and collided to different fixed nuclear targets. ^{16}O and ^{28}Si at energy of 14.5 GeV/N are used to bombard various nuclei at AGS of BNL. In this year Pb and Au will be accelerated to the energy of 170 GeV/N and 10 GeV/N by SPS and AGS, respectively. In 1997 a relativistic heavy ion collider (RHIC) at BNL will accomplish $Au + Au$ colliding at c.m. energy of $\sqrt{S_{NN}} = 200\ GeV/N$. When the large hadron collider (LHC) is constructed at the end of this century at CERN $Pb + Pb$ colliding at $\sqrt{S_{NN}} = 6.3\ TeV/N$ would become possible.

There are mainly two kinds of works going on in this field:

Many works are to analyze whether relativistic heavy ion collisions can reach the predicted phase transition conditions. For example, to estimate the reached energy density, to follow the space-time development of the colliding system finding out whether the volume and life-time are enough for approaching equilibrium. Theoretically various methods are applied in the analyses. In the hadronic phase, the geometric multiple collision models and the hydrodynamic models are both widely applied. The Monte Carlo simulations of the collision processes based on transport theory are also developed to follow the space-time development of each event. At the same time, considering collisions between partons, models based on the above mentioned physical pictures have also started to develop in the quark-gluon phase. Various global observables are evaluated based on these models and compared to experimental data. These analyses not only provide the test to different models, but also give us reliable knowledge about the space-time development of the energy density, the volume and lifetime of the system and tell us whether the critical conditions for QGP phase transition are fulfilled during the collision.

Another kind of works is concentrated on the QGP signals. Even if QGP is formed at the early stage of relativistic heavy ion collisions how can one be sure of it by measuring the final observables? Various possible signals of QGP formation have been predicted theoretically. Some of them have been confirmed by experimental data. However, other explanations based on a dense and hot hadronic phase seem to work equally well. To finally confirm or rule out any signals more complete data sets, including reference data of $p - p$ and $h - A$ processes, are needed. Besides, unified explanation of all data is really a challenge to any proposed theoretical suggestions.

II. The conditions reached in relativistic heavy ion collisions

Here is a general picture of relativistic heavy ion collisions: Two Lorentz contracted nuclei collide at an impact parameter b. The nucleons within the overlapping region of the two colliding nuclei are called "participant nucleons" which are strongly excited due to the violent collisions, while those outside this region are passing by and not very much influenced by the collision process. After the collision many particles are produced, most of which are pions. The transverse motion of these produced particles reflects the excitation of the system. On the longitudinal direction the information of the original parton's motion is still partly kept. Therefore, the descriptions of the two different directions are usually based on different variables.

2.1 Global observables

The following global observables are the most often applied ones:

* multiplicity n, charged one n_c, negatively charged one n_-,

* energy ϵ_i measured in a small solid angle cell i at the polar angle θ_i,

* transverse energy $E_T = \sum_i \epsilon_i \, sin\theta_i$ where the summation is performed in a certain observation area. Experimentally, one finds that the average transverse energy $< E_T >$ is well proportional to the average multiplicity $< n >$. This indicates that the transverse excitation of each produced particle seems in the same level.

* zero-degree energy E_{ZD} measured by a zero-degree calorimenter set in the forward direction with an open angle $\Delta\theta \sim 0.3^0$. This energy measures the unaffected part of the incident energy.

For individual particle, except for its mass and charge, its momentum is expressed in its two components:

* transverse momentum $p_\perp$ (correspondingly, the transverse mass $m_\perp = \sqrt{p_\perp^2 + m^2}$)

* rapidity y which is related to the longitudinal momentum as

$$y = \frac{1}{2} \, ln \, \frac{E + p_\parallel}{E - p_\parallel} \ . \tag{4}$$

Rapidity is widely applied for its special properties. Under a Lorentz transformation y simply shifts by $\Delta = \frac{1}{2} \, ln \, \frac{1+\beta}{1-\beta}$, where β is the velocity between the two Lorentz frames. For small y, it behaviors like velocity, while for large y, when the mass of the particle can be neglected, $y \sim \eta = -ln \, tg \, \theta/2$, where the pseudorapidity η is related to the angle θ, a quantity easily to be measured.

2.2 Energy density ϵ

Now let us look at the energy density reached in relativistic heavy ion collisions. There is a widely applied formula by Bjorken[1] to estimate the energy density at a certain time τ_0:

$$\epsilon_{Bj} = \frac{1}{\pi R^2 \tau_0}\left(\frac{dE_T}{dy}\right)_{max} , \tag{5}$$

where R is the radius of the smaller one of the two colliding nuclei. This formula is based on a very simple idea that ϵ is the ratio of the excitation energy dE_T measured in a small volume to the size of the volume $\pi R^2 \tau_0 \, dy$. Using the Stefan-Boltzmann relation the temperature at τ_0 can also be estimated from

$$\epsilon = (47.5\pi^2/30)T^4 , \tag{6}$$

where 47.5 is the total number of degree of freedom in the QGP phase. The estimations based on eqs.(5) and (6) show that for the two presently operating accelerators, namely AGS at BNL and SPS at CERN, the reached ϵ and T are already enough to realize the critical values for QGP formation, predicted by Lattice calculation. However, when c.m. energy increases, ϵ increases rather slowly. For an increase of 2-3 orders of magnitude of c.m. energy, ϵ increases only up to~one order of magnitude[2].

Related to the analysis of the reached energy density the following points are of interests:

2.3 Collision centrality

To reach a high energy density it is important to choose central events with small impact parameter b. Different models have been proposed to provide quantitative criteria of the collision centrality based on geometric picture. The total observed transverse energy E_T (or multiplicity n) is provided by many independent sources. The distribution of the source number can be evaluated based on the geometric picture. The individual sources could be participant nucleons, or individual N-N collisions, or individual N-A collisions, depending on the model assumptions. All these models give satisfactory overall explanation of the data when choosing the fitting parameter properly. In this sense they provide a kind of criteria for choosing central events. However, they have little predicting ability except for some extrapolation to higher incident energy region in terms of phenomenological parameters. Based on the participant-spectator picture it is assumed in our analysis[3] that the transverse energy E_T is mainly provided by decaying products of the participant nucleons, while the spectator nucleons are almost not effected and go through providing the energy measured in zero-degree calorimeter, E_{ZD}. Data show clearly an anti-correlation of E_T and E_{ZD}. Experimentally, a high E_T-cut or a low E_{ZD}- cut are usually applied to choose central events.

2.4 Collision time

Now, I would like to give a comment on Bjorken's formula. There is one assumption behind it: the two Lorentz contracted nuclei are very thin, so the $A - A$ collision is supposed to be completed at once. Then the system starts to expand from a thickness $\Delta z = 0$. However, the Lorentz contracted nuclei still have a finite thickness which gives a finite collision time. In fact, the system starts to expand and cooling during the collision. Therefore, the real energy density never reaches Bjorken's estimation. Our Monte Carlo simulation based on a hadronic cascading picture shows the reached $\epsilon \sim \frac{1}{2}\epsilon_{Bj}$[4]. Fortunately, when c.m. energy $\sqrt{s}$ increases, the colliding nuclei become thinner due to a stronger Lorentz contraction. So, with decreasing collision time the reached $\epsilon \to \epsilon_{Bj}$.

2.5 Stopping power

The next effect which influences the reached ϵ is the stopping power. It determines the energy fraction stored in the collision region. There are two interesting aspects about stopping power.

First, is there enough energy stored during the collision to form a highly excited system? Data at SPS and AGS show that the nuclear stopping power changes with the incident energy considerably. At AGS energy ($\sim$ 14.5 GeV/N) the E_T- distributions saturate when the target is heavier than Ag. This means that the incident energy has completely stopped in Ag target. Even the target becomes heavier there is no more energy available to excite additional nucleons and provide more E_T. While at SPS the incident energy is much higher ($\sim 200\ GeV/N$), and the data of E_T-distributions do not show such saturation up to the heaviest target. When we look at the E_{ZD}-distribution for the central collision events at AGS energy it has a sharp peak at $E_{ZD} \sim 0$, while at SPS energy the probability of $E_{ZD} = 0$ is zero, i.e. there is always some energy left in the forward direction. It shows that from $\sim 10\ GeV/N$ to $\sim 200 GeV/N$ the target nucleus changes from black to gray and the energy fraction deposited in the collision region becomes smaller. Because of the change of the stopping power the reached energy density does not increase much from AGS to SPS, although their incident energy differs by one order of magnitude. Fortunately, the estimated energy density for both AGS and SPS are enough for the predicted phase transition to happen. When the incident energy increases further the stopped energy fraction becomes even smaller and the colliding nuclei are almost transparent to the incident nucleons.

On the other hand, since the transparency of the colliding nuclei become higher and all incident nucleons are passing through for higher incident energy one may reach another interesting situation where a baryon-free central system may form after the collision. This system provides a direct check of the present Lattice calculation based on zero baryon number, $n_B = 0$. It is also a possible candidate of the excited vacuum which is a very important issue in particle physics. In hadron-nucleus collisions the typical average rapidity shift of the incident hadron is $\Delta y \sim 2$, while in A-A collisions data indicate a larger shift of $\Delta y \sim 3.5$ due to the existence of more secondary collisions. Based on these two values of Δy the pure baryon distribution for different incident energies is estimated in ref.[2]. The result shows that for $\Delta y \sim 2$ a baryon-free central region may be formed at RHIC energy, however with $\Delta y \sim 3.5$ only LHC can provide such conditions. Therefore, the baryon-free QGP can be obtained at least at energies higher than RHIC.

2.6 Space-time evolution

Up to now we have only looked at a static picture of relativistic heavy ion collisions. How about the space-time evolution of the system? This evolution could be separated in several steps according to different proper time $\tau = \sqrt{z^2 - t^2}$.

At the end of the pre-equilibrium stage $\tau = \tau_0$ the volume of the system for a small rapidity interval dy can be estimated as $V_0 \sim \pi R_A^2 \tau_0 \, dy$ when no transverse expansion is included for $\tau < \tau_0$. The energy density at $\tau = \tau_0$ could be estimated by Bjorken's formula:

$$\epsilon_0 = \frac{1}{\pi R_A^2 \tau_0} \frac{dE_T}{dy} \quad . \tag{7}$$

If ϵ_0 is larger than the critical value ϵ_c for QGP formation the system would enter QGP-state at $\tau = \tau_0$. Then the system starts to expand and the energy density decreases. Up to $\tau = \tau_c$ and $\epsilon = \epsilon_c$ the phase transition from QGP to hadronic phase starts. Between $\tau_c < \tau < \tau_h$ the system is in a mixed phase if we assume a first order phase transition. After τ_h there is an interacting hadronic system continuing expanding and finally reaching the freeze out point τ_f when all hadrons cease interacting and fly out freely.

However, one should keep in mind that the real situation is much more complicated than the picture shown above. First, the finite collision time would give a finite time interval for energy restoration. The space-time distribution of the energy density in the system has also a complicated development. Second, the phase transition is not an ideal first order transition. Recent analyses show[5] that at $T \leq T_c$ the properties of hadrons may change considerably, e.g. the hadron masses may reduce and influence the property of the phase transition. On the other hand, at $T \geq T_c$ the remaining binding effect between quarks or gluons make the transition far from an ideal one.

It would be nice if the information of the size and life time of the system could be extracted from experimental data. In the next section a brief description of π-interferometry is given, which does provide some information of the source size and life time.

III Bose-Einstein correlation

In 1954 it was proposed by Hanbury Brown and Twiss[6] (HBT) that based on the Bose-Einstein correlation between identical Bosons the momentum correlation between the emitted photons from a distant star could be used to measure the size of the star, which is otherwise very difficult due to the very small open angle of the star in the sky.

Similarly it is possible to extract information of the size of the $A - A$ collision system at its freeze-out point from the momentum correlation between emitted identical pions (or kaons). This method is named as π-interferometry. The momentum correlation function between two pions with 4-momenta k_1 and k_2 can be expressed as

$$C(q) = \frac{p_2(k_1 k_2)}{p_1(k_1)p_1(k_2)} \tag{8}$$

where $q = k_1 - k_2$. It can be proved that, for a source distribution independent of the momentum, $C(q)$ is related to the Fourian transformation of the π-source space-time distribution $\rho(x)$ by

$$C(q) = 1 + \lambda \int e^{iqx} \rho(x) \quad , \tag{9}$$

where λ is a parameter expressing the chaoticity of the source. For a Gaussian form source distribution, $C(q)$ can also be expressed as a Gaussian and the widths of the correlation at small q are just the size and life time of the source.

Experimental analyses show that when $\sqrt{S_{NN}} < 5 \ GeV$ the measured source size $R_f \sim R_B$, where R_B is the radius of the smaller colliding nucleus, while $\sqrt{s} \geq 20 \ GeV$

one obtains $R_f \sim 2R_B$ i.e. the measured freeze- out size indeed shows the expansion of the system.

Recent experiments obtained an empirical relation between the measured R_f and $\frac{dn}{dy}$ [5]. The freeze out size of the system at different $\frac{dn}{dy}$ (or $\sqrt{s}$) can be predicted based on this relation. Freeze out means that there are no more interactions between the produced pions. There are two different ways to estimate it. If one assumes that freeze out happens when the mean free path of the pions reaches the source size, $\lambda_\pi \sim R_f$, one obtains

$$R_f \propto \left(\frac{dn}{dy}\right)^{1/2} \; .$$

However, assuming the pion's density $\rho_\pi = \rho_f$, being a constant at the freeze out point would lead to

$$R_f \propto \left(\frac{dn}{dy}\right)^{1/3} \; .$$

At present data seem to favor the later one. The expansion of the system could be an indication of the phase transition from QGP to hadronic phase. Consider the entropy of the two different phases, sV, where V is the volume, s is the entropy density proportional to the degree of freedom of the system. Since the degree of freedom changes from 47.5 in QGP phase to 3 in pion phase (or a little larger if including other resonances), i.e. $s_Q \gg s_h$, the volume of the system must increase drastically to keep the total entropy increasing during the phase transition: $(sV)_h \geq (sV)_Q$. Besides, the life time measurement could provide information of the property of the phase transition. For a first order phase transition from QGP the mixed phase must exist quite long to release the latent heat and complete the transition. This would lead to a long life time of the source. However, the present data statistics is too low to give a reliable estimation of the life time. Usually a zero life time $\tau_f = 0$ is assumed for estimating the source size R_f in the analysis. However, it is shown in our work[7] that there are strong correlations between the parameters R_f, τ_f and λ. Set $\tau_f = 0$

usually leads to an overestimation of R_f. The system formed in relativistic heavy ion collisions is much more complicated than a static star. Here the π-emitting source is expanding, having a finite life time. Besides, the space distribution of the source is correlated to the momentum of the emitting pions; the resonance decay influences the property of the parameter λ. Therefore, one has to start from the basic formulation including these additional effects. There are still a lot of works in this field both in experimental and theoretical aspects.

In the following two sections the two most probable candidates of QGP signatures in relativistic heavy ion collisions will be discussed: J/ψ suppression and strangeness enhancement.

IV. J/ψ suppression

The J/ψ suppression in QGP phase was first predicted by Matsui and Satz in 1986[8]. In 1987 the result of NA38 at CERN confirmed this prediction. Since then there have been plenty of new experimental data and detailed theoretical analyses. The main experimental results from NA38 can be briefly summarized as follows[9]: The ratio of the J/ψ signal to the Drell-Yan continuum in the dimuon mass spectrum is reduced with increasing transverse energy E_T. The E_T-dependence of this ratio can be expressed as a scaling function of $E_T/A^{2/3}$. Later, more detailed data show that the J/ψ suppression decreases rapidly with increasing p_T.

If QGP is formed the large amount of existing quarks, antiquarks and gluons produce a strong screening effect on the binding force between the $c\bar{c}$ quark pair, preventing the formation of J/ψ. For events with higher E_T, which means higher collision centralities, a higher energy density may be reached in a larger space region. All these are favorable to the formation of QGP and to a stronger J/ψ suppression. The finite size and existing time of QGP allow only relatively slow $c\bar{c}$ to form J/ψ

inside the QGP. However, fast $c\bar{c}$ pairs, especially those with large p_T would easily escape from the QGP region and forming J/ψ without suffering any screening effect, therefore with no suppression for J/ψ at high p_T.

However, in the hadronic phase the final state interactions between the produced J/ψ and other hadrons would suppress the J/ψ products. Higher E_T means more central collisions, which would provide a higher hadronic density and give stronger final state interactions and a stronger J/ψ absorption. To explain the p_T-dependence the initial state interaction of the partons which later produce the $c\bar{c}$ pairs must be introduced. After these initial state interactions the $c\bar{c}$ would be produced at higher p_T, therefore the J/ψ suppression at high p_T would be weakened.

It is for sure that J/ψ suppression can happen only at a high density. But, which of the above two mechanisms is more likely the real one? One has to find definite criteria and to confirm it in future experiments.

Similar suppression of J/ψ or Υ have also been observed in $h - A$ collisions[10], where there is no room for QGP formation. These reference data, therefore, could be used to analyze the background. A nice analysis[11] was reported at $QM'91$ that with one value of absorption cross section of $\sigma \sim 6$ mb all J/ψ suppression data in $h - A$ and $A - A$ collisions could be fitted. Of course, the real situation is much more complicated. Many works[12] have been done later to analyze the effect of gluon shadowing, comover absorption, intrinsic heavy-quark component, etc. . We have done an analysis[13] based on absorption mechanism including the gluon shadowing effect. The result shows that when gluon shadowing is included the fitted absorption cross section in $h - A$ process is considerably smaller than the one in $A - A$ collisions. To finally confirm if J/ψ suppression in relativistic heavy ion collisions can provide a signal for QGP formation one has to work out a model unifying all effects and to analyze all data both for $h - A$ and $A - A$ processes.

V. Strangeness enhancement

People have expected that the stringeness production will be enhanced in QGP phase, because of the lower threshold for $s\bar{s}$ production in it, which is about $2m_s \sim$ 350 MeV, comparing to the one of 700 MeV for $\pi\pi \to K\bar{K}$ reaction in the hadronic phase. The partially restored chiral symmetry would decrease the threshold of $s\bar{s}$ production even further. This is favorable to the enhancement of strangeness production. Particularly, in baryon- rich QGP there are always plenty of u and d quarks from the original baryons, which would suppress the $u\bar{u}$ and $d\bar{d}$ pair production. Furthermore, it is much easier for a produced $\bar{s}$ quark to find a u quark forming K^+ than an s to find a $\bar{u}$ forming K^-. Besides, K^- could easily be absorbed through processes like $K^- + N \to \pi^- + \Lambda$, while for K^+ it is very difficult to be absorbed due to the requirement of strangeness conservation. This makes the enhancement of K^+ much stronger than the one of K^- in the baryon-rich QGP.

Experimentally, strangeness enhancement was first observed as K^+/π^+ ratio enhancement from $\sim 5\%$ in $p-p$ process to $\sim 20\%$ for $Si+Au$ central collision at AGS[14]. Generally a factor of 2 enhancement has been obtained for various strangeness production processes in relativistic heavy ion collisions[15].

However, the enhancement could also be due to various secondary collisions existing in the dense hadronic gas, such as $NN \to NYK$, $\pi N \to KY$, $\pi\pi \to K\bar{K}$, $\cdots$. There are models introducing an equilibrium hadronic fireball during the collision. With high enough initial density and large cross sections of strangeness production in secondary collisions the experimentally obtained enhancement could be explained[16]. However, the results change considerably with different choices of these parameters which could not be determined from other experiments. Besides, the equilibrium condition is rather strong and the freeze-out point is not well defined. Another kind of models are based on secondary collisions without introducing the equilibrium condition[17]. Here, the formation time of the secondaries give strong effects on the results. The interaction cross sections between various hadrons and resonances are not well known. These also make the results not very decisive. Therefore, it seems still too early to make the final conclusion.

Recently, people are interested in the change of the hadron's properties in dense and hot hadronic matter[5]. If most hadrons would reduce their mass at high density or high temperature the particle production ratios would change drastically. With increasing data on strangeness production, especially on multi-strange anti-baryon production[15] a unified explanation of all data with one set of parameters either based on QGP phase or on hadronic phase is necessary to fully understand the property of the formed high density matter during relativistic heavy ion collisions.

Relativistic heavy ion collisions and QGP is still a new field. What is formed during the collision, a QGP or a dense hot hadronic matter, is still an open question. This is also a field which needs wide cooperation of particle physics, nuclear physics, cosmology, statistics, hydrodynamics, transport theory and computer science.

Experimentally both SPS and AGS are going to accelerate heavier ions using present accelerators. In the long run, RHIC and LHC would provide heavy ion colliders with much higher energies. With huge nuclei colliding at such high energies we are looking forward to exploring the "new physics" in the formed dense and hot "hadronic" matter or QGP.

References

[1] J. D. Bjorken, Phys. Rev. D27(1983)140.

[2] H. Satz, Proc. of Joint Inter. Lepton-Photon Symposium and Europhys. Conf. on High Energy Phys., Geneva, 7/25-8/1, 1991.

[3] Chao Wei-qin, Liu Bo, Z. Phys. C42(1989)337;
Chao Wei-qin, Ren Jiang-long, Europhys. Lett. 8(1989)123;
Miao Bi-xia, Chao Wei-qin, Nucl. Phys. A494(1989)620.

[4] Chao Wei-qin, S. P. Sorensen, High Energy Phys. and Nucl. Phys. 17(1993)476.

[5] R. Stock, Nucl. Phys. A544(1992)405c.

[6] R. Hanbury-Brown, R. Q. Twiss, Philos. Mag. 45(1954)633.

[7] Zhang Qinghui, Chao Wei-qin, Preprint BIHEP-TH-9327.

[8] T. Matsui, H. Satz, Phys. lett. 178B(1986)416.

[9] C. Baglin, et al., Phys. lett. B220(1989)471, B255(1991)459.

[10] M. J. Leitch, Nucl. Phys. A544(1991)197c.

[11] C. Gerschel, J. Hüfner, Z. Phys. C56(1992)171.

[12] S. Gupta, H. Satz, Phys. Lett. B283(1992)439;
S. J. Brodsky, Nucl. Phys. A544(1991)223c.

[13] Chao Wei-qin, Liu Bo, Preprint BIHEP-TH-9312.

[14] Y. Miake et al., (E802 Collaboration), Z. Phys. C38(1988)135.

[15] J. Rafelski, Nucl. Phys. A544(1991)279c.

[16] B. L. Friman, Nucl. Phys. A498(1989)161C;
C. M. Ko, L. Xia, Nucl. Phys. A498(1989)561C.

[17] R. Mattiello et al., Phys. Rev. lett. 63(1989)1459;
Chao Wei-qin, Gao Chong-shou, Zhu Yun-lun, Nucl. Phys. A514(1990)734.

Why the central values of $m_{\nu_e}^2$ obtained experimentally with the magnetic spectrometers by various laboratories over the world are most negative? *

Ching Chengrui and Ho Tsohsiu

(Center of Theoretical Physics,CCAST(World Laboratory) and
Institute of Theoretical Physics, Academia Sinica
P.O. Box 2735, Beijing 100080)

Liang Dongqi, Mao Yajun, Chen Shiping and Sun Hancheng

(China Institute of Atomic Energy
P. O. Box 275(10), Beijing 102413)

ABSTRACT By comparing the results obtained using the same experimental data of CIAE but different theoretical formula fits it is pointed out that the negative value of m_ν^2 is most likely linked to the inaccuracy of the theoretical formula of the β-spectrum. With the use of a theoretical formula with up to second order energy sum rule included, the experimental data of CIAE are re-fitted, and the problem with the negativeness of the central value of m_ν^2 has alleviated significantly.

In the previous paper[1] we have reported on the ν_e -mass limit obtained by the CIAE(China Institute of Atomic Energy) group as $m_\nu < 12.4$ eV (95% C.L.). It can be compared with other results reported since 1986 by several other laboratories, namely

$$
\begin{array}{lllll}
m_\nu & < & 12.4\,\text{eV} & \text{CIAE} & (1992)\,[1] \\
      & < & 11.0\,\text{eV} & \text{Zurich} & (1992)\,[2] \\
      & < & 9.3\,\text{eV} & \text{LANL} & (1991)\,[3] \\
      & < & 13.0\,\text{eV} & \text{INS} & (1991)\,[4] \\
      & < & 18.0\,\text{eV} & \text{Zurich} & (1986)\,[5]
\end{array}
$$

and also the hotly disputed result by ITEP

$$17eV < m_\nu < 40eV\,(1987)[6]$$

. However, a striking feature is that all the central values of m_ν^2 are negative, and it is hard to explain it merely by the experimental uncertainties(Tab. 1)

*Supported by the National Natural Science Foundation of China

67

V. de Sabbata and H. Tso-Hsiu (eds.), Cosmology and Particle Physics, 67–77.
© 1994 Kluwer Academic Publishers. Printed in the Netherlands.

Later, the Particle Data Group (PDG) has combined the results of Ref. [2-5] and set a new world average value as

$$m_\nu < 7.3eV(90\%C.L.)[7],$$

and pointed at the same time that "Caution is urged in interpreting this result, because the m_ν^2 average is dominated by the Robertson' 91 result, which is nearly 2 σ negative." This value is also shown in the last line of Table 1.

Recently a more stringint limit of m_ν was reported by Mainz group[8] as

$$m_\nu{}^2 = -39 \pm 34 \pm 15(eV)^2$$

and

$$m_\nu < 7.2eV.$$

However, in obtaining the quoted value, the authors stressed that a specific energy interval of 137 eV was selected, and if a wider energy range is used instead, a more negative $m_\nu{}^2$ was resulted, so that the negativeness of the $m_\nu{}^2$ appears once again. Moreover, it is also interesting to note that in the experiments of Ref.[2-4] both the statistical and systematicl errors are approximately the same, nevertheless the deduced mass limit differs by as large as 3.7 eV. The origin lies in that in determining the mass limit the following assumptions are made: 1, the normal distribution is assumed to be centered at the measured central value with σ as the deviation; 2, the region with $m_\nu^2 < 0$ is abandoned as unphysical region; 3, the remaining part is normalized to unit, and 95 % of this area with the corresponding m_ν^2 is taken to be the m_ν - limit at 95 % C.L..

Then the questions are 1, why the region with $m_\nu^2 < 0$ can be eliminated and 2, why the remaining area should be normalized to unit and the confidence level is defined according to this area ? All these are not well-founded. And the inevitable consequence of this analysis is the farther the measured central value m_ν^2 is below zero, the lower the upper limit for m_ν is. Or in other words, the most improbable event determines with the highest weight the neutrino mass limit now we have.

However, we have two arguments that disfavour the results presented in Table 1. In what follows we shall show that 1, the negative central value of m_ν^2 should not be regarded as unbiased measurement resulted from statistical fluctuation and consequently, 2, the normal distribution should not be taken as centered at this biased value. Now let us discuss these two problems in turn.

The theoretical β spectrum shape for the experimental data fit can be written down as

$$N_{th}(E) = AF(Z, E)pE_t \sum_n W_n(E_0 - E - E_{fn}) \times [(E_0 - E_{fn} - E)^2 - m_\nu^2]^{1/2} \quad (1)$$

where, A is the normalization constant; $F(Z, E)$ is the Fermi function, Z is the daughter nuclear charge; p, E and E_t are the momentum, kinetic and total energy of β rays, respectively; W_n and E_{fn} are, respectively, the relative probability and the excitation energy of the final state with $E_{f0} = 0$ in our definition; E_0 is the end point of the β-spectrum. For simplicity we have omitted in Eq.(1) all corrections resulting from a given experiment.

As is well-known, W_n and E_{fn} must be calculated based on a specific molecular model, which should be chosen to be as faithful as possible to reproduce the source used in a given experiment. The radioactive source of CIAE experiment is ^3T - labelled PAD($C_{14} H_{15} T_6 O_2 N_3$) with tritium sitting in the C-H covalent bond of the molecule. Therefore a realistic approach is to approximate this big molecule to covalent bond such as CH_3T, $CH_3\text{-}CH_2T$, or $CH_3\text{-}CHT\text{-}CH_3$. Such approach was adopted, for instance, in Ref.[2] and Ref.[4]. For comparison, we have also tried the T_2 molecule, the T-atom, the T-nucleus as well as Valine 2, where a theoretical calculation for the latter is available[9]. All the fitted results are shown in Table 2, and the corresponding mass limits are deduced following the recipe we have just outlined. One can, however, argue in advance that it is highly unlike that ^3T in PAD can be mimiced by a bare nucleus or an ^3T atom. In accordance with this conjecture, the least square fits presented in the last two lines of Table 2 clearly show that the bare nucleus model yields the negative m_ν^2 as big as about 3σ away from zero, (and amazingly the smallest mass limit!) and the highest χ^2 value. And the atom model also does no good. Both these two models can be rejected on the ground of the more negative m_ν^2 and higher χ^2 - values. As for the results of other molecular models, the CH_3T model with 7 levels, and the C_3H_7T with 20 levels give the smallest negative value of m_ν^2 as well as the smallest χ^2 - value. Thus the results present in Table 2 suggest strongly that there is a correlation between the negativeness of m_ν^2 and the precision of the theoretical formulas and the corresponding model as well. This has led us to question the precision of the theoretical formulas that were used in getting the results of Table 2.

From the theoretical point of view, and for a many-electron system, the ground state wave functions of the parent and daughter molecules can be calculated, and are calculated with better precision for most models listed in Table 2. However, it is not always the case for E_{fn} and W_n when n lies highly above the ground state . It is difficult even for simple two-electrons'molecular system such as T_2 and $(HeT)^+$. In order to see this point let us recall that there is a theoretically rigorous criterion–the sum rule, which should be fulfilled in any theoretical calculation. Therefore what we have done is to construct the first and second order energy sum rules using the latest calculations of different models, where the spectra are given on the one hand, and calculate the same quantities using the wave functions of the parent molecules on the other hand, and then to see the difference. It is appropriate to notice here that, in principle, the higher order energy sum rules can also be constructed, but as we shall see later, the first and the second order energy sum rules enter the β -spectrum shape formula explicitly, we therefore concern only these two sum rules. In Table 3, Table 4, and Table 5 we have summarized respectively, the best calculated branching ratios and the excitation energies of 7 levels of $CH_3T \rightarrow CH_3He^+$, 20 levels of $C_3H_7T \rightarrow C_3H_7He^+$, and 12 levels of $T_2(TH) \rightarrow THe^+(HHe^+)$ known to us.

Now according to the definitions, the first and the second order energy sum rules can be written down as follows:

$$< \Psi_i \mid \Delta H \mid \Psi_i >= \sum W_n(\Delta E_{fn} + E_{f0} - E_{i0}) \qquad (2)$$

and

$$< \Psi_i \mid (\Delta H)^2 \mid \Psi_i >= \sum W_n(\Delta E_{fn} + E_{f0} - E_{i0})^2 \qquad (3)$$

with

$$\sum W_n = 1 \qquad (4)$$

where ΔH is the difference of the Hamiltonians of the initial and final molecular systems RT and RHe^+. Ψ_i is the ground state wave function of RT, ΔE_{fn} is the excitation energy of the nth -state with respect to the ground state of the daughter molecule RHe^+, so that it is identical with E_{fn} in Eq. (1) and in Tables 3 - 5. E_{i0} and E_{f0} are the ground state binding energies of RT and RHe^+ respectively.

The average excitation energy $\overline{\Delta E^*}$ is defined according to the following equation:

$$\overline{\Delta E^*} = \sum W_n \Delta E_{fn} \qquad (5)$$

Then with the use of Eq. (2-5) one can construct the energy dispersion function σ^2 as the following:

$$\sigma^2 \equiv < \Psi_i \mid (\Delta H)^2 \mid \Psi_i > -(< \Psi_i \mid \Delta H \mid \Psi_i >)^2 = \overline{\Delta E^{*2}} - (\overline{\Delta E^*})^2 \qquad (6)$$

with

$$\overline{\Delta E^{*2}} = \sum W_n(\Delta E_{fn})^2 \qquad (7)$$

We have calculated directly the quantities $\overline{\Delta E^*}$ and σ^2 using the spectras presented in Tables (3-5) and compared them with that obtained from the definitions using the initial wave functions of the parent molecules. The results are presented in Tables (6-8).

It should be noted that since in practical calculations $\sum W_n$ does not equal to 1 exactly, this leads to a small correction term in σ^2. Therefore σ^2 in Tables (6-8) is calculated, if needed, using the following formula

$$\sigma^2 = \overline{\Delta E^{*2}} - (\overline{\Delta E^*})^2 + (E_{i0} - E_{f0})^2 \Delta W_n \sum W_n - 2(E_{i0} - E_{f0})\overline{\Delta E^*} \times \Delta W_n$$

with

$$\Delta W_n = 1 - \sum W_n$$

From Tables 6-8 it can be seen that although $\sum W_n$ in all calculations are very close to unit, and if the precision achieved in $\overline{\Delta E^*}$ computation is tolerable–less than a few percent–the precision in the calculated σ^2 is very poor, and it does not exceed $\approx 30\%$ to 40%, and hence is not acceptable. This shows clearly that it is really very hard to achieve reasonably high accuracy even for HT-HHe^+ or T_2-THe^+ molecular systems. In fact, the best and the most accurate calculation for T_2 and THe^+ so far was reported in Ref.[17] by W. Kolos et. al., and the 12 levels' formula was extracted based on this calculation. But if one is looking beyond the ground state, one immediately sees that the excited states were calculated not as precise as the former. The similar argument is also expressed in Ref.[15].

Then the question is: if it is sufficient to remain satisfied, as claimed in Ref.[10], with the first order energy sum rule , which is calculated with an accuracy of a few percent? Unfortunately it is not the case for β-spectrum shape. It is obvious by recalling that the β-spectrum including the final state interaction is proportional to the following expression:

$$\sum W_n(E_0 - E - \Delta E_{fn})((E_0 - E - \Delta E_{fn})^2 - m_\nu^2)^{1/2}$$

Therefore the average spectrum shape is related directly to the first and second order energy sum rule

$$N(E) \propto ((E_0 - E)^2 - 2(E_0 - E)\overline{\Delta E^*} + \overline{\Delta E^{*2}})$$

for $m_\nu = 0$. This point has been unfortunately neglected in most of the literatures, and the theoretical formula with first and second order energy sum rules included was derived for small neutrino mass in Ref. [16] in 1982 and later was rewritten in Ref.[14] in 1984. Now since the second order energy sum rule is reproduced with rather poor precision for all models listed above, it is natual to inquire the reliability of the theoretical formulas used in the present data analysis concerning the m_ν determination.

In order to re-analyse the β-spectrum properly our principle is the following: since the ground state wave functions are calculated with highest precision, we therefore rely only on the ground state branching ratio, and the first and the second energy sum rules, which are evaluated using only the initial wave functions. The theoretical β-spectral shape is given as the following:

$$
\begin{aligned}
N(E) =\ & AF(Z,E)pE_t\{W_1(E_0 - E)[(E_0 - E)^2 - m_\nu^2]^{1/2}\theta(E_0 - E - m_\nu) \\
& +(1 - W_1) \times [(E_0+ < \Delta H >_1 -E)^2+ < \Delta H^2 >_1 -< \Delta H >_1{}^2 \\
& -m_\nu^2/2] \times \theta(E_0+ < \Delta H >_1 -E - m_\nu)\}
\end{aligned}
$$

(8)

where

$$
< \Delta H >_1= \overline{\Delta E^*}/(1 - W_1)
$$

and

$$
< \Delta H^2 >_1= \overline{(\Delta E^*)^2}/(1 - W_1)
$$

Eq.(8) is the so-called two-levels' formula with the ground state transition treated exactly while the contribution from all excited state transitions is estimated using the first and second energy sum rules. For a formula with more transitions treated exactly we refer the reader to Ref.[14].

Now using formula Eq.(8) the CIAE data are re-fitted. The results are shown in Table 9. Two remarkable features from Table 9 can be seen, namely, 1, in all models without exception the least square fits with closure formula (8) lead to smaller negative values of m_ν^2 and smaller χ^2 values, particularly for C_3H_7T, the m_ν^2 is $+4(eV)^2$ with the χ^2 equal to 1.091, the smallest one; and 2, all the obtained m_ν^2 values using formula (8) are compatible with zero within one standard deviation. This result clearly demonstrates that the quality of the fitting as well as the m_ν^2 value itself rely heavily on the precision of the theoretical spectrum. and the negativeness of m_ν^2 seems at least alleviated. It is also interesting and appropriate to note that the sum rule approach for both C_3H_7T and T_2 models here leads to very similar results. This is related to the fact that incidentally these two models have roughly the same values of W_1 and $\overline{\Delta E^*}$, as well as σ^2.

From the result we obtained, we may conclude, that the negative central value of m_ν^2 is most likely stemmed from our incomplete knowledge of the radiactive source. And it is more reasonable to consider the negative m_ν^2 as a systematic error in theory, which has not been included in the previous data processing.

REFERENCES

[1] Sun Hancheng, Liang Dongqi, Chen Shiping et al.,
To appear in Chinese Journ.Nucl.Phys.

[2] E. Holzschuh, M. Fritschi and W. Kundig, *Phys. Lett.*, B 287 (1992) 381.

[3] R.G.H. Robertson, T.J. Bowles, G.J. Stephenson et al., *Phys. Rev. Lett.*,
67(1991) 957.

[4] H. Kawakami, S. Kato, T. Ohshima et al., *Phys. Lett.*, B 256 (1991) 105.

[5] E. Holzschuh, M. Fritschi, W. Kundig et al., *Phys. Lett.*,
B 137 (1986) 485.

[6] S. Boris, A. Golutvin, L. Laptin et al., *Phys. Rev. Lett.*, 58 (1987) 2019.

[7] Particle Data Group, *Phys. Rev.* D 45 No. 11 (1992) Part 2, P. VI.6.

[8] Ch. Weinheimer et. al., *Phys. Lett.*, B 300 (1993) 210,

[9] I.G. Kaplan, G.V. Smelov and V.N. Smutny, *Phys. Lett.*,
B 161 (1985) 389.

[10] S. Schafroth, *Doctoral Thesis*, Ab Initio Calculations on ^{3}T Substituted
Molecules, p.69.

[11] R.L. Martin and J.S. Cohen, *Phys. Lett.*, A 110 (1985) 95.

[12] I.G. Kaplan and G.V. Smelov, *Nuclear β Decay and neutrinos*
Proc. Intern. Symp., Osaka, Japan, June 1986. Editors:
T. Kotani et.al., World Scientific, p.354.

[13] J. Arafune et.al., *J. phys. Soc. Japan*, 55 (1986) 3806

[14] Ching Chengrui and Ho Tsohsiu, *Phys. Reports*, 112, No.1 (1984) 1.
see also X.L.Zhao,C.R.Ching,*Commun. Theor.Phys.*,7,(1987) 195.

[15] R.L. Martin, Presentation at 11th Intern. Workshop on Weak
Interactions, Santa Fe, NM (1987)

[16]C.R. Ching and T.H. Ho, *Commun. Theor. Phys.*, 1 (1981) 11,
C.R. Ching, T.H. Ho and H.L. Zhao, *ibid*, 1 (1982) 267.

[17]W. Kolos et.al., *Phys. Rev.*, A 31(1985)551,
O. Fackler et.al., *Phys. Rev. Lett.*, 55 (1985) 1388.

Table 1: m_ν^2 and m_ν - **upper limit**

Reference		m_ν^2 (eV^2)	Error (eV)^2		Upper limit
			stat.	syst.	for m_ν(95%C.L.)
CIAE -92	[1]	-31	± 75	± 48	12.4 eV
Zurich-92	[2]	-24	± 48	± 61	11.0 eV
LANL -91	[3]	-147	± 68	± 41	9.3 eV
INS -91	[4]	-65	± 85	± 65	13.0 eV
Zurich-86	[5]	-11	± 63	± 178	18.0 eV
ITEP -87	[6]	919	± 60	± 150	17 < m_ν < 40
PDG -92	[7]	-107	± 60		7.3(90%C.L.)

Table 2: m_ν^2, m_ν-**limit and** χ^2-**value for CIAE data fit[1]**

Model	No. of level	m_ν^2	m_ν	χ^2	E_0-18500
		(eV)^2	(eV)		(eV)
CH$_3$T	7	-31± 75	12.4	1.141	78.3
CH$_2$=CHT	2	-51± 75	12.0	1.145	79.7
CH$_3$-CHT-CH$_3$	2	-43± 75	12.2	1.144	79.9
CH$_3$-CHT-CH$_3$	20	-9± 75	12.9	1.134	79.4
VALINE II	2	-141± 75	10.3	1.140	78.9
T-molecule	2	-68± 75	11.7	1.148	79.9
T-molecule	12	-177± 75	9.7	1.145	77.6
T-atom	2	-191± 75	9.4	1.146	75.2
T-nuclei	1	-237± 75	8.9	1.158	67.2

Table 3: **The** W_N **and** E_{FN} **in the 7 levels transition of** CH_3T
molecule.[9]

W_n	$E_{fn}(\text{eV})$	W_n	$E_{fn}(\text{eV})$
0.6056	0.00	0.017	57.50
0.084	22.50	0.075	72.50
0.141	32.50	0.044	91.33
0.033	47.50		

Table 4: **The** W_N **and** E_{FN} **in the 20 levels transition of** C_3H_7T
molecule.[10]

W_n	$E_{fn}(\text{eV})$	W_n	$E_{fn}(\text{eV})$
.571036	0.00	.045685	63.740
.117594	23.105	.012837	65.565
.073401	35.655	.007596	73.222
.012191	38.874	.053237	78.616
.008831	42.572	.005876	82.119
.020535	44.795	.002271	86.099
.007183	48.285	.001970	92.813
.012122	51.384	.003415	97.807
.008235	55.890	.011496	106.532
.015856	58.777	.001496	120.988

Table 5: **The W_N and E_{FN} in the 12 levels transition of T_2 molecule.[11]**

W_n	$E_{fn}(eV)$	W_n	$E_{fn}(eV)$
.5822	0.00	.0089	41.75
.1675	27.29	.0143	46.03
.0787	33.89	.0166	51.71
.0081	37.96	.0789	65.28
.0001	38.82	.0297	75.45
.0092	39.38	.0061	88.07

Table 6: **A comparison of $\overline{\Delta E^*}$ and σ^2 between the direct calculation and the sum rule approach for 7 levels in CH_3T model.**

	Direct calcul.	Sum rule	Deviation
W_n	1	1	very small
$\overline{\Delta E^*}(eV)$	18.51	18.98[12]	2.5%
$\sigma^2(eV)^2$	744.07	1207.6[12]	$\approx 40\%$

Table 7: **A comparison of $\overline{\Delta E^*}$ and σ^2 between the direct calculation and the sum rule approach for 20 levels in C_3H_7T model.**

	Direct calcul.	Sum rule	Deviation
W_n	.9929	1.00	.7%
$\overline{\Delta E^*}(eV)$	20.56	19.1±.4[3,10]	7.5%
$\sigma^2(eV)^2$	795.69	1231.14[13]	$\approx 30\%$

Table 8: A comparison of $\overline{\Delta E^*}$ and σ^2 between the direct calculation and the sum rule approach for 12 levels in T_2 model.

	Direct calcul.	Sum rule(HT)	Sum rule(T_2)	Deviation
W_n	0.9997	1.00	1.00	.3%
$\overline{\Delta E^*}$(eV)	17.67	18.62[12]	18.80[14]	5-6%
$\sigma^2 (eV)^2$	566.50	1109.5[12]	1045.9[14]	43-50%

Table 9: Re-fitted m_ν^2 using the closure formula (8) and the comparison with that using formula (1).

Model	No. of levels	Formula(1) or(8)	W_1	E_0-18500 (eV)	m_ν^2 (eV)^2	χ^2
CH_3T	2	(8)	0.6056	79.7	-21	1.128
CH_3T	7	(1)	0.6056	78.3	-31	1.141
CH_3-CHT-CH_3	20	(1)	0.5710	79.4	-9	1.134
CH_3-CHT-CH_3	2	(8)	0.5710	82.2	+4	1.091
T-molecule	2	(8)	0.5820	78.2	1	1.098
T-molecule	12	(1)	0.5822	77.6	-177	1.145

ULTRACRYOGENIC GRAVITATIONAL WAVE EXPERIMENTS: NAUTILUS

E. Coccia
Dipartimento di Fisica, Universita' di Roma "Tor Vergata" and INFN
Via della Ricerca Scientifica, 00133 Roma, Italy

Abstract

The challenging problem in the gravitational wave research is the very small size of the effect one wishes to measure. The advantages emerging from cooling a resonant antenna to very low temperatures are discussed in the first part of the paper, together with the main requirements and features of cryogenic systems for gravitational wave experiments.
In the second part we report on the ultralow temperature detector NAUTILUS. The goal of this new generation antenna is to detect burst of gravitational radiation from sources located at distances up to the Virgo Cluster of galaxies. NAUTILUS is installed in the Frascati INFN Laboratories and it should start operating at the end of 1993.

1. INTRODUCTION

The goals of a gravitational wave (g.w.) experiment are to verify directly the existence of gravitational radiation, to study its features and to use this new tool for astronomical observations.

The detector first developed by Weber [1] and subsequently used by other experimental groups consists of a carefully suspended resonant mass object, usually a cylindrical bar, whose vibrational normal modes having the appropriate symmetry should be excited by g.w. bursts of extraterrestrial origin.

The problem of the detection has been clarified in its main aspects since many years [2]. The detector is assumed to be a single large mass quadrupole, and the energy absorbed E_a due to gravitational waves, can be calculated by means of the cross section Σ, which for the most favourable polarization, direction of propagation and antenna mode of vibration (the first longitudinal one), takes the form [3]:

$$\Sigma = (8GM/\pi c)(v_s/c)^2 \qquad (1)$$

where v_s is the sound velocity and M the mass of the bar.

The energy absorbed can be written as:

V. de Sabbata and H. Tso-Hsiu (eds.), Cosmology and Particle Physics, 79–95.
© 1994 *Kluwer Academic Publishers. Printed in the Netherlands.*

(Rome,LSU, Stanford) have set a new upper limit on the intensity of g.w. bathing the Earth [5]. In 1989 The Rome antenna EXPLORER, cooled at 2.0 K with superfluid helium, reached the record sensitivity of $h \sim 7\ 10^{-19}$ [5], soon followed by the LSU detector ALLEGRO. These detectors can observe collapses taking place up to the Large Magellanic Cloud. The expected rate is of the order of 1 per 10 years.

Third generation experiments, making use of ultralow temperature tecniques, will operate soon with the goal of increasing the expected rate of events to several per year, reaching an instrumental sensitivity such as to observe burst sources (like supernovae and coalescing close binaries) occurring at distances of the order of the Mpc, approaching the Virgo Cluster (2500 galaxies at ~19Mpc). In order to reach this goal, the experimental parameters which determine the detector sensitivity must be pushed at the extreme limit of the existing technology.

We report in tab. 1 the main features of the three generations of resonant antennas, with the minimum detectable value of the g.w. amplitude h for a conventional g.w. burst of 1 ms duration. In the last column it is reported the distance d^* from the Earth to the collapse in which $10^{-2}\ M_o$, isotropically converted in g.w., would give the indicated value of h.

Table 1
Main features of three generations of resonant g.w. detector

Generation	T(K)	ß	Q	T_{eff} (K)	h_c	d*(Kpc)
I	300	10^{-4}	10^5	10	10^{-17}	5
II	4.2	10^{-3}	10^6	10^{-2}	10^{-19}	50
III	0.1	10^{-2}	10^7	10^{-7}	10^{-21}	10^4

The strategy of the Rome group to obtain a strain sensitivity of the order of 10^{-21} (which means an energy captured by the antenna as low as 10^{-31} J) implies the use of the highest Q aluminium alloy as antenna material (Al 5056), the use of a resonant capacitive transducer [7] with $\beta \sim 10^{-2}$, the development of a dc SQUID amplifier [8] with noise temperature near to the quantum limit ($T_n \sim h\omega_o/k \sim 10^{-7}$ K) and the cooling of the bar below 0.1 K temperatures [9].

$$E_a = f(\upsilon_R) \, \Sigma \qquad (2)$$

where $f(\upsilon_R)$ is the spectrum energy density at the resonant frequency.

The energy absorbed by the antenna must be large compared to the energy fluctuations in the detector. In a simple form the energy noise can be written as [4]:

$$E_n = kT/\beta Q + 2kT_n \qquad (3)$$

where T is the thermodynamic temperature of the antenna, Q the quality factor of the mode of vibration, β is the ratio of the electromagnetic energy in the transducer to the total energy of the antenna and T_n is the noise temperature of the amplifier connected to the transducer.

For instance a typical room temperature cylindrical bar (T=300 K, Q=10^5) equipped with a resonant capacitive transducer ($\beta=10^{-3}$) and a FET amplifier (T_n=1K) can observe with SNR=1 a spectral energy density at the Earth of the order of 1 J m^{-2} Hz^{-1}. This value corresponds to a collapse occurring at a distance of 5 Kpc in which 10^{-2} M_0 are converted in a millisecond burst. The g.w. amplitude h, which is of the order of the relative antenna displacement $\Delta L/L$, is in this case $h \sim \Delta L/L \sim 10^{-17}$. The discouraging expected rate of such collapses is of the order of 1 per 100 years. Thus the achievement of a reasonable rate of events implies to extend the observing range beyond our Galaxy and the Local Group of galaxies.

It is straightforward from the above formulae that the low temperature is necessary to reduce the random thermal fluctuations kT in the antenna itself. Moreover at low temperature it is possible to take advantage of the steep increase of the quality factor Q for aluminium alloys, and also to use properties of superconductors to build very low noise parametric amplifiers based on the Josephson effect (SQUIDS).

The thermodynamic temperature of the antenna has been used to classify in different generations the g.w. experiments: the low cost and high reliability room temperature antennae developed by Weber in the '60s are said to be of the first generation. The 4.2 K cryogenic detectors developed in the '70s and beginning to be operationals in the '80s are of the second generation. In 1986 three of such cryogenic detectors

In the years 1982-1984 a feasibility study was conducted to establish the technical possibility of the cooling of a multiton Al 5056 bar to milliKelvin temperatures [9]. In 1986 INFN financed the Rome group project of an ultralow temperature antenna, NAUTILUS. In 1989/91 the detector has been assembled and tested to ultralow temperatures in the CERN laboratories [10]. In 1992 NAUTILUS has been installed in its operating site at LNF (INFN Laboratori Nazionali di Frascati) and is now ready to start a first period of observations. Two more ultralow temperature detectors are in the assembling or testing stage: one, AURIGA, at LNL (INFN Laboratori Nazionali di Legnaro), and another at Stanford University.

In the next section we report some fundamental considerations about cryogenics and gravitational wave experiments. In sec. 3 we present the results of the feasibility study for the third generation ultralow temperature detectors. In sec. 4 the NAUTILUS detector is described.

2. CRYOGENICS AND GRAVITATIONAL WAVE EXPERIMENTS

The need of the low temperatures in the gravitational wave research was clear since the '60s, when Joe Weber was beginning the operation of a separated pair of room temperature g.w. antennas in coincidence.

The intriguing results of these measurements and the fascination of this field of research stimulated, among the others, the birth of the g.w. groups of Stanford, Baton Rouge (Louisiana) and Rome. From the very beginning (1970) the plan of these groups was to build cryogenic detectors, making use of the major improvements of the low temperature technique. Bill Fairbank, leader of the Stanford group, presented at that time an ambitious project for cooling at 3 mK a large antenna by combining the use of a ^{3}He-^{4}He dilution refrigerator and of the adiabatic demagnetization of a CMN salt.

The low temperatures made it also possible the use of the properties of superconductors to make very low noise parametric amplifier based on the Josephson effect (SQUID), whose noise temperature can in principal approach the quantum limit ($T_n = h\omega/k = 10^{-7}$ K for frequencies operation around 1 KHz). Morover after few years the Rome group first observed the steep increase in the quality factor Q of some aluminium alloys at low temperatures. In particular Al 5056 presents Q~ 10^7 -10^8 below 10 K.

In 1971 Weber cooled for the first time a 1.5 ton aluminium cylinder to liquid helium temperatures and.."*Immediately after cooling, a very large amount of noise was observed. Some of the excess noise appeared to be associated with internal structural relaxation of the cylinder. Some noise was due to acoustic coupling of high intensity noise associated with the liquid nitrogen and liquid helium system*" [11].

These considerations reflected the difficulties of operating a g.w. antenna at low temperatures, demonstrated by the twenty years efforts of several researchers in 4 continents. The problem was that for the first time a high Q resonant mass of various tons had to be cooled to 4.2 K or below, being free to move and isolated so well from the rest of the world to allow the detection of a displacement of the order of ΔL~10^{-18} m.

In a low temperature system for a g.w. antenna, cryogenics and acoustic isolation requirements are strictly joined; they can be summarized as follows:

a) ensure a long operation time, with rare and short interruptions in the data taking for cryogenic manteinance;

b) ensure a constant and uniform temperature of the antenna; a stationary gaussian distribution of the amplitude of the bar vibrations, in absence of signals, is an important condition for a reliable antenna.

c) do not add extra mechanical noise, in order not to excite the vibrational modes at a detectable level;

d) preserve the inherent high mechanical quality factor Q of the bar.

The peculiar problem of a cryogenic g.w. experiment is to put a large resonant mass at the same time in good thermal contact but in very poor mechanical contact with an effective and long autonomy cooling source.

At 4.2 K a liquid helium bath surrounding the antenna vacuum chamber serves as heat sink and some helium exchange gas is used to rapidly thermalize the detector. The gas is then pumped out before the data taking. The antenna temperature remains about 4.2 K as long as the vacuum chamber it is completely surrounded by the liquid helium. It is found that the source of mechanical noise constituted by the evaporating liquid helium must be attenuated by a factor 10^{-7} to not disturb the detector. This attenuation is provided by the suspension system, which is carefully designed to act as an efficient mechanical filter.

Temperatures as low as 0.8 K or 0.25 K can be reached by pumping on a bath of liquid ^{4}He or ^{3}He, repectively. In the first case the superfluidity of the liquid may help to fulfill the cryogenics and acoustic requirements, as will be shown later.

Using the adiabatic demagnetization of a suitable paramagnetic salt much lower temperatures are possible. With cerium magnesium nitrate (CMN) 2 mK can be reached. Another method allowing to reach 1 mK is Pomeranchuk cooling, i.e. cooling by adiabatic compression of a liquid-solid mixture of ^{3}He [12]. These methods have a drawback: they are "one shot" rather than continuous methods of cooling.

The only method able to mantain temperatures as low as 4-10 mK continuously and in presence of large thermal inputs is ^{3}He-^{4}He dilution refrigeration [12].

As the temperature decreases below 1 K, the only effective cold transfer mechanism between the antenna and the cooling source becomes the conduction by solid. This makes the mechanical isolation from the cooling source of the ultralow temperature antennae a much more difficult task than for the antennae above 1K.

3. FEASIBILITY STUDY FOR MILLIKELVIN DETECTORS

A refrigerator coupled to a g.w. detector must have enough cooling power to absorb the residual heat leak below 0.1 K and to cool the bar reasonably quickly. The thermal contact with the refrigerator will be on the central section of the bar, where requirements c) and d) are easier to fulfill.

3.1 Thermal inputs. We have considered the thermal inputs on our antenna cooled at 0.1 K and surrounded by a shield at 1.0 K. Table 2 resumes the various contributions.

We have neglected all the time dependent heat leaks due to relaxation fenomena, mainly present in dielectric materials and depending strongly on magnetic fields.

Table 2

Estimated thermal inputs on the antenna at 100 mK if the antenna is supposed surrounded by a shield at temperature about 1 K, at a residual pressure of 10^{-9} mbar, and suspended to it by a central section cable of titanium.

Conduction by solid	1 μW
Conduction by residual gas	20 μW
Radiation	30 nW
Cosmic rays	20 nW

It turns out that the main thermal inputs come from the residual gases, which at a pressure of the order of 10^{-9} mbar may give a heat leak of the order of 0.3 nW/cm^2 (20 μW total).

There are no systematic studies of the residual heat leak in ultra-low temperature apparatuses, the knowledge of them being mainly based on conclusions from trial-and-error improvement work. Some considerations present in the specialised literature [13] indicate an upper limit of 1 nW/cm^2 for the residual heat leak below 0.1 K, which is compatible with our estimates. 1 nW/cm^2 means 60 μW of total thermal input on the antenna. This figure determines the minimum cooling power of the refrigerator at those temperatures.

A further reduction is obtainable surrounding the antenna with shields cooled below 1K.

3.2 Thermal gradients.

The antenna material Al 5056 becomes superconducting at the critical temperature T_c=0.925 K [9]. Consequently the thermal conductivity k becomes very low at millikelvin temperatures. The experimental data follow below 1 K the law:

$$k \approx 2.5 \, T^{-3} \quad [\, W \, m^{-1}K^{-4} \,]$$

Because of the low thermal conductivity, at very low temperatures the residual heat leak may cause a very large thermal gradient along the antenna, which does not depend on the performance of the refrigerator. The mixing chamber of the dilution refrigerator will be in fact in contact with the central section of the cylindrical bar, keeping it at a certain temperature, while the ends temperature (which is also the transducer temperature) will stabilize at some warmer temperature, depending on the amount of thermal inputs on the antenna. The uniform thermalization of the antenna at 100 mK, within 20%, is compatible with a maximum heat leak of 10 µW. Uniform cooling at 50 mK would require an heat leak not larger then 1 µW. These heat leaks also express the required values of the refrigerator cooling power at those temperatures.

3.3 Cooling times, The thermal time costant of a bar of mass M, length L, cross section A, density ρ, specific heat c and thermal conductivity κ, cooled from one end, is:

$$\tau_c = \frac{Mc}{\kappa A/L} = L^2 \rho \frac{c}{\kappa}$$

For a 3 m long Al 5056 bar in the temperature range 4.2K - 0.05K τ_c is of the order of a few hours, indicating a reasonable cooling time.

Using the measured thermal properties of the antenna material [13] and numerically integrating the heat diffusion equation, it can be computed the exact time dependence of the temperature of a 2300 kg Al 5056 bar, 3 m long, 0.6 m of diameter, thermally anchored to a dilution refrigerator at the middle section [14]. The antenna can be supposed initially thermalized at 1 K, and put, at the time t = 0, into contact with 0.1 K thermal ground. The results of these calculations have been reported in refs. 9 and 14: we recall here that six hours are necessary to reach 0.110 K and the heat flux is, apart for an initial interval of some tens of minutes, well below the cooling power reachable by commercial dilution refrigerators.

Following these estimates, the uniform cooling below 0.1 K of a large antenna is possible. The cryogenic apparatus must be consequently designed with the main goal of reducing the total heat leak to the antenna at the level of few µW. A configuration of various intermediate shields around the antenna, at temperatures below 1 K, is useful to cryopump the residual gases. As a consequence of the high thermal insulation, the required cooling power of the dilution refrigerator is quite low (as low as the residual heat leak). It is clear that the refrigerator has to be specially designed to constitute part of a large cryostat and cool intermediate shields.

3.4 Thermal contact and mechanical isolation. The possibility of measuring g.w. amplitudes of the order of 10^{-21} (which means an antenna displacement of the order of 10^{-21} m) imposes severe limits on the sources of mechanical vibration present in the experimental apparatus. For instance the boiling of liquid helium in the cryostat container of the 4.2 K antennae must be attenuated of at least a factor 10^{-7} by mechanical filters placed between the helium container and the antenna. In the new milliKelvin criostats this noise source is suppressed by the use of several shields below 1 K, which constitute intermediate masses for the needed mechanical filters between the noise source and the antenna. Only the new source constituted by the refrigerator should become relevant.

An operating dilution refrigerator works with a circulation of almost pure ^{3}He gas. The circulation is sustained by a room temperature leak tight mechanical pump. Both the mechanical noise of the pump, transmitted along cables and pipes, and the noise due to the ^{3}He circulation are carried to the mixing chamber, which is the coldest part and have to be in good thermal contact with the aluminium bar.

The main experimental problem is to provide at the same time an excellent vibration isolation and a good thermal link between antenna and mixing chamber.

The simplest design is that of suspending the bar with the traditional central section cable whose ends are thermally anchored to the mixing chamber of the dilution refrigerator (see fig.2).

Let us find a precise requirement for the refrigerator mechanical noise source S_R[m/$\sqrt{Hz}$], applied at the upper ends of the suspension cable. The resulting excitation of the antenna mode of vibration is[15]:

$$x_0 = S_R \, Q \, A \, \lambda \, \sqrt{\Delta\nu}$$

where x_0 is the amplitude of vibration of the antenna ends, Q is the quality factor and $\Delta\nu$ the bandwidth of the antenna resonance, A is the attenuation provided by the suspension and λ is a factor which expresses the ability of the external vibration to excite that mode of vibration and depends on the geometry of the contact between suspension and antenna. For instance if want to observe amplitudes x_0 of the order of 10^{-20} m, it turns out that S_R has to be kept below 10^{-19} m/$\sqrt{Hz}$ around 1 KHz.

This appears as the major experimental problem of the third generation detectors.

In order to minimize S_R, the flow of ^{3}He in the heat exchangers and in the mixing chamber must be minimized. Low flow causes low cooling power. The resulting indication is to use a dilution refrigerator with high cooling power (say of the order of 1 mW at 100mK) when ^{3}He is fully circulating (generally some mmoles/s), so that it still has the needed few μW when set in a minimum flow condition. Moreover the vibrations trasmission can be dumped using soft multiwires plait as thermal paths between the refrigerator and the antenna.

3.5 Feasibility study conclusions

i) The cooling of a 2.3 ton Al 5056 bar, 3 meters long, below 0.1 K temperatures is possible using a ^{3}He-^{4}He dilution refrigerator: a cooling power of the order of 100μW should be sufficient to thermalize the bar at 0.1 K in half a day time.

ii) A multishield configuration at temperatures below 1 K is necessary in order to:

a) reduce the heat leak to few μW for obtaining the thermalization of the whole bar uniformly at temperatures below 0.1K.

b) have a sufficient number of intermediate masses for variuos mechanical filters needed to isolate the antenna from the noise of the boiling liquid helium.

c) allow the refrigerator to work with a minimum ^{3}He flow, in order to reduce the influence on the antenna of the acoustic noise coming from the heat exchangers and the mixing chamber.

These results determined the design of the NAUTILUS antenna.

4. NAUTILUS

4.1 The cryogenics. The general layout of the cryogenic apparatus is shown in fig.1 [10]. We recall here that the relevant feature of the cryostat is its central section, which is shorter than the cylindrical bar antenna itself. This section contains two helium gas cooled shields, the liquid helium (LHe) reservoir (2000 liters of capacity), three OFHC copper massive rings and, through the top central access, a special ^{3}He-^{4}He dilution refrigerator [16]. End caps are fastened at each stage of the cryostat to complete the seven shields surrounding the bar. The shields are suspended to each other by means of titanium rods and constitute a cascade of low pass mechanical filters. The overall mechanical vibration

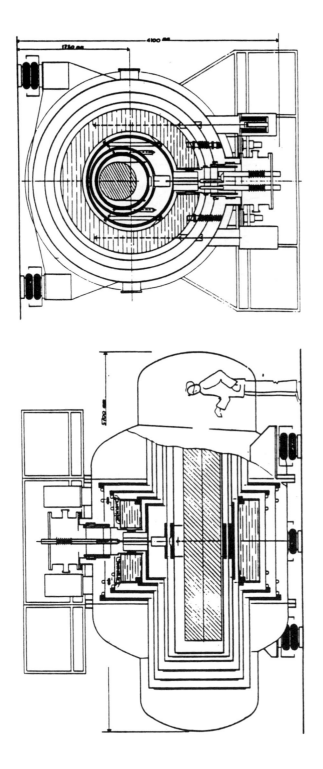

Fig. 1. Layout of the Nautilus antenna.

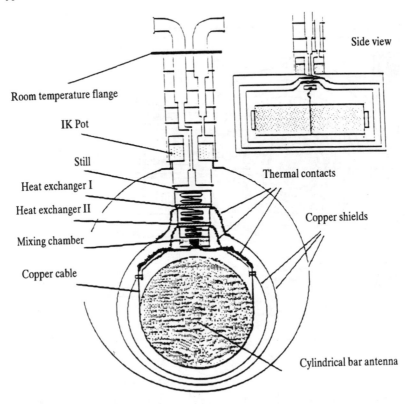

Room temperature flange

IK Pot

Still

Heat exchanger I

Heat exchanger II

Mixing chamber

Copper cable

Side view

Thermal contacts

Copper shields

Cylindrical bar antenna

Fig. 2. Schematic layout of the dilution refrigerator and its connections to the cylindrical bar and to the three copper shields. The shields are suspended each other and are cooled by different stages of the refrigerator. The external shield (2090 Kg) is cooled to 1.3 K by the 1K pot, the heat exchanger I cools to about 350 mK the intermediate shield (860 Kg) and the heat exchanger II cools to about 180 mK the internal shield (800 Kg) sorrounding the bar, which (2350 Kg) is cooled to below 100 mK by the mixing chamber, via the copper cable suspension.

isolation at the bar resonant frequency (about 900 Hz) is of the order of -260 dB.

The first copper shield is thermally anchored to the 1 K pot of the refrigerator. The intermediate and inner shields are in thermal contact with two step heat exchangers of the dilution refrigerator; the mixing chamber [17] cools the bar by means of an OFHC copper rod wrapped around the bar central section. The thermal path in these cases is constituted by soft multiwire copper braids, in order to minimize the transmission of mechanical vibrations to the bar (see fig.2). We recall here that the bar has a mass of 2350 kg, length of 3 m and diameter of 0.6 m.

Fig. 3 shows the bar temperature during the first cool-down. About three weeks were needed to reach 77 K, using 8000 liters of liquid nitrogen, and about one week to achieve 4.2 K, using about 5000 liters of LHe. We then kept the bar temperature in the range 4.2-8 K for about two weeks, to perform various tests.

When we started the ultralow temperature cooling, the initial temperatures of the bar and of the three copper shield were about 8 K. We filled the 1 K pot with LHe at low pressure and started to condense and circulate the ^{3}He-^{4}He mixture in the dilution refrigerator. After three days the calibrated Ge thermometers indicated a temperature of 95 mK on the bar end face and of 63 mK on the mixing chamber. As far as we know, it was the first time that such massive bodies were cooled at these very low temperatures.

The observed features of the cooling agree with the earlier model [14]. From the measured thermal gradient between the mixing chamber and the bar end (about 30 mK) we deduce an upper limit of 10μW for the antenna heat leak (corresponding to 1.7 μWm^{-2}).

The overall LHe evaporation rate at regime was 50 litres/day.

We remark that in this run we could not optimize the ^{3}He flow, because of an electrical short in the still heater, so that all of the above results were obtained with a reduced refrigerator cooling power. We think that a bar temperature of about 50 mK could be possible.

4.2 *Readout and sensitivity.* NAUTILUS has been moved in the summer 1992 to LNF.

In this new site (41.5N, 12.4E) the detector has been placed on a special platform which can rotate for the proper orientation of the

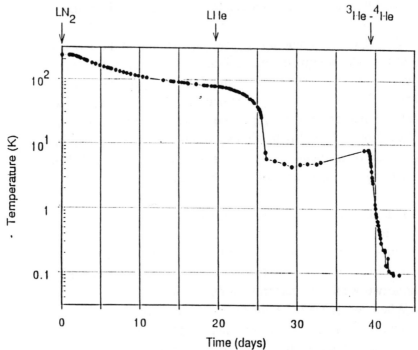

Fig. 3. Temperature of the cylindrical bar versus time. The arrows indicate the main cryogenic operations, described in the text.

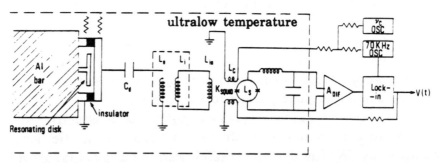

Fig.4. Electrical scheme of the experimental apparatus. The vibration of the bar are converted into electrical signals by a capacitive transducer, resonating at the antenna frequency in order to improve the energy transfer from the bar to the electronics. Bar and transducer form a system of two coupled oscillators. The signals are applied to the input coil of a dc SQUID amplifier by means of a superconducting transformer, which provides the required impedance matching. The output signal from the SQUID instrumentation contains the informations on the vibrational state of the antenna and can be properly processed.

detector with respect to either an array of detectors or even a fixed source.

Nautilus is equipped with a resonant capacitive transducer and a dc SQUID preamplifier (see fig. 4), following the electrical measurement configuration developed for the EXPLORER detector [6], in operation at CERN.

We recall that the vibrations of the bar are converted into electrical sygnals by the capacitive transducer, resonating at the antenna frequency in order to improve the energy transfer from the bar to the electronics. Bar and transducer form a system of two coupled oscillators. The signals are applied to the input coil of a dc SQUID amplifier by means of a superconducting transformer, which provides the required impedance matching. The output signal from the SQUID instrumentation contains the informations on the vibrational state of the antenna and can be properly processed.

The detector is completed with seismic, electromagnetic and cosmic rays veto systems.

The necessity of a cosmic rays detector is due to the fact that extensive air showers or energetic single particles (muons or hadrons) interacting in the antenna may produce signals with rates which increases with the increasing sensitivity of the antenna to g.w.. For instance 1 cosmic ray event per day is estimated for NATILUS having $T_{eff} \sim 1$ mK. This rate increase to 10^3 when $T_{eff} \sim 1$ μK and to 10^5 if the quantum limit $T_{eff} \sim h\omega_0/k \sim 0.1$ μK. is reached.

The veto system consists of two layers of streamer tubes for a total of 102 counters. The first layer is located on the top of the cryostat, the dimensions are 6x6 m^2. The second layer is under the cryostat, the dimensions are 6x2.5 m^2 . The system detects about 50% of the single track events leaving in the antenna more than 10 Gev (corresponding to $T_{eff} \sim 1.5$ μK). For extensive air shower the efficiency is almost 100%.

The effective temperature in the first run, end 1993, should be of the order of 800 μK.

In fig.5 the pulse sensitivity of the detector at the quantum limit is shown.

It is crucial for the unequivocal detection of gravitational wave bursts that various antennae of comparable sensitivity and bandwidth operate in continuous and well coordinated coincidence. It is natural to plan future coincidences with the similar ultralow temperature antenna AURIGA in preparation at LNL.

In order to push the NAUTILUS sensitivity towards and possibly beyond the quantum limit, new non linear schemes of amplification and electromechanical transduction are currently under investigation.

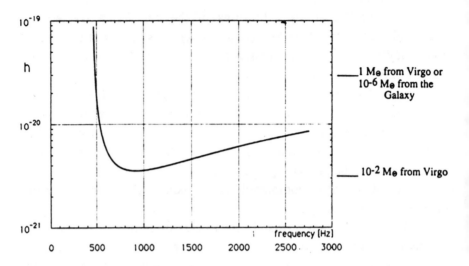

Fig. 5 Planned sensitivity of NAUTILUS to g.w. pulses. It is assumed that the pulse consists of one sinusoidal cycle with the frequency reported in the abscissa. On the right we report the expected g.w. amplitudes at the Earth corresponding to the conversion of solar masses into g.w. isotropically emitted from the indicated distances.

REFERENCES

[1] J. Weber, Phys. Rev. **117**, 306 (1960).

[2] See for instance the Proceedings of the Fourth Marcel Grossmann Meeting on General Relativity, Rome 1985. North Holland Pub. Co. 1986. or K.S. Thorne in "Three hundred years of gravitation" S.W. Hawking and W Israel editors, Cambridge University Press 1987.

[3] M. Rees, R. Ruffini and J.A. Wheeler: "Black holes, Gravitational Waves and Cosmology", Gordon and Beach, New York, 1974.

[4] R.P. Giffard, Phys. Rev. D 14, 2478 (1976).

[5] E Amaldi. et al. Astron. Astrophys **216**, 325 (1989).

[6] E. Amaldi et al. Europhysics letters, 12, 5 (1990)

[7] P. Rapagnani, Nuovo Cimento C **5**, 385 (1982).

[8] C. Cosmelli et al. IEEE Trans. on Mag. 23, 454 (1987)

[9] E. Coccia and T.O. Niinikoski, J. of Phys. E: Sci. Instr. 16, 695 (1983).

[10] P. Astone et al. Europhysics Letters, 16, 231 (1991).

[11] J Weber, in "Experimental Gravitation", Proceeding of the Int. Symposium held in Pavia, Acc. Naz. dei Lincei, 1977.

[12] O.V. Lounasmaa, "Experimental Principles and Methods Below 1K", Academic Press, 1974.

[13] T.O. Niinikoski, Proc. ICEC 10, Helsinki 1984.

[14] E. Coccia and T. Pomentale, CERN Internal Report DD/82/16, (1982).

[15] E. Coccia, Rev. Sci. Instr. 55, 1980 (1984).

[16] The dilution refrigerator has been designed and assembled in collaboration with Oxford Instruments Ltd. For details see: M.Bassan, E.Coccia, I.Modena, G.Pizzella, P.Rapagnani, F.Ricci: "Features of the Cooling at Ultra-Low Temperatures of a Resonant Gravitational-Wave Antenna", Proceedings of the 5th Marcel Grossmann Meeting on General Relativity. Perth, Western Australia 1988, Blair, Buckingham and Ruffini editors, World Scientific 1989.

[17] E. Coccia and I. Modena, Cryogenics 31, 712 (1991).

SPIN AND TORSION IN THE EARLY UNIVERSE

Venzo de Sabbata
World Laboratory, Lausanne, Switzerland
Dept of Physics, University of Bologna and Ferrara, Italy
and
Istituto di Fisica Nucleare, sezione di Ferrara, Italy

Abstract
It is shown the importance of torsion, which is a necessary geometrical feature of the space-time, when one consider the early universe and some property of string theory, and when one like to find a way toward the quantization of gravity. It is also shown that through the time-temperature uncertainty relation it is possible to arrive at the notion of a minimum acceleration that can be important in connection with observations involving galaxies. At the end a brief comment is given on the problem of cosmological constant and inflationary expansion when torsion is considered.

V. de Sabbata and H. Tso-Hsiu (eds.), Cosmology and Particle Physics, 97–128.

1. Introduction

In the last years many progresses are been made on the theoretical aspects of the torsion and results are achieved in the context of the influence that the spin, inserted through torsion, that is through a geometrical property of the space-time, has on the early universe and in the field of elementary particle physics.

One of the main result is that torsion seems to constitute a way toward quantization of gravity.

As is well known, torsion is a slight modification of the Einstein theory of relativity (proposed in the 1922-23 by Cartan [1]), but is a generalization that appears to be necessary when one try to conciliate general relativity with quantum theory.

Briefly the only modification is to introduce an asymmetric affine connection, as Cartan did [1], where the antisymmetric part is the torsion tensor:

$$Q_{ij.}^{\ \ k} = \Gamma_{[ij]}^{\ \ k} \qquad (1)$$

In the presence of torsion the space-time is called a Riemann-Cartan manifold and is denoted by U_4 (usual Riemann space-time is denoted by V_4).

One of the most important geometrical property of torsion is that a closed contour in an U_4 manifold becomes in general a non-closed contour in the flat space-time V_4. As we will see, this non-closure property that is the fact that the integral

$$1^\alpha = \oint Q_{\beta\gamma}^{\ \ \alpha} \, dA^{\beta\gamma} \neq 0 \qquad (2)$$

over a closed infinitesimal contour is different from zero,

can be treated as defects in space-time in analogy to the geometrical description of dislocations (defects) in crystals, and this can constitute a way to go toward the quantization of gravity that means quantization of the space-time itself.

Gravity in fact, according to Einstein, is not a force but is the curvature of . space-time and then we cannot quantize gravity like other forces but we must try to quantize the space-time itself.

We know that one of the major difficulty in the contemporary physics both in quantum field theory and in general relativity is the presence of singularities and of infinities. In general this depends on the fact that we are used to considering point particles so that we are faced with the divergence of self energy integrals that go to infinity, or when in general relativity we have to do with a collapsing body (i.e. a black hole) we are faced with the singularities. These difficulties may disappear if, with torsion, we will have a minimal unit of length and time: that is if we have to do with a discretized space-time. **In fact if we like to quantize gravity we cannot follow the procedure of quantization that we use to quantize for instance other forces like weak, strong or electromagnetic. In fact gravity is not a force but is the curvature (and torsion!) of the space-time and we must try to quantize the space-time itself.**

But first of all let me consider the connection between torsion and strings.

2. Torsion and string tension

In string theories, the coupling is characterized by the so called string tension T, which is the energy or mass per unit length. At the Planck scale (M_{Pl} = $(\hbar c/G_N)^{1/2}$), T is given by:

$$T_{Pl} = c^2/G_N \simeq 1.6 \cdot 10^{28} g \; cm^{-1} \qquad (3)$$

(G_N is the Newtonian gravitational constant). Thus T has units of G^{-1} (c = 1), i.e. T $\propto$ 1/G. At any other energy scale, M, T is given as

$$T = T_{Pl}(M/M_{Pl})^2 \qquad (4)$$

T scales as the mass squared, i.e. the energy at which the symmetry breaking occurs. So the maximum value of the string tension would be at the Planck mass, i.e. $T_{Pl}= c^2/G = T_{max}$, (M $\leq$ M_{Pl}). At this scale the string length is the Planck length $L_{Pl}=(\hbar G/c^3)^{1/2}$ so that the string mass is $T_{Pl}L_{Pl} \Longrightarrow M_{Pl}$.

However in the usual picture it is not very clear how the string tension arises (for instance at the Planck scale) from the interaction between point particles. Then we explore a possible connection with the spin-torsion interaction between spinning constituents and to account for its scaling with energy. Now as is well known, the torsion tensor $Q_{ij}{}^k$ is related to the spin density tensor $J_{ij}{}^k$ (which acts as the source term for torsion, analogous to matter density being source term for curvature) through the equation:

$$Q_{ij}{}^k = (8\pi G/c^3)(J_{ij}{}^k - (1/2)\delta_i^k J_{bj}{}^b - (1/2)\delta_j^k J_{ib}{}^b) \qquad (5)$$

Thus for N aligned spins per unit volume, the torsion is given as [2,3]:

$$\bar{Q} = (4\pi G/c^3)\bar{J}, \qquad (6)$$

where $J = N\hbar/2$

Consider now a string of Planck length $L_{Pl} = (\hbar G/c^3)^{1/2}$ with the interacting spins having a net spin of $\hbar/2$. Then the spin density is $J = (\hbar/2)/[(\hbar G/c^3)^{3/2}(4\pi/3)]$. The spin-torsion energy is then $cQ \cdot \hbar/2$ and is thus given by:

$$E_{ST} = (4\pi G/c^2)(\hbar/2)(\hbar/2)/[(4\pi/3)(\hbar G/c^3)^{3/2}] \qquad (7)$$

Assuming the energy of the string is solely given by this spin-torsion field, then the energy or mass per unit length (i.e. the tension) is given by:

$$T = E_{ST}/L_{Pl}c^2 = (3/4)\left[\hbar^2/(\hbar G/c^3)^2\right](G/c^4) \approx (3/4)c^2/G \qquad (8)$$

which is just the tension in the string which is assumed in the superstring models with unification at the Planck scale. This shows the underlying importance of spin or torsion in determining the tension.

If the string length were now $L_{Pl} \cdot (M_{Pl}/M)$, i.e. the energy scale at which the strings are produced is $M \Longrightarrow$ ($M < M_{Pl}$), then the tension would evidently be reduced by the ratio of $(M/M_{Pl})^2$, i.e. it would now be:

$$T = (c^2/G)(M/M_{Pl})^2 = T_{Pl}(M/M_{Pl})^2 \qquad (9)$$

in agreement with eq.(2). We have shown elsewhere [4] that the spin-torsion coupling constant G depends on energy in the early universe as E^{-2}, i.e. as M^{-2}, i.e. it can pass through the values of the coupling constants of the fundamental interactions during the evolution of the universe, enabling the understanding of the ratios of the different coupling strengths. Thus we had the relations:

$$G_N = G_w (M_w/M_{Pl})^2 \qquad (10)$$

(where $G_w \implies G_F (c/\hbar)^2$ where G_F is the Fermi beta-decay constant and M_w the intermediate weak boson mass) and also:

$$(G_f/G_N) = (M_{Pl}/M_f)^2 \approx 10^{38} \qquad (11)$$

and

$$(G_f/G_F) = (M_w/M_f)^2 \qquad (12)$$

where G_f is the strong gravity constant characterizing the strength of strong interactions, and $M_f \approx 1$ Gev is the f-meson mass. Eqs. (10), (11) and (12) enable us to understand the coupling strengths of strong, weak and gravitational interactions.

The above relations would enable us to write the string tension as:

$$T = c^2/G_{eff} \qquad (13)$$

where G_{eff} is the value of G at the scale M,

$$G_{eff} = G_N (M_{Pl}/M)^2 \qquad (14)$$

and we have just the relation .

$$T = (c^2/G_N)(M/M_{Pl})^2 \qquad (15)$$

or

$$T = T_{Pl}(M/M_{Pl})^2 \qquad (16)$$

For the hadronic string describing the Regge trajectories of hadrons, the tension is given by the appropriate strong gravity constant G_f, i.e. [5]:

$$T_{strong} = c^2/G_f \approx 10^{-11} g\ cm^{-1} \approx 10^{-38} c^2/G_N \qquad (17)$$

3. Quantum gravity and torsion

One thing that I would like to say is that torsion may constitute a way toward quantization of gravity. In fact we will show that introducing torsion in General Relativity, that is considering the effect of the spin and linking the torsion to defects in space-time topology, we can have a minimal unit of length and a minimal unit of time.

We know that one of the major difficulty in the contemporary physics both in quantum field theory and in general relativity is the presence of singularities and of infinities. In general this depends on the fact that we are used to considering point particles so that we are faced with the divergence of self energy integrals that go to infinity, or when in general relativity we have to do with a collapsing body (i.e. a black hole) we are faced with the singularities.

These difficulties may disappear if, with torsion, we will have a minimal unit of length and time: that is if we have to do with a discretized space-time. In fact if we like to quantize gravity we cannot follow the procedure of quantization that we use to quantize for instance other forces like weak, strong or electromagnetic. In fact gravity is not a force but is the curvature (and torsion!) of the space-time and we must try to quantize the space-time itself.

Now the torsion gives rise to defects in space-time topology: we know that in the geometrical description of crystal dislocations and defects, torsion plays the role of defect density and in this context one can consider space-time as an elastic deformable medium in the sense of

Sakharov. But before to go on in this way, we like to show that without torsion we are faced with some difficulties.

In fact in a recent paper Treder and Borzeskowski [6] have pointed out that the Landau-Peierls type of uncertainty relations

$$\Delta F \ (c\Delta t)^2 \ \geq \ (\hbar c)^{1/2} \qquad (18)$$

(F denotes the magnitude of the field strength F_{ik} and ΔF is measurement inaccuracy Δt is the time interval between two field measurements) could correspond to commutation rules of the form:

$$i \ [F_{ik} \ , \ dx^i \wedge dx^k] \ = \ \sqrt{\hbar c} \ , \qquad (19)$$

In this case inequality (18) would have the meaning of quantum uncertainty relations, that is it would imply, via Rosenfeld's analysis, the uncertainty inequalities for quantum general relativity as:

$$\Delta g_{ik} \Delta x^i \Delta x^k \ \geq \ \hbar G/c^3 \qquad (20)$$

and

$$\Delta \Gamma_{ikl} \Delta x^i \Delta x^k \Delta x^l \ \geq \ \hbar G/c^3 \qquad (21)$$

(Δg and $\Delta \Gamma$ are the corresponding inaccuracies). However they note that in general relativity the "field strength" Γ^i_{kl} (which are the usual Christoffel symbols) are not tensorial quantities. So distances cannot be defined independently of the field quantities g_{ik}.

In fact, as demonstrated in [6], one should not expect to find commutation rules corresponding to Rosenfeld's inequality relations

$$g_{ik} \, L_o^{\,2} \;\geq\; \hbar G/c^3$$

$$\Gamma^i_{\;kl} L_o^{\,3} \;\geq\; \hbar G/c^3 \tag{22}$$

(L_o denotes the dimension of the spatial region over which the average value of g_{ik} and $\Gamma^i_{\;kl}$ is measured) because the quantities "field strength" and "length" appearing in (22) cannot be defined independently of each other. The independent definition of those quantities for which commutation rules are to be formed is however necessary to require. Therefore, the inequalities (22) do not have the status of uncertainty relations like (20) and (21); they are rather relations describing the inaccuracy for the measurement of the single quantities g_{ik}, $\Gamma^i_{\;kl}$, etc. As a consequence of the dependence of distances on the metric g_{ik}, the inequalities (22) say of course also something about limitations on distance measurements. Indeed, assuming e.g. a coordinate system in which $|g_{oo}| = 1$ is satisfied, with the condition $|\Delta g_{oo}| < |g_{oo}|$ one is led from the first of (22) to

$$L_o^{\,2} \;>\; |\Delta g_{oo}| L_o^{\,2} \;\geq\; \hbar G/c^3 \tag{23}$$

showing that for distances of the order of magnitude of Planck's length the notion "length" loses its physical meaning. But this statement about limitations on field and length measurements does not coincide with the statement implied by uncertainty relations of type (18) and valid for canonically conjugate quantities.

Therefore, the conclusion drawn in Ref.[6] was that the inequalities (22) establish limitations on the validity of the relativistic quantum field conception. And it was stressed that such limitations should only occur in quantized General Relativity Theory but not in any other theory of quantum gravity.

This problem may not arise if torsion is considered, since in this case the asymmetric part of the connection i.e. $\Gamma^i_{[kl]}$, i.e.the torsion tensor Q^i_{kl} is a true tensorial quantity. In fact, as we have said, with torsion one can define distances in this sense: if we consider a small closed circuit and write $l^i = \oint Q^i_{kl} dA^{kl}$ where $dA^{kl} = dx^k \wedge dx^l$ is the area element enclosed by the loop, then l^α represents the so called "closure failure", i.e. torsion has an intrinsic geometric meaning: it represents the failure of the loop to close, l^i having the dimension of length. (Q^i_{kl} has dimension of inverse length, dA is area).

In the above relation, we know that torsion can be related to the intrinsic spin $\hbar$, and, as the spin is quantized, we can say that the defect in space- time topology should occur in multiples of the Planck length, $(\hbar G/c^3)^{1/2}$ i.e.

$$\oint Q^i_{kl} dx^k \wedge dx^l = n \ (\hbar G/c^3)^{1/2} \qquad (24)$$

n is an integer. This is analogous to the well known $\oint pdq = n\hbar$, i.e. the Bohr-Sommerfeld relation. As Q^i_{kl} plays the role of the field strength (analogous to F_{ik} for electromagnetism), relation (24) is analogous to that of eq.(19) (with $\sqrt{\hbar}$ on the right hand side). So distance has been defined independently of g_{ik}. In fact (24) would define a minimal fundamental length, i.e. the Planck length entering through the minimal unit of spin or action $\hbar$. So $\hbar$ has to deal with the intrinsic defect built into torsion structure of space time through [7] $1^{\alpha} = \oint Q^{\alpha}_{\beta\gamma} dx^{\beta} \wedge dx^{\gamma}$, i.e. we have

$$[Q^{\alpha}_{\beta\gamma}, dx^{\beta} \wedge dx^{\gamma}] \geq (\hbar G/c^3)^{1/2} \qquad (25)$$

i.e. $\hbar$ is related to a quantized timelike vector with dimensions of length related to an intrinsic geometrical structure with torsion (after all we known that intrinsic spin $\hbar$ is a universal attribute of all matter). So torsion must enter into geometry.

In conclusion [8] on, has a classical background metric g_{ik} (saying, for instance, what $dx^i \wedge dx^k$ means) and an (nearly) independent tensor field Q^i_{kl} quantized via condition (24):

$$\oint Q^i_{kl} dx^k \wedge dx^l = n (\hbar G/c^3)^{1/2} . \qquad (26)$$

This quantization leads to the commutation rules (25):

$$[Q^i_{kl}, dx^k \wedge dx^l] = (\hbar G/c^3)^{1/2} \qquad (27)$$

and to corresponding uncertainty relations:

$$\Delta Q^i_{kl} (\Delta x^k \wedge \Delta x^l) \geq (\hbar G/c^3)^{1/2} \qquad (28)$$

Therefore, the Einstein-Cartan theory of gravitation should in contrast to Einstein's General Relativity Theory, provide genuine quantum-gravity effects.

We can also observe that from relation (24), considering the fourth component, we can defined time in the quantum geometric level through torsion as

$$t = (1/c) \oint Q \, dA = n(\hbar G/c^5)^{1/2} \qquad (29)$$

so torsion is essential to have a minimum unit of time ≠ 0!

This in fact would give us the smallest definable unit of time as $(\hbar G/c^5)^{1/2} \cong 10^{-43}$s. In the limit of $\hbar \Rightarrow 0$ (classical geometry of general relativity) or $c \Rightarrow \infty$ (Newtonian case), we would recover the unphysical $t \Rightarrow 0$ of classical cosmology or physics. So both $\hbar$ and c must be finite to give a geometric unit for time (i.e. $\hbar \Rightarrow 0$ and $c \Rightarrow \infty$ are equivalent). The fact that $\hbar$ is related to a quantized timelike vector, discretize time. This quantum of time or minimal unit of time also correspondingly implies a limiting frequancy of $f_{max} \approx (c^5/\hbar G)^{1/2}$. This would have consequences even for perturbative QED, in estimating self energies of electrons and other particles, i.e. the self energy integral (in momentum space) taken over the momenta of all virtual photons. To make the integral converge Feynman in his paper on QED [9], multiplied the photon propagator, k^{-2}, by the ad hoc factor: $-f^2/(k^2 - f^2)$, where k is the frequency (momentum) of the virtual photon. This convergence factor although it preserves relativistic invariance, is objectionable because of its ad hoc character without any theoretical justification. Feynman considers f to be arbitrarily large without definite theoretical basis. Here the presence of space-time defects associated with the torsion due to the intrinsic spin would give a natural basis for the maximal value for f_{max}^2 as (from eq.(29)) $\approx c^5/G\hbar \approx 10^{96}$(and extremely large as required by Feynman), giving finite result (instead of ∞) for the self energy. This makes f_{max} another fundamental constant for particle physics serving as a high frequency cut off which is not arbitrary.

But we can go a little further, trying to see if there are conjugate variables for which we can write commutation relations. We will see that R and Q can be such conjugate variables. Consider in fact the equation of geodesic deviation of a free particle with non-zero spin (in which coupling of spin to the space-time curvature occurs):

$$\frac{d^2 x^\mu}{ds^2} + \begin{Bmatrix} \mu \\ \alpha\beta \end{Bmatrix} \frac{dx^\alpha}{ds} \frac{dx^\beta}{ds} = Q^\mu_{\alpha\beta} \frac{dx^\alpha}{ds} \frac{dx^\beta}{ds} \tag{30}$$

Following the above discussion, in the quantum picture the right hand side is proportional to multiples of the Planck length. As the left hand side of eq.(30) is obtained from variation of $\int g_{\alpha\beta} (dx^\alpha/ds)(dx^\beta/ds)$, we would simply have

$$\Delta\Gamma^\mu_{\alpha\beta} \, dx^\alpha \wedge dx^\beta = [Q^\mu_{\alpha\beta} \, , \, dx^\alpha \wedge dx^\beta] \geq (\hbar G/c^3)^{1/2} \tag{31}$$

In the limit of vanishing torsion, we have the usual geodesic equation. So torsion in eq.(30) can be identified with the quantum correction (in multiples of $\hbar$) to the classical equation of motion.

Unlike electromagnetism where a single particle suffices to measure the components of the field tensor, in the case of gravitation we require the relative acceleration between two particles to specify the components of the curvature tensor, i.e. $R^\alpha_{\beta\gamma\delta}$ is defined through the acceleration:

$$d^2 x^\alpha/ds^2 = R^\alpha_{\beta\gamma\delta} (dx^\beta/ds)(dx^\gamma/ds) n^\delta \tag{32}$$

(n^δ = unit vector in direction of the line joining the two particles). As many approaches lead to the notion of "maximal acceleration", for instance in the case of gravitation, this is of $\sim m_{Pl} c^3/\hbar = c^{7/2}/(\hbar G)^{1/2}$ (m_{Pl} = Planck mass), then the maximal value for d^2x^α/ds^2 would also imply a maximal curvature which turns out to be $\approx c^3/\hbar G \approx 10^{66} cm^{-2}$. In the classical limit, i.e. $\hbar \Rightarrow 0$, we have the usual curvature singularity, i.e. the maximal value is infinite, the relation between acceleration and curvature being of the form: $a = c^2\sqrt{R}$, R = curvature scalar. This would imply a relation of the type

$$\Delta R_{\alpha\beta\gamma\delta} (\Delta x^\alpha \wedge \Delta x^\beta)(\Delta x^\gamma \wedge \Delta x^\delta) \geq \hbar G/c^3. \tag{33}$$

So a general relativistic upper limit on acceleration due to quantum effects (an operational way of measuring curvature tensor) would also imply a maximal curvature and hence uncertainty relation involving Planck length.

Thus in order to arrive at a quantum theory of gravitation for which no limitations but only genuine uncertainty relations occur one has to decouple metrical background and quantized (gravitational) field, i.e. to modify General Relativity Theory.

The gravitational theories with torsion discussed here are of this nature. There one has a classical background metric $g_{\alpha\beta}$ (saying, for instance, what $dx^\alpha \wedge dx^\beta$ means) and an independent tensor field $Q^\alpha{}_{\beta\gamma}$ quantized via condition (24)

$$\oint Q^i{}_{kl} dx^k \wedge dx^l = n \ (\hbar G/c^3)^{1/2} \tag{34}$$

This quantization leads to the commutation rules (25):

$$[Q^i_{kl} , dx^k \wedge dx^l] = (\hbar G/c^3)^{1/2} \qquad (35)$$

and to corresponding uncertainty relations:

$$\Delta Q^i_{kl} (\Delta x^k \wedge \Delta x^l) \geq (\hbar G/c^3)^{1/2} \qquad (36)$$

In that way we have quantized torsion field but not the background. We believe that there is the possibility to include also the metrical background (the curvature).

We can start from these considerations: if we wish to connect the initial and final positions of one and the same particle we cannot avoid uncertainty associated with torsion, i.e. for a sufficiently small area element dS, uncertainty in distance between initial and final position would be $\Delta l^\mu = Q^\mu dS$ and this would induce fluctuations in distance in the metric through $\Delta l = \sqrt{g_{\mu\nu} dx^\mu dx^\nu}$. So what is important is not the point themselves but the "fluctuations" in their position, i.e. the interval between them caused by a deformation of space itself through torsion. Note that plastic deformations are induced by torsion and are different from elastic deformations considered by Sacharov (which depend only on curvature). With quantized values, these fluctuations would also manifest as metric fluctuations. Since curvature causes relative acceleration between neighbouring test particles we have momentum uncertainty related to curvature as:

$$ma^\mu dS = \Delta p^\mu = m R^\mu_{\alpha\beta\gamma} \frac{dx^\alpha}{ds} dx^\beta n^\gamma = m R \frac{dS}{ds} n^\mu \qquad (37)$$

So as position fluctuations are given by torsion, momentum

fluctuation are due to curvature and we can interpret quantum effects (and then uncertainty principle) as consequences of space-time deformation i.e.

$$\Delta p^{\mu} \cdot \Delta x_{\mu} = \hbar \qquad (38)$$

where

$$\Delta x^{\mu} = Q \ c \ dS \ n^{\mu} \quad \text{and} \quad \Delta p^{\mu} = m \ R \ \frac{dS}{dS} \ n^{\mu} \qquad (39)$$

(i.e. uncertainty in initial and final positions and relative acceleration between them due to space-time deformations), Q (torsion) and curvature (R) thus both playing simultaneous roles as conjugate variables of the geometry (gravitational field), thus enabling us to write commutation relations between curvature and torsion (analogous to $[x,p] = i\hbar$) as:

$$\left[Q \ , \ R \right] \ = \ (\hbar G/c^{3})^{-3/2} \qquad (40)$$

or, more explicity, as:

$$\left[Q^{\alpha}_{\ \mu\nu} \ dx^{\mu} \wedge dx^{\nu} \ , \ R^{\alpha}_{\ \mu\nu\rho} \ dx^{\mu} \wedge dx^{\rho} \wedge n^{\nu} \right] \ = \ i\hbar G/c^{3} \qquad (41)$$

or

$$\Delta Q \ \Delta R \ \geq \ L_{Pl}^{-3} \qquad (42)$$

where L_{Pl} is the Planck length.

Thus in a sense, we have written quantum commutation relations for the observables of the background geometry, rather than for the gravitational field in a fixed background as in the usual picture; thus partially removing the inconsistency between quantizing the gravitational field and not the geometry, gravity being the geometry itself!

In fact in the usual methods of quantizing the gravitational field one has the field on a fixed background, which gives rise to an inconsistency, because the gravitational field is just geometry so that how can we separate the geometry and the field? Here, on the contrary, we had not only quantized torsion on a fixed background but we have considered both torsion and background curvature which appear as conjugate variables.

Curvature and torsion in this context appear to be conjugate variables for which we can write commutation relations like (40).

Another result, that seems to be relevant as regards the problem of the dark matter, is that through torsion we arrive at the definition of a minimal acceleration which appear to be:

$$a_{min} = c\, H_o$$

In order to see the question of minimal acceleration in some details we will start from time-temperature uncertainty relations.

4. Torsion and Time-Temperature Uncertainty relation

Now consider again the relation (29) where time is defined in the quantum geometric level through torsion as

$$t = (1/c) \oint Q\, dA = n(\hbar G/c^5)^{1/2} \qquad (43)$$

so, as we have seen, torsion is essential to have a minimum unit of time $\neq 0$ where the smallest definible unit of time is $(\hbar G/c^5)^{1/2} \cong 10^{-43}$ s. This quantum of time or minimal unit of time, as we have seen, implies a limiting frequency of $f_{max} \approx (c^5/\hbar G)^{1/2}$.

Now in a recent paper [10] it has been pictured

particles with rest mass m as vortices, which would give them a life time related through torsion as (i.e. analogous to charge being connected to $\int B \, dA$ we had mass related to $\int Q \, dA$ exploiting analogy between torsion and magnetism [11]) (see also [12]):

$$t \cong (1/c)\int Q \, dA \cong \hbar/mc^2 \qquad (44)$$

which is just the lifetime of quantum particles.

We understand the connection between eqs.(43) and (44) having an energy-dependent G! This is clearly stated in ref. [4]. We have in fact:

$$G \approx \hbar c/m^2 \qquad (45)$$

For $m = M_{Pl}$, we have $G = G_N$. For strong gravity with $m = m_p$, we have $G = G_f$ and so on. Substituting (45) in (43), we arrive at eq.(44) (or, more generally, at eq.(24)) i.e. we have space-time defect lengths l^α scaling with energy as E^{-1}, where the strength of interaction is fixed by the value of G_{eff} (see eq.(14)) related to l^α through eqs.(43), (44) and (45).

The main point that must be stressed is that the existence of the different defect lengths l^α is primary and is due to the introduction of torsion that physically means the introduction of spin, and then is related to $\hbar$, the unit of spin.

So the absence of torsion implies absence of defects in space-time and, consequently, absence of masses, charges and of the very existnece of time!

Now ℏ is related to time through the energy, that is we have:

$$\text{time} \implies \hbar/\text{energy} \quad \text{i.e. we have } \hbar\nu = \text{energy} = E \quad (46)$$

We have also the thermodynamic definition

$$E = k_B T \quad (47)$$

where k_B is the Boltzmann constant (the unit of entropy: see for instance [13,14]) and T is the temperature. Eqs. (46) and (47) imply the relation:

$$\text{time } \times \text{ temperature } = \hbar/k_B = \text{constant} \quad (48)$$

i.e. we find

$$\Delta t \, \Delta T = \hbar/k_B = 10^{-27}/1.3 \cdot 10^{-16} \approx 10^{-11} \quad (49)$$

Eq. (49) is universally valid as can be seen by several examples. In the early universe, for $\Delta t = 10^{-43}$s, we have $\Delta T \approx 10^{32}$K (from eq. (49)). This is just the temperature $T \approx (\hbar c^5/G_N)^{1/2}(1/k_B) \approx 10^{32}$K, at the Planck epoch. At the hadron era, $\Delta t \approx 10^{-23}$s, $\Delta T \approx 10^{12}$K, the hadronic temperature $(m_p c^2/k_B \approx 10^{12}$K) etc. $\approx (\hbar c^5/G_f)^{1/2}(1/k_B)$. Also for a typical electromagnetic interaction time scale $\approx 10^{-16}$s, we have the corresponding temperature from eq. (49) as $T = 10^5$ K, which is just that corresponding to the Rydberg theory (≈ 13 ev) for ionized atoms.

So we can say that the existence of these times and temperatures in the early universe (as implied by eq. (49)) fixes the strength of the dominating interactions at the different epochs, G_N at t_{Pl}, G_f at t_{hadron} etc. i.e. G is automatically determined, once eq.(49) is assumed (through eqs.(43), (44) and (45)).

The time-temperature uncertainty relation eq.(49) is valid in also curved space-time. In general in curved space-time, the temperature T is modified as [12]:

$$T(g_{oo})^{1/2} = \text{const.} \qquad (50)$$

But we also know that frequencies are redshifted in a gravitational field i.e.

$$\nu(g_{oo})^{1/2} = \text{const.} \qquad (51)$$

This is dilated in a gravitational field as $t(g_{oo})^{-1/2}$ (in general $\Delta\nu\Delta t = 1$!). So we have:

$$\Delta T(g_{oo})^{1/2} \cdot \Delta t(g_{oo})^{-1/2} = \Delta T \Delta t = \hbar/k_B = \text{constant} \qquad (52)$$

So the time-temperature uncertainty relation also holds in curved space and $\hbar$ and k_B are universal constants even in the presence of gravitation.

As regards the minimal temperature it can be related to minimal possible energy allowed in the cosmological context. Also from time-temperature uncertainty relation $\Delta t \Delta T \approx \hbar/k_B$ we have that the maximal conceivable time is related to the Hubble H_o constant as:

$$\Delta t_{max} \approx 1/H_o \qquad (53)$$

then

$$\Delta T_{min} \approx \hbar H_o/k_B \approx 10^{-27} \cdot 10^{-18}/10^{-16} \qquad (54)$$

So the smallest possible operationally definable temperature
is

$$\Delta T_{min} \approx 10^{-29} \text{ K} \qquad (55)$$

Here $\hbar H_o$ can have the interpretation as the minimum amount of
energy that can be operationally defined in a closed
universe. In general one cannot reach absolute zero. One can
come arbitrarily close (at present we have reached $\approx 10^{-6}$ K).
It is remarkable that this minimal temperature can also be
arrived at by considering a black hole of maximum possible
mass, i.e. mass of the universe $\approx 10^{55} g \approx 10^{60} M_{Pl}$. Since
temperature of a black hole scales inversely of mass: $T \approx$
$\hbar c^3/G K_B M$, maximal possible mass M_{max} will give minimum
possible operationally definable temperature which is
consistent with (55)!

Also from entropy considerations we have the maximal
possible entropy of $\approx 10^{120} k_B$ [13,14]. This implies minimum
temperature of $\approx 10^{-29} K$.

We have seen the relation between acceleration and
background curvature through the equation of geodesic
deviation, but we like to show that we can define not only
maximal acceleration but also minimal acceleration a_{min}.

In fact we have seen how the general time-temperature
relation valid in curved space, and explicity found if
torsion is considered, gives rise to a minimal operationally
definable temperature as (see eq.(54)):

$$T_{min} = \hbar c/k_B R_H \qquad (56)$$

where R_H , the maximal background scale, is the Hubble
radius.

To sum up eq.(56) comes when one consider time-temperature uncertainty relation $\Delta t \Delta T = \hbar/k_B$ (see ref.[12] where k_B is the Boltzmann constant). This last relation is universally valid and in correspondence of t_{min} and t_{max} defined, as we have seen, through torsion, we have T_{max} and T_{min}; in ref.[12] we also shown that this uncertainty relation is valid also in curved space-time.

Now we have another familiar result that an observer in a state of uniform acceleration a, finds himself immersed in a thermal bath of temperature.

$$T = \hbar a/k_B c \qquad (57)$$

Thus T_{min} as given by eq.(57) implies a minimal acceleration a_{min} as:

$$a_{min} = k_B T_{min} c/\hbar \qquad (58)$$

and substituting for $T_{min} = \hbar c/k_B R_H$ (cf eq.(57) and using $R_H = c/H_0$ then gives:

$$a_{min} = c H_0 \qquad (59)$$

We can see better how the minimum temperature as given by (54) is derived from torsion.

In fact in a recent paper [18] we have noted that the maximal acceleration exists given by $a_{max} = c^{7/2}/G^{1/2}\hbar^{1/2}$. This originates as a quantum effect due to torsion, and would give rise to a maximal force, as we shall explain, which can be operationally defined as $F_{max} = c^4/G$.

Briefly we remember that the spin-torsion interaction energy between spin S and torsion Q has been shown to be [19]

$$E = - \overline{S} \cdot \overline{Q} \qquad (60)$$

and the formal analogy of this expression with the interaction energy of a magnetic dipole $\bar{\mu}$ in constant magnetic field $\bar{H}$ i.e. $\bar{\mu} \cdot \bar{H}$, shows that spin in spaces with torsion behaves like a dipole in a magnetic field, i.e. will orient predominantly along the torsion direction. Then it is easy to show that we are led to a magnetic field

$$B = (8\pi/3c)(2\alpha G)^{1/2}\sigma \tag{61}$$

where α is the fine structure constant and σ the spin density. In that way [20] we obtain a maximum primeval field at the Plank epoch of 10^{58}G.

Subsequently flux conservation would have made the field behave as $Bt = BT^{-2} = $ const. with expansion time t and temperature T. Such a magnetic field would have accelerated charged particles in the early universe. For a particle with charge e, the relativistic Larmor frequency is given by $\omega_L = (ecB/\hbar)^{1/2}$ implying a magnetic energy $\hbar\omega \sim (e\hbar cB)^{1/2}$. It is known that this would imply a critical magnetic field when $(e\hbar cB)^{1/2}$ equals the rest mass energy mc^2 of the particles i.e. when the gyroradius r/G becomes smaller than the Compton length. Thus quantum considerations impose a critical magnetic field strength of $B_c = m^2c^3/e\hbar$. At the Planck epoch when $m = m_{Pl} = (\hbar c/G)^{1/2}$, this implies a B_{max} of $c^4/eG \approx 10^{58}$G; now, as $\omega_L = (ecB/\hbar)^{1/2}$ implies a circular (or helical) acceleration in the magnetic field of the charged particle with $a \cong \omega_L^2 r_G$ we will have $a \sim (ecB/\hbar)r_G$. If we substitute $B_{max} \approx c^4/Ge$ and the corresponding $r_G = (\hbar G/c^3)^{1/2}$ at the Planck epoch we have the expression for maximal acceleration as given before i.e. $a_{max} = c^{7/2}/(\hbar G)^{1/2} =$

$m_{Pl} c^3/\hbar = 5 \cdot 10^{53} cm \ s^{-2}$. So this expression for maximal acceleration is originated as a quantum effect due to torsion and one can also note that in this expression of maximal acceleration for a charged particle the electric charge is not involved : instead there is the Planck constant which is connected with spin. The minimal length L_{min}, also dictated by torsion [8], implies a maximal mass or energy.

Also here one can remember that this minimal length can be arrived at by considering the non-closure property of a contour in the space-time with torsion, that is:

$$l^\alpha = \oint Q^\alpha_{\mu\nu} \ dA^{\mu\nu} \neq 0 \qquad\qquad (62)$$

(where $dA^{\mu\nu} = dx^\mu \wedge dx^\nu$ is the area element enclosed by the loop) over a closed infinitesimal contour is different from zero and we have seen that as torsion is related to the intrinsic spin, if we connect torsion to the fundamental unit of intrinsic spin $\hbar$, we find that the defect in space-time topology should occur in multiplies of the Planck length, so that we can write $\oint Q \ dA = n(\hbar G/c^3)^{1/2}$ and for the same reason considering the fourth component we have also an analogous situation as regards the time, namely $t = (1/c)\oint Q \ dA = n(\hbar G/c^5)^{1/2}$ which gives a minimum unit of time $\neq 0$ (for n = 1). So the minimal length and the minimal time are dictated by torsion; this implies a maximal mass or energy. For a quantum state we have $M_{max} = (\hbar/L_{min} c) = (\hbar c/G)^{1/2}$. This would give rise to a maximal force as:

$$F_{max} = M_{max} \, a_{max} = c^4/G \qquad (63)$$

This expression can also be arrived at by noting that given a mass M, general relativistic considerations imply that it cannot be localized to a distance R_{min} less than $G \, M/c^2$ (for the Schwarzschild case it is $2MG/c^2$ and for maximal rotation [of the type $J = GM^2/c$ which we have earlier shown [13] is to be expected for object forming black holes) it is GM/c^2).

The gravitational self force $F = GM^2/R^2$ is thus maximized for R_{min} giving $F_{max} = GM^2/R_{min}^2 = c^4/G$, that is (63). Note that F_{max} is independent of M showing that it is a universal quantity. Then the maximal mass operationally definible is F_{max}/a_{min}.

We shall now indicate how a_{min} can be defined: as clearly shown in ref.[18], we have the formula for the acceleration due to torsion in an exact solution of Einstein-Cartan theory as

$$\ddot{R} = G^2 s^2/c^4 R^5 \qquad (64)$$

We had applied this formula in the case of black hole evaporation. Here we shall indicate how the above relation can be used to obtain both a_{max} and a_{min}. For a_{max} using $S \approx \hbar$ and $R_{min} = (\hbar G/c^3)^{1/2}$ (as explained earlier), we get

$$(\ddot{R})_{max} = a_{max} = G^2 \hbar^2/c^4 (\hbar G/c^3)^{5/2} = c^{7/2}/(\hbar G)^{1/2} \qquad (65)$$

which agrees with what was stated above!

For a_{min} we have to invoke cosmological parameters in the above equation. For S we take the total spin of universe which, as was shown earlier [21,22], was given by a cosmological solution involving torsion as:

$$S_U = 10^{120}\hbar \tag{66}$$

which for R, the maximal radius i.e. $R = R_H$, the Hubble

radius, we have

$$\ddot{R} = a_{min} = G^2 S_U^2 / c^4 R_H^5 = \left(c^{7/2} / (\hbar G)^{1/2}\right) \left(S_U/\hbar\right)^2 \left(L_{Pl}/R_H\right)^5$$

$$= a_{max} \left(S_U/\hbar\right)^2 \left(L_{Pl}/R_H\right)^5 \simeq 10^{-8} cm \ sec^{-2} \tag{67}$$

which agrees with what is assumed in MOND.

So we see that both a_{max} and a_{min} follow as consequence of an exact solution of Einstein-Cartan theory based on torsion.

We will now proceed to estimate the maximal and minimal temperature associated with a_{max} and a_{min} respectively. First of all, as we have shown [12], we notice that the minimal operationally definable temperature is (as written in (54))

$$T_{min} = \hbar c / k_B R_H \tag{68}$$

where R_H , the maximal background scale, is the Hubble radius.

Now we shall proceed to show that this minimal temperature follow from the minimal acceleration a_{min} as given above by the effect of torsion. In fact another familiar result that an observer in a state of uniform acceleration, a, finds himself in a thermal bath of temperature [18]

$$T = \hbar a / k_B c \tag{69}$$

Thus a_{min} as given above implies a minimal temperature of

$$T_{min} = \hbar a_{min} / k_B c \tag{70}$$

Substituting for a_{min} as found above through torsion, we find finally (after substituting also for S_U etc.)

$$T_{min} = \hbar c/k_B R_H \qquad (71)$$

which is the same as equation (68). From eq.(69) we can express a_{min} as

$$a_{min} = k_B T_{min} c/\hbar \qquad (72)$$

Substituting equation (71) for T_{min} and using $R_H = c/H_0$ then gives

$$a_{min} = c^2/R_H = c \, H_0 \qquad (73)$$

which is the value assumed in MOND (Modification of Newtonian Dynamics) [16,17] in order to explain a wide variety of observations involving galaxies and clusters such as well known relationship involving luminosity-velocity -rotation- dispersion etc. However in MOND the introduction of a_{min} given by (73) is ad hoc while here the existence of a fundamental minimal acceleration (or temperature) can be considered a consequence of the introduction of torsion. In this case, for instance, we do not need dark matter in order to explain the flat rotation curves of spiral galaxies, i.e. the discrepancies between the observable and newtonian dynamical mass (see [23]).

At last we like to mention that considering the torsion in the early universe we are led to an inflationary expansion (see [24]). In fact the simplest E-C generalization of standard big-bang cosmology is obtained by considering the universe filled with unpolarized spinning fluid and solving for the modified Einstein equations [15,25]:

$$G^{\alpha\beta}\left(\left\{\begin{array}{c}\end{array}\right\}\right) = \chi \, \vartheta^{\alpha\beta} \qquad (74)$$

(the brackets { } indicates that $G^{\alpha\beta}$ is made through Christoffel symbols)

where
$$\vartheta^{\alpha\beta} = <T^{\alpha\beta}> + <\tau^{\alpha\beta}>$$

$$= (\rho + p - (1/2)\chi\sigma^2)u^\alpha u^\beta - (p - (1/4)\chi\sigma^2)g^{\alpha\beta}$$

$\hspace{10cm}$ (75)

$T^{\alpha\beta}$ is the usual energy-momentum tensor and $\tau^{\alpha\beta}$ can be considered as representing the contribution of an effective spin-spin interaction. $\sigma^2 = (1/2)<S_{\alpha\beta}S^{\alpha\beta}>$ where $S_{\alpha\beta}$ is the spin density associated with the quantum mechanical spins of elementary particles; ρ, p and σ depend only on time. In the comoving frame $u^\mu \equiv (0,0,0,1)$, we get the following modified field equations of the Robertson-Walker universe, which in general for $k \neq 0$ and $\Lambda \neq 0$ is of the form [15,25]:

$$\dot{R}^2/R^2 = (8\pi G/3)\left[\rho - (2/3)\pi G\sigma^2/c^4\right] + \Lambda c^2/3 - kc^2/R^2 \qquad (76)$$

We immediately notice that the torsion term in eq.(76), (the second term within the square brackets), is of _opposite_ sign to that of the cosmological constant term. This raises the possibility that a sufficiently large spin-torsion term in early universe might cancel a correspondingly large cosmological torsion term. We shall see that this is indeed the case. For instance consider the universe at the Planck epoch when as we noted in [15] (see also [25]) the Λ term was $\approx 10^{66} cm^{-2}$, implying $\Lambda_{Pl} c^2 \approx 10^{87}$ in eq.(76). At $t_{Pl} \approx 10^{-43}s$ the universe had a density of $c^5/G^2\hbar \approx 10^{93}g\ cm^{-3}$, and as the particle masses were $\approx 10^{19}Gev \approx 10^{-5}g$, ⸱the particle number density was $n_{Pl} \approx 10^{98}cm^{-3}$, so that σ, the spin density, was $\sigma_{Pl} \approx 10^{98} \cdot 10^{-27}$ (i.e. $n_{Pl}\hbar$) $\approx 10^{71}$. This gives for the term $- (8\pi G/3)(2/3)\pi G\sigma_{Pl}^2/c^4$ (i.e. the torsion term in eq.(76)) the value of $\approx - 10^{87}$, which is exactly equal and of opposite sign to that of the cosmological term which is $\approx + 10^{87}$, so that the two terms would have cancelled each other in the early universe at the Planck epoch.

We can also see that they would continue to cancel at later epochs in the early universe. The Λ-term would evolve with temperature T as $\Lambda \propto T^2$ [26]. σ being the spin density (i.e.proportional to number density of spins) would scale as T^3 ($n \propto T^3$ for relativistic particles). So σ^2 would scale as T^6. Now in earlier work [4] it was argued that the spin-torsion coupling in the early universe was energy dependent and scaled as $G \propto T^{-2}$. So $G^2 \propto T^{-4}$, so that $G^2\sigma^2$ term would scale as T^2, which has the same dependence on T as the Λ-term so that if they cancel each other at the Planck epoch, they would also cancel at later epochs in the early universe.

Thus, in short, we have a more natural mechanism for the understanding of a vanishing Λ-term by the simple incorporation of spin-effects (a universal property of particles!) in general relativity.

As regards the inflation, we note that the quantity within the square brackets of eq.[76] corresponds to an effective density of the form:

$$\rho_{eff.} = [\rho - (2/3)\pi G \sigma^2/c^4] \tag{77}$$

We have seen that the torsion term (i.e.the second term

within the square brackets) was at the Planck epoch equal and opposite in sign to the cosmological term $\Lambda_{Pl} c^2 \approx 10^{87}$ at that epoch, i.e.we had $- (8\pi G/3)(2/3)\pi \, G \, \sigma^2/c^4 \approx - 10^{87}$. The $8\pi G\rho/3$ term is also of comparable magnitude (with ρ Planck density). This raises the question of whether ρ_{eff} can become negative just around or before the Planck epoch. If ρ_{eff} and consequently (as in general, from kinetic considerations $p_{eff} \approx k\rho_{eff}c^2$, where k is a numerical coefficient $\approx$ 1) the pressure p_{eff} becomes negative, we have the required condition for inflation, as negative pressure drives inflation. So inflation would follow as a natural consequence of a spin-dominated (i.e.torsion-dominated) phase in the very early universe, spin being a basic property besides the mass (for more details see [24].

References

[1]   H.E.Cartan - Compt.Rend.174, 437, 593 (1922); Ann.Ecole
      Normale 41, 1 (1924)

[2]   V.de Sabbata and M.Gasperini - Il Nuovo Cimento 27, 289
      (1980);

[3]   V.de Sabbata and C.Sivaram - Il Nuovo Cimento 101A, 273
      (1989)

[4]   V.de Sabbata and C.Sivaram - Astrophys.Spc.Sci.158, 347
      (1989)

[5]   V.de Sabbata and C.Sivaram - Ann.der Physik 47,
      419 (1991)

[6]   H.-H.v.Borzeszkowski and H.-J.Treder - Ann.der Physik
      46, 315 (1989); see also "On Quantum Gravity" PRE-EL
      88-05, Potsdam 1988, and H.-H.v.Borzeszkowski and
      H.-J.Treder, The Meaning of Quantum gravity, Dordrecht,
      Reidel Publ.Comp.1988.

[7]   V.de Sabbata and C.Sivaram - "Torsion and Quantum effects"
      in "Modern Problems of Theoretical Physics", Festschrift
      for Professor D.Ivanenko, ed.by P.I.Pronin and N.Obukhov,
      World Sci.Singapore 1991, pag.143

[8]   V.de Sabbata, C.Sivaram, H.-H.v.Borzeszkowski and H.-J.
      Treder - Ann.der Physik 48, 197 (1991)

[9]   P.A.M.Dirac - "New directions in Physics" J.Whiley, N.Y.
      1978

[10] V.de Sabbata and C.Sivaram - Found.of Physics Letters
      6, 201 (1993)

[11] V.de Sabbata and M.Gasperini - Lett.Nuovo Cimento 27,
      133 (1980)

[12] V.de Sabbata and C.Sivaram - Found.of Phyisics Letters
      5, 183 (1992)

[13] V.de Sabbata and C.Sivaram - "The central role of spin
      in black hole evaporation" in 'Black hole physics'
      Kluwer Academic Publ. NATO ASI Series, Vol.364, 225
      (1992)

[14] V.de Sabbata, C.Sivaram and D.Wang - Ann.der Physik 47,
      508 (1990)

[15] V.de Sabbata and C.Sivaram - "Torsion, Quantum Effects
and the problem of cosmological constant" in
"Gravitation and Modern Cosmology " devoted to 75th
jubilee of Peter Gabriel Bergmann - ed. by A.Zichichi,
V.de Sabbata and N.Sanchez, Plenum Press, vol.56, 19
(1992)

[16] D.Lindley and D.Nature 359, 583 (1992)

[17] M.Milgrom and R.H.Sanders - Nature 362, 25 (1993)

[18] V.de Sabbata and C.Sivaram - Astrophys.Space Sci.176,145
(1991)

[19] V.de Sabbata and M.Gasperini - Lett.Nuovo Cimento 27,138
(1980)

[20] V.de Sabbata and C.Sivaram - Il Nuovo Cimento 102B, 107
(1988)

[21] V.de Sabbata - "On the relation between fundamental
constants" in 'Gravitational Measurements, Fundamental
Metrology and Constants' Kluwer Academic Publ. NATO ASI
Series, Vol 230, 115 (1989)

[22] V.de Sabbata and M.Gasperini - Lett.Nuovo Cimento 25,
489 (1979)

[23] V.de Sabbata and C.Sivaram - "On limiting field
strengths in Gravitation" accepted in Found.Phys.Letters

[24] V.de Sabbata and C.Sivaram - Astrophys.Space Sci. 176,
141 (1991)

[25] V.de Sabbata and C.Sivaram - Astrophys.Space Sci. 165,
51 (1990)

[26] C.Sivaram, K.Shina and E.C.G.Sudarshan - Found.Phys.
6, 717 (1976).

Four-Dimensional String/String Duality [†]

M. J. Duff and Ramzi R. Khuri[*]

Center for Theoretical Physics
Texas A&M University
College Station, TX 77843

We present supersymmetric soliton solutions of the four-dimensional heterotic string corresponding to monopoles, strings and domain walls. These solutions admit the $D = 10$ interpretation of a fivebrane wrapped around 5, 4 or 3 of the 6 toroidally compactified dimensions and are arguably exact to all orders in α'. The solitonic string solution exhibits an $SL(2, Z)$ strong/weak coupling duality which however corresponds to an $SL(2, Z)$ target space duality of the fundamental string.

[†]   Talk given by MJD. Work supported in part by NSF grant PHY-9106593.

[*]   Supported by a World Laboratory Fellowship.

V. de Sabbata and H. Tso-Hsiu (eds.), Cosmology and Particle Physics, 129–143.

1. Introduction

A major problem in string theory is to go beyond a weak-coupling perturbation expansion. A possible approach to this problem is provided by the string/fivebrane duality conjecture [1,2], which states that, in their critical spacetime dimension $D = 10$, superstrings (extended objects with one spatial dimension) are dual to superfivebranes (extended objects with five spatial dimensions). There is now a good deal of evidence in favor of this idea, which may be divided into: Poincare duality [1], strong/weak coupling duality [2–8], singularity structure duality [9] and classical/quantum duality [10,11]. Most of these discussions have focused on the $D = 10$ heterotic string and its dual counterpart the $D = 10$ heterotic fivebrane, but in this paper we wish to examine the four-dimensional consequences.

That the field theory limit of the $D = 10$ heterotic string admits as a soliton a heterotic fivebrane [1] was first pointed out by Strominger [2]. He went on to suggest a strong/weak coupling duality between the string and the fivebrane in analogy with the Montonen-Olive strong/weak coupling conjecture in four-dimensional super Yang-Mills theories [12]. This strong/weak coupling was subsequently confirmed from the point of view of Poincare duality in [3]. There it was shown that just as the string loop expansion parameter is given by

$$g(\text{string}) = e^{\phi_0}, \tag{1.1}$$

where ϕ_0 is the dilaton VEV, so the analogous fivebrane parameter is given by

$$g(\text{fivebrane}) = e^{-\phi_0/3}, \tag{1.2}$$

and hence

$$g(\text{fivebrane}) = g(\text{string})^{-1/3}. \tag{1.3}$$

In the same paper [2], Strominger pointed out that after toroidal compactification to four dimensions, the fivebrane would appear as either a 0-brane, a 1-brane or a 2-brane, depending on how the fivebrane wraps around the compactified directions [13–15]. Thus it ought to be possible to find soliton solutions directly from the four-dimensional string corresponding to monopoles (0-branes), strings (1-branes) and domain walls (2-branes). Such fivebrane-inspired supersymmetric monopoles have already been found [16–18], and here we discuss the string and domain wall. (The possibility of monopole solutions in this context was anticipated in [19], where a monopole-like solution in the massless fields sector of the bosonic string was discovered.)

To find these multi-string and multi-domain wall solutions we shall follow the procedure outlined in [16,17] for monopoles, where it was argued that exact solutions of the heterotic string could be obtained by modifying the 't Hooft ansatz for the Yang-Mills instanton. We shall present them from both the $D = 10$ and $D = 4$ points of view. As with the fivebrane and the monopole, there are three types of string and domain wall solutions: neutral, gauge and symmetric. Following the arguments of [5,6], all symmetric solutions correspond to $(4,4)$ supersymmetry on the worldsheet of the fundamental string and are thus presumably exact to all orders in α'.

Of particular interest is the solitonic string, since its couplings to the background fields of supergravity compared to those of the fundamental string are such that the dilaton/axion field S is replaced by the modulus field T. Thus under string/fivebrane duality, the $SL(2,Z)$ strong/weak coupling duality trades places with the $SL(2,Z)$ target space duality, in accordance with recent observations of Schwarz and Sen [20] and Binétruy [21].

2. The General Ansatz

We first summarize the 't Hooft ansatz for the Yang-Mills instanton. Consider the four-dimensional Euclidean action

$$S = -\frac{1}{2g^2} \int d^4x \, \mathrm{Tr} F_{\mu\nu} F^{\mu\nu}, \qquad \mu,\nu = 1,2,3,4. \qquad (2.1)$$

For gauge group $SU(2)$, the fields may be written as $A_\mu = (g/2i)\sigma^a A_\mu^a$ and $F_{\mu\nu} = (g/2i)\sigma^a F_{\mu\nu}^a$ (where σ^a, $a = 1,2,3$ are the 2×2 Pauli matrices). A self-dual solution (but not the most general one) to the equation of motion of this action is given by the 't Hooft ansatz

$$A_\mu = i\overline{\Sigma}_{\mu\nu}\partial_\nu \ln f, \qquad (2.2)$$

where $\overline{\Sigma}_{\mu\nu} = \overline{\eta}^{i\mu\nu}(\sigma^i/2)$ for $i = 1,2,3$, where

$$\overline{\eta}^{i\mu\nu} = -\overline{\eta}^{i\nu\mu} = \epsilon^{i\mu\nu}, \qquad \mu,\nu = 1,2,3,$$
$$= -\delta^{i\mu}, \qquad \nu = 4 \qquad (2.3)$$

and where $f^{-1}\Box f = 0$. The ansatz for the anti-self-dual solution is similar, with the δ-term in (2.3) changing sign. From this ansatz, depending on how many of the four coordinates f is allowed to depend and depending on whether we compactify, we shall obtain $D = 10$ multi-fivebrane and $D = 4$ multi-monopole, multi-string and multi-domain wall solutions.

We will discuss these four cases in the next section. In this section, we do not specify the precise form of f or the dilaton function, but show that the derivation of the solution and most of the arguments used to demonstrate the exactness of the heterotic solution are equally valid for any f satisfying $f^{-1}\Box f = 0$.

It turns out that there is an analog to the 't Hooft ansatz for the Yang-Mills instanton in the gravitational sector of the string, namely the axionic instanton [22]. In its simplest form, this instanton appears as a solution for the massless fields of the bosonic string [19]. The identical instanton structure arises in all supersymmetric multi-fivebrane solutions [4,5], in particular in the tree-level neutral solution [4]:

$$g_{\mu\nu} = e^{2\phi}\delta_{\mu\nu} \qquad \mu,\nu = 1,2,3,4,$$
$$g_{ab} = \eta_{ab} \qquad a,b = 0,5,...,9, \qquad (2.4)$$
$$H_{\mu\nu\lambda} = \pm 2\epsilon_{\mu\nu\lambda\sigma}\partial^{\sigma}\phi \qquad \mu,\nu,\lambda,\sigma = 1,2,3,4,$$

with $e^{-2\phi}\Box\, e^{2\phi} = 0$. The D'Alembertian refers to the four-dimensional subspace $\mu,\nu,\lambda,\sigma = 1,2,3,4$ and ϕ is taken to be independent of $(x^0, x^5, x^6, x^7, x^8, x^9)$. For zero background fermionic fields the above solution breaks half the spacetime supersymmetries.

The generalized curvature of this solution was shown [19,23] to possess (anti) self-dual structure similar to that of the 't Hooft ansatz. To see this we define a generalized curvature $\hat{R}^{\mu}{}_{\nu\rho\sigma}$ in terms of the standard curvature $R^{\mu}{}_{\nu\rho\sigma}$ and $H_{\mu\alpha\beta}$:

$$\hat{R}^{\mu}{}_{\nu\rho\sigma} = R^{\mu}{}_{\nu\rho\sigma} + \frac{1}{2}\left(\nabla_{\sigma}H^{\mu}{}_{\nu\rho} - \nabla_{\rho}H^{\mu}{}_{\nu\sigma}\right) + \frac{1}{4}\left(H^{\lambda}{}_{\nu\rho}H^{\mu}{}_{\sigma\lambda} - H^{\lambda}{}_{\nu\sigma}H^{\mu}{}_{\rho\lambda}\right). \qquad (2.5)$$

One can also define $\hat{R}^{\mu}{}_{\nu\rho\sigma}$ as the Riemann tensor generated by the generalized Christoffel symbols $\hat{\Gamma}^{\mu}_{\alpha\beta}$, where $\hat{\Gamma}^{\mu}_{\alpha\beta} = \Gamma^{\mu}_{\alpha\beta} - (1/2)H^{\mu}{}_{\alpha\beta}$. The crucial observation for obtaining higher-loop and even exact solutions is the following. For any solution given by (2.4), we can express the generalized curvature in terms of the dilaton field as [19]

$$\hat{R}^{\mu}{}_{\nu\rho\sigma} = \delta_{\mu\sigma}\nabla_{\rho}\nabla_{\nu}\phi - \delta_{\mu\rho}\nabla_{\sigma}\nabla_{\nu}\phi + \delta_{\nu\rho}\nabla_{\sigma}\nabla_{\mu}\phi - \delta_{\nu\sigma}\nabla_{\rho}\nabla_{\mu}\phi$$
$$\pm \epsilon_{\mu\nu\rho\lambda}\nabla_{\sigma}\nabla_{\lambda}\phi \mp \epsilon_{\mu\nu\sigma\lambda}\nabla_{\rho}\nabla_{\lambda}\phi. \qquad (2.6)$$

It easily follows that

$$\hat{R}^{\mu}{}_{\nu\rho\sigma} = \mp\frac{1}{2}\epsilon_{\rho\sigma}{}^{\lambda\gamma}\hat{R}^{\mu}{}_{\nu\lambda\gamma}. \qquad (2.7)$$

So the (anti) self-duality appears in the gravitational sector of the string in terms of its generalized curvature.

We now turn to the exact heterotic solution. The tree-level supersymmetric vacuum equations for the heterotic string are given by

$$\delta\psi_M = \left(\nabla_M - \tfrac{1}{4}H_{MAB}\Gamma^{AB}\right)\epsilon = 0,$$
$$\delta\lambda = \left(\Gamma^A\partial_A\phi - \tfrac{1}{6}H_{ABC}\Gamma^{ABC}\right)\epsilon = 0, \qquad (2.8)$$
$$\delta\chi = F_{AB}\Gamma^{AB}\epsilon = 0,$$

where $A, B, C, M = 0, 1, 2, ..., 9$ and where ψ_M, λ and χ are the gravitino, dilatino and gaugino fields. The Bianchi identity is given by

$$dH = \frac{\alpha'}{4}\left(\text{tr}R \wedge R - \frac{1}{30}\text{Tr}F \wedge F\right). \qquad (2.9)$$

The $(9+1)$-dimensional Majorana-Weyl fermions decompose into chiral spinors according to $SO(9,1) \supset SO(5,1) \otimes SO(4)$ for the $M^{9,1} \rightarrow M^{5,1} \times M^4$ decomposition. Then (2.4) with arbitrary dilaton and with constant chiral spinors $\epsilon_\pm$ solves the supersymmetry equations with zero background fermi fields provided the YM gauge field satisfies the instanton (anti) self-duality condition [2]

$$F_{\mu\nu} = \pm\frac{1}{2}\epsilon_{\mu\nu}{}^{\lambda\sigma}F_{\lambda\sigma}. \qquad (2.10)$$

In the absence of a gauge sector, the multi-fivebrane solution is identical to the "neutral" tree-level solution shown in (2.4). A perturbative "gauge" fivebrane solution was found in [2]. An exact solution is obtained as follows. Define a generalized connection by

$$\Omega^{AB}_{\pm M} = \omega^{AB}_M \pm H^{AB}_M \qquad (2.11)$$

in an $SU(2)$ subgroup of the gauge group, and equate it to the gauge connection A_μ [24] so that the corresponding curvature $R(\Omega_\pm)$ cancels against the Yang-Mills field strength F and $dH = 0$. For $e^{-2\phi}\Box\, e^{2\phi} = 0$ (or $e^{2\phi} = e^{2\phi_0}f$) the curvature of the generalized connection can be written in terms of the dilaton as in (2.6) from which it follows that both F and R are (anti) self-dual. This solution becomes exact since $A_\mu = \Omega_{\pm\mu}$ implies that all the higher order corrections vanish [5]. The self-dual solution for the gauge connection is then given by the 't Hooft ansatz. So the heterotic solution combines a YM instanton in the gauge sector with an axionic instanton in the gravity sector. In addition, the heterotic solution has finite action. Further arguments supporting the exactness of this solution based on $(4,4)$ worldsheet supersymmetry are shown in [5]. Note that at no point in this discussion do we refer to the specific form of f, so that all of the above arguments apply for an arbitrary solution of $f^{-1}\Box\, f = 0$.

3. Monopoles, Strings and Domain Walls

We now go back to the 't Hooft ansatz (2.1)-(2.3) and solve the equation $f^{-1}\Box f = 0$. If we take f to depend on all four coordinates we obtain a multi-instanton solution

$$f_I = 1 + \sum_{i=1}^{N} \frac{\rho_i^2}{|\vec{x} - \vec{a}_i|^2}, \tag{3.1}$$

where ρ_i^2 is the instanton scale size and $\vec{a}_i$ the location in four-space of the ith instanton. For $e^{2\phi} = e^{2\phi_0} f_I$, and assuming no dimensions are compactified, we obtain from (2.4) the neutral fivebrane of [4] and the exact heterotic fivebrane of [5,6] in $D = 10$. The solitonic fivebrane tension $\widetilde{T_6}$ is related to the fundamental string tension T_2 $(= 1/2\pi\alpha')$ by the Dirac quantization condition [3]

$$\kappa_{10}^2 \widetilde{T_6} T_2 = n\pi, \tag{3.2}$$

where n is an integer and where κ_{10}^2 is the $D = 10$ gravitational constant. This implies $\rho_i^2 = e^{-2\phi_0} n_i \alpha'$, where n_i are integers. Near each source the solution is described by an exact conformal field theory [22,19,5].

Instead, let us single out a direction in the transverse four-space (say x^4) and assume all fields are independent of this coordinate. Since all fields are already independent of x^5, x^6, x^7, x^8, x^9, we may consistently assume the $x^4, x^5, x^6, x^7, x^8, x^9$ are compactified on a six-dimensional torus, where we shall take the x^4 circle to have circumference $Le^{-\phi_0}$ and the rest to have circumference L, so that $\kappa_4^2 = \kappa_{10}^2 e^{\phi_0}/L^6$. Then the solution for f satisfying $f^{-1}\Box f = 0$ has multi-monopole structure

$$f_M = 1 + \sum_{i=1}^{N} \frac{m_i}{|\vec{x} - \vec{a}_i|}, \tag{3.3}$$

where m_i is proportional to the charge and $\vec{a}_i$ the location in the three-space (123) of the ith monopole. If we make the identification $\Phi \equiv A_4$ then the lagrangian density may be rewritten as

$$F_{\mu\nu}^a F_{\mu\nu}^a = F_{jk}^a F_{jk}^a + 2F_{k4}^a F_{k4}^a = F_{jk}^a F_{jk}^a + 2D_k \Phi^a D_k \Phi^a, \tag{3.4}$$

where $j, k = 1, 2, 3$. We now go to $3 + 1$ space (0123) with the Lagrangian density

$$\mathcal{L} = -\frac{1}{4} G_{\alpha\beta}^a G^{\alpha\beta a} - \frac{1}{2} D_\alpha \Phi^a D^\alpha \Phi^a, \tag{3.5}$$

where $\alpha, \beta = 0, 1, 2, 3$. It follows that the above multi-monopole ansatz is a static solution with $A_0^a = 0$ and all time derivatives vanish. The solution in $3+1$ dimensions has the form

$$\Phi^a = \mp \frac{1}{g} \delta^{aj} \partial_j \omega,$$
$$A_k^a = \frac{1}{g} \epsilon^{akj} \partial_j \omega,$$

(3.6)

where $j, k = 1, 2, 3$ and where $\omega \equiv \ln f$ and g is the YM coupling constant. This solution represents a multi-monopole configuration with sources at $\vec{a}_i, i = 1, 2 ... N$ [16,17]. For $e^{2\phi} = e^{2\phi_0} f_M$, we obtain from (2.4) a neutral monopole solution and the exact heterotic monopole solution of [16,17]. The monopole strength is given by $\tilde{g} = \sqrt{2} \kappa_4 \widetilde{T_1}$, where $\widetilde{T_1} = \widetilde{T_6} L^5$ obeys, from (3.2), the quantization condition

$$e^{-\phi_0} \kappa_4^2 T_2 \widetilde{T_1} = \frac{n\pi}{L}.$$

(3.7)

This implies $m_i = e^{-\phi_0} n_i \pi \alpha' / L$. Similarly the "electric" charge of the fundamental string is $e = \sqrt{2} \kappa_4 T_1$, where $T_1 = T_2 L e^{-\phi_0}$, and hence

$$e\tilde{g} = 2\pi n$$

(3.8)

as expected. Unlike for the instanton, in the monopole case we cannot identify the explicit coset conformal field theory near each source. A noteworthy feature of this solution is that the divergences from both gauge and gravitational sectors cancel to yield a finite lagrangian, and finite soliton mass.

It is straightforward to reduce the multi-monopole solution to an explicit solution in the four-dimensional space (0123). The gauge field reduction is exactly as above, i.e. we replace A_4 with the scalar field Φ. In the gravitational sector, the reduction from ten to five dimensions is trivial, as the metric is flat in the subspace (56789). In going from five to four dimensions, one follows the usual Kaluza-Klein procedure of replacing g_{44} with a scalar field $e^{-2\sigma}$. The tree-level effective action reduces in four dimensions to

$$S_4 = \frac{1}{2\kappa_4^2} \int d^4x \sqrt{-g} e^{-2\phi-\sigma} \left(R + 4(\partial\phi)^2 + 4\partial\sigma \cdot \partial\phi - e^{2\sigma} \frac{M_{\alpha\beta} M^{\alpha\beta}}{4} \right),$$

(3.9)

where $\alpha, \beta = 0, 1, 2, 3$, where $M_{\alpha\beta} = H_{\alpha\beta} = \partial_\alpha B_\beta - \partial_\beta B_\alpha$, and where $B_\alpha = B_{\alpha 4}$. The four-dimensional monopole solution for this reduced action is then given by

$$e^{2\phi} = e^{-2\sigma} = e^{2\phi_0} \left(1 + \sum_{i=1}^{N} \frac{m_i}{|\vec{x} - \vec{a}_i|} \right),$$
$$ds^2 = -dt^2 + e^{2\phi} \left(dx_1^2 + dx_2^2 + dx_3^2 \right),$$
$$M_{ij} = \pm \epsilon_{ijk} \partial_k e^{2\phi}, \qquad i, j, k = 1, 2, 3.$$

(3.10)

Since the tree-level solution is exact, we need not reduce the higher order corrections to the action.

We now modify the solution of the 't Hooft ansatz even further and choose two directions in the four-space (1234) (say x^3 and x^4) and assume all fields are independent of both of these coordinates. We may now consistently assume that $x^3, x^4, x^6, x^7, x^8, x^9$ are compactified on a six-dimensional torus, where we shall take the x^3 and x^4 circles to have circumference $Le^{-\phi_0}$ and the remainder to have circumference L, so that $\kappa_4^2 = \kappa_{10}^2 e^{2\phi_0}/L^6$. Then the solution for f satisfying $f^{-1}\Box f = 0$ has multi-string structure

$$f_S = 1 - \sum_{i=1}^{N} \lambda_i \ln |\vec{x} - \vec{a}_i|, \tag{3.11}$$

where λ_i is the charge per unit length and $\vec{a}_i$ the location in the two-space (12) of the ith string. If we make the identification $\Phi \equiv A_4$ and $\Psi \equiv A_3$ then the lagrangian density for the above ansatz can be rewritten as

$$F_{\mu\nu}^a F_{\mu\nu}^a = F_{jk}^a F_{jk}^a + 2D_k\Phi^a D_k\Phi^a + 2D_k\Psi^a D_k\Psi^a, \tag{3.12}$$

where $j, k = 1, 2$. We now go to the $3 + 1$ space (0125) with the lagrangian density

$$\mathcal{L} = -\frac{1}{4}G_{\rho\sigma}^a G^{\rho\sigma a} - \frac{1}{2}D_\rho\Phi^a D^\rho\Phi^a - \frac{1}{2}D_\rho\Psi^a D^\rho\Psi^a, \tag{3.13}$$

where $\rho, \sigma = 0, 1, 2, 5$. It follows that the multi-string ansatz is a static solution with $A_0^a = 0$ and all time derivatives vanish. The solution in $3 + 1$ dimensions has the form

$$\Phi^a = \mp\frac{1}{g}\delta^{aj}\partial_j\omega,$$

$$\Psi^k = \frac{1}{g}\epsilon^{kj}\partial_j\omega, \tag{3.14}$$

$$A_k^a = -\delta^{a3}\frac{1}{g}\epsilon^{kj}\partial_j\omega,$$

where $j, k = 1, 2$ and where $\omega \equiv \ln f$. This solution represents a multi-string configuration with sources at $\vec{a}_i, i = 1, 2...N$. By setting $e^{2\phi} = e^{2\phi_0}f_S$, we obtain from (2.4) a neutral multi-string solution and an exact heterotic multi-string solution. The solitonic string tension $\widetilde{T_2}$ is given by $\widetilde{T_6}L^4$ and from (3.2) is related to the fundamental string tension T_2 by

$$e^{-2\phi_0}\kappa_4^2 T_2 \widetilde{T_2} = \frac{n\pi}{L^2}. \tag{3.15}$$

This implies $\lambda_i = n_i 2\pi\alpha'/L^2$. Like the monopole and unlike the instanton, we cannot identify an explicit coset conformal field theory near each source. Also like the monopole, the lagrangian per unit length for the string solution is finite as a result of the cancellation of divergences between the gauge and gravitational sectors.

As in the multi-monopole case, it is straightforward to reduce the multi-string solution to a solution in the four-dimensional space (0125). The gauge field reduction is done in (3.14). In the gravitational sector, the reduction from ten to six dimensions is trivial, as the metric is flat in the subspace (6789). In going from six to four dimensions, we compactify the x_3 and x_4 directions and again follow the Kaluza-Klein procedure by replacing g_{33} and g_{44} with a scalar field $e^{-2\sigma}$. The tree-level effective action reduces in four dimensions to

$$S_4 = \frac{1}{2\kappa_4^2} \int d^4x \sqrt{-g} e^{-2\phi-2\sigma} \left(R + 4(\partial\phi)^2 + 8\partial\sigma \cdot \partial\phi + 2(\partial\sigma)^2 - e^{4\sigma}\frac{N_\rho N^\rho}{2} \right), \quad (3.16)$$

where $\rho = 0,1,2,5$, where $N_\rho = H_{\rho 34} = \partial_\rho B$, and where $B = B_{34}$. The four-dimensional string soliton solution for this reduced action is then given by

$$e^{2\phi} = e^{-2\sigma} = e^{2\phi_0} \left(1 - \sum_{i=1}^{N} \lambda_i \ln|\vec{x} - \vec{a}_i| \right),$$
$$ds^2 = -dt^2 + dx_5^2 + e^{2\phi} \left(dx_1^2 + dx_2^2 \right), \qquad (3.17)$$
$$N_i = \pm\epsilon_{ij}\partial_j e^{2\phi}.$$

Again since the tree-level solution is exact, we do not bother to reduce the higher order corrections to the action.

We complete the family of solitons that can be obtained from the solutions of the 't Hooft ansatz by demanding that f depend on only one coordinate, say x^1. We may now consistently assume that $x^2, x^3, x^4, x^7, x^8, x^9$ are compactified on a six-dimensional torus, where we shall take the x^2, x^3 and x^4 circles to have circumference $Le^{-\phi_0}$ and the rest to have circumference L, so that $\kappa_4^2 = \kappa_{10}^2 e^{3\phi_0}/L^6$. Then the solution of $f^{-1}\Box f = 0$ has domain wall structure with the "confining potential"

$$f_D = 1 + \sum_{i=1}^{N} \Lambda_i |x_1 - a_i|, \qquad (3.18)$$

where Λ_i are constants. By setting $e^{2\phi} = e^{2\phi_0} f_D$, we obtain from (2.4) a neutral domain wall solution and an exact heterotic domain wall solution. The solitonic domain wall tension $\widetilde{T_3}$ is given by $\widetilde{T_6}L^3$ and from (3.2) is related to the fundamental string tension T_2 by

$$e^{-3\phi_0}\kappa_4^2 T_2 \widetilde{T_3} = \frac{n\pi}{L^3}. \tag{3.19}$$

This implies $\Lambda_i = e^{\phi_0}n_i(2\pi)^2\alpha'/L^3$. Like the monopole and string we cannot identify an explicit coset conformal field theory near each source. Again the reduction to $D = 4$ is straightforward. In the gauge sector, the action reduces to YM + three scalar fields Φ, Ψ and Π. For the spacetime (0156) the solution for the fields is given by

$$
\begin{aligned}
\Phi^1 &= \mp\frac{\Lambda}{g(1+\Lambda|x_1|)}, \\
\Psi^3 &= \frac{\Lambda}{g(1+\Lambda|x_1|)}, \\
\Pi^2 &= -\frac{\Lambda}{g(1+\Lambda|x_1|)}, \\
A_\mu &= 0,
\end{aligned}
\tag{3.20}
$$

where $\mu = 0,1,5,6$. In the gravitational sector the tree-level effective action in $D = 4$ has the form

$$S_4 = \frac{1}{2\kappa^2}\int d^4x\sqrt{-g}\,e^{-2\phi-3\sigma}\left(R + 4(\partial\phi)^2 + 12\partial\sigma\cdot\partial\phi + 6(\partial\sigma)^2 - e^{6\sigma}\frac{P^2}{2}\right), \tag{3.21}$$

where $P = H_{234}$. The four-dimensional domain wall solution for this reduced action is then given by

$$
\begin{aligned}
e^{2\phi} = e^{-2\sigma} &= e^{2\phi_0}\left(1 + \Lambda|x_1|\right), \\
ds^2 &= -dt^2 + dx_5^2 + dx_6^2 + e^{2\phi}dx_1^2, \\
P &= \Lambda\left(\Theta(x_1) - \Theta(-x_1)\right).
\end{aligned}
\tag{3.22}
$$

Again since the tree-level solution is exact, we do not bother to reduce the higher order corrections to the action. A trivial change of coordinates reveals that the spacetime is, in fact, flat. Dilaton domain walls with a flat spacetime have recently been discussed in a somewhat different context in [25,26].

As for the fivebrane in $D = 10$, the mass of the monopole, the mass per unit length of the string and the mass per unit area of the domain wall saturate a Bogomol'nyi bound with the topological charge. (In the case of the string and domain, wall, however, we must follow [27] and extrapolate the meaning of the ADM mass to non-asymptotically flat spacetimes.)

4. String/String Duality

Let us focus on the solitonic string configuration (3.17) in the case of a single source. In terms of the complex field

$$T = T_1 + iT_2$$
$$= B_{34} + ie^{-2\sigma} \tag{4.1}$$
$$= B_{34} + i\sqrt{\det g^S_{mn}} \qquad m, n = 3, 4,$$

where g^S_{MN} is the string σ-model metric, the solution takes the form (with $z = x_1 + x_2$)

$$T = \frac{1}{2\pi i} \ln \frac{z}{r_0},$$
$$ds^2 = -dt^2 + dx_5^2 - \frac{1}{2\pi} \ln \frac{r}{r_0} dz d\bar{z}, \tag{4.2}$$

whereas both the four-dimensional (shifted) dilaton $\eta = \phi + \sigma$ and the four-dimensional two-form $B_{\mu\nu}$ are zero. In terms of the canonical metric $g_{\mu\nu}$, T_1 and T_2, the relevant part of the action takes the form

$$S_4 = \frac{1}{2\kappa_4^2} \int d^4x \sqrt{-g} \left(R - \frac{1}{2T_2^2} g^{\mu\nu} \partial_\mu T \partial_\nu \bar{T} \right) \tag{4.3}$$

and is invariant under the $SL(2, R)$ transformation

$$T \rightarrow \frac{aT + b}{cT + d}, \qquad ad - bc = 1. \tag{4.4}$$

The discrete subgroup $SL(2, Z)$, for which a, b, c and d are integers, is just a subgroup of the $O(6, 22; Z)$ target space duality, which can be shown to be an exact symmetry of the compactified string theory at each order of the string loop perturbation expansion.

This $SL(2, Z)$ is to be contrasted with the $SL(2, Z)$ symmetry of the elementary four-dimensional solution of Dabholkar et al. [27]. In their solution T_1 and T_2 are zero, but η and $B_{\mu\nu}$ are non-zero. The relevant part of the action is

$$S_4 = \frac{1}{2\kappa_4^2} \int d^4x \sqrt{-g} \left(R - 2g^{\mu\nu} \partial_\mu \eta \partial_\nu \eta - \frac{1}{12} e^{-4\eta} H_{\mu\nu\rho} H^{\mu\nu\rho} \right). \tag{4.5}$$

The equations of motion of this theory also display an $SL(2, R)$ symmetry, but this becomes manifest only after dualizing and introducing the axion field a via

$$\sqrt{-g} g^{\mu\nu} \partial_\nu a = \frac{1}{3!} \epsilon^{\mu\nu\rho\sigma} H_{\nu\rho\sigma} e^{-4\eta}. \tag{4.6}$$

Then in terms of the complex field

$$S = S_1 + iS_2$$
$$= a + ie^{-2\eta} \tag{4.7}$$

the Dabholkar *et al.* fundamental string solution may be written

$$S = \frac{1}{2\pi i} \ln \frac{z}{r_0},$$
$$ds^2 = -dt^2 + dx_5^2 - \frac{1}{2\pi} \ln \frac{r}{r_0} dz d\bar{z}. \tag{4.8}$$

Thus (4.2) and (4.8) are the same with the replacement $T \leftrightarrow S$. It has been conjectured that this second $SL(2, Z)$ symmetry may also be a symmetry of string theory [28,29,30], but this is far from obvious order by order in the string loop expansion since it involves a strong/weak coupling duality $\eta \to -\eta$. What interpretation are we to give to these two $SL(2, Z)$ symmetries: one an obvious symmetry of the fundamental string and the other an obscure symmetry of the fundamental string?

While the present work was in progress, we became aware of recent interesting papers by Sen [31], Schwarz and Sen [20] and Binétruy [21]. In particular, Sen draws attention to the Dabholkar *et al.* string solution (4.8) and its associated $SL(2, Z)$ symmetry as supporting evidence in favor of the conjecture that $SL(2, Z)$ invariance may indeed be an exact symmetry of string theory. He also notes that the spectrum of electric and magnetic charges is consistent with the proposed $SL(2, Z)$ symmetry [31].[†]

All of these observations fall into place if one accepts the proposal of Schwarz and Sen [20]: *under string/fivebrane duality the roles of the target-space duality and the strong/weak coupling duality are interchanged !* This proposal is entirely consistent with an earlier one that under string/fivebrane duality the roles of the σ-model loop expansion and the string loop expansion are interchanged [10]. In this light, the two $SL(2, Z)$ symmetries discussed above are just what one expects. From the string point of view, the T-field $SL(2, Z)$ is an obvious target space symmetry, manifest order by order in string loops whereas the S-field $SL(2, Z)$ is an obscure strong/weak coupling symmetry. From the fivebrane point of view, it is the T-field $SL(2, Z)$ which is obscure while the S-field $SL(2, Z)$ is an "obvious" target

---

[†] Sen also discusses the concept of a "dual string", but for him this is obtained from the fundamental string by an $SL(2, Z)$ transform. For us, a dual string is obtained by the replacement $S \leftrightarrow T$.

space symmetry. (This has not yet been proved except at the level of the low-energy field theory, however. It would be interesting to have a proof starting from the worldvolume of the fivebrane.) This interchange in the roles of the S and T field in going from the string to the fivebrane has also been noted by Binétruy [21]. It is made more explicit when S is expressed in terms of the variables appearing naturally in the fivebrane version

$$S = S_1 + iS_2$$
$$= a_{346789} + ie^{-2\eta}, \tag{4.9}$$
$$= a_{346789} + i\sqrt{\det g^F_{mn}}, \qquad m, n = 3, 4, 6, 7, 8, 9,$$

where $g^F_{MN} = e^{-2\phi/3} g^S_{MN}$ is the fivebrane σ-model metric [3] and a_{MNPQRS} is the 6-form which couples to the 6-dimensional worldvolume of the fivebrane, in complete analogy with (4.1).

Note, however, that unlike the Dabholkar *et al.* solution, our symmetric solution (3.14) also involves the non-abelian gauge fields A_ρ, Φ, Ψ whose interactions appear to destroy the $SL(2, Z)$. This remains a puzzle (A generalization of the $D = 4$ Dabholkar *et al.* solution involving gauge fields may also be possible by obtaining it as a soliton of the fivebrane theory. This would involve a $D = 4$ analogue of the $D = 10$ solution discussed in [7].)

5. Discussion

It may at first sight seem strange that a string can be dual to another string in $D = 4$. After all, the usual formula relating the dimension of an extended object, d, to that of the dual object, $\tilde{d}$, is $\tilde{d} = D - d - 2$. So one might expect string/string duality only in $D = 6$ [10]. However, when we compactify n dimensions and allow the dual object to wrap around $m \leq \tilde{d}-1$ of the compactified directions we find $\tilde{d}_{\text{effective}} = \tilde{d}-m = D_{\text{effective}}-d-2+(n-m)$, where $D_{\text{effective}} = D - n$. In particular for $D_{\text{effective}} = 4$, $d = 2$, $n = 6$ and $m = 4$, we find $\tilde{d}_{\text{effective}} = 2$.

Thus the whole string/fivebrane duality conjecture is put in a different light when viewed from four dimensions. After all, our understanding of the quantum theory of five-branes in $D = 10$ is rather poor, whereas the quantum theory of strings in $D = 4$ is comparatively well-understood (although we still have to worry about the monopoles and domain walls). In particular, the dual string will presumably exhibit the normal kind of mass spectrum with linearly rising Regge trajectories, since the classical ($\hbar$-independent) string expression $\widetilde{T}_6 L^4 \times$ (angular momentum) has dimensions of (mass)^2, whereas the analogous classical expression for an uncompactified fivebrane is $(\widetilde{T}_6)^{1/5} \times$ (angular momentum) which has dimensions (mass)$^{6/5}$ [1]. Indeed, together with the observation that the $SL(2, Z)$ strong/weak coupling duality appears only after compactifying at least 6 dimensions, it is tempting to revive the earlier conjecture [1,15] that the internal consistency of the fivebrane may actually *require* compactification.

Acknowledgements

We would like to thank Pierre Binétruy, Ruben Minasian, Joachim Rahmfeld, John Schwarz and Ashoke Sen for helpful discussions.

References

[1] M. J. Duff, Class. Quant. Grav. **5** (1988).

[2] A. Strominger, Nucl. Phys. **B343** (1990) 167.

[3] M. J. Duff and J. X. Lu, Nucl. Phys. **B354** (1991) 129.

[4] M. J. Duff and J. X. Lu, Nucl. Phys. **B354** (1991) 141.

[5] C. G. Callan, J. A. Harvey and A. Strominger, Nucl. Phys. **B359** (1991) 611.

[6] C. G. Callan, J. A. Harvey and A. Strominger, Nucl. Phys. **B367** (1991) 60.

[7] M. J. Duff and J. X. Lu, Phys. Rev. Lett. **66** (1991) 1402.

[8] M. J. Duff and J. X. Lu, Class. Quant. Gravity **9** (1992) 1.

[9] M. J. Duff, R. R. Khuri and J. X. Lu, Nucl. Phys. **B377** (1992) 281.

[10] M. J. Duff and J. X. Lu, Nucl. Phys. **B357** (1991) 534.

[11] J. A. Dixon, M. J. Duff and J. C. Plefka, Phys. Rev. Lett. **69** (1992) 3009.

[12] C. Montonen and D. Olive, Phys. Lett. **B72** (1977) 117.

[13] M. J. Duff, P. Howe, T. Inami and K. S. Stelle, Phys. Lett. **B191** (1987) 70.

[14] M. J. Duff, T. Inami, C. N. Pope, E. Sezgin and K. S. Stelle, Nucl. Phys. **B297** (1988) 515.

[15] K. Fujikawa and J. Kubo, Nucl. Phys. **B356** (1991) 208.

[16] R. R. Khuri, Phys. Lett. **B294** (1992) 325.

[17] R. R. Khuri, Nucl. Phys. **B387** (1992) 315.

[18] J. P. Gauntlett, J. A. Harvey and J. T. Liu, EFI-92-67, IFP-434-UNC.

[19] R. R. Khuri, Phys. Lett. **B259** (1991) 261.

[20] J. H. Schwarz and A. Sen, NSF-ITP-93-46, CALT-68-1863, TIFR-TH-93-19.

[21] P. Binétruy, NSF-ITP-93-60.

[22] S. J. Rey, Phys. Rev. **D43** (1991) 526.

[23] R. R. Khuri, Phys. Rev. **D46** (1992) 4526.

[24] J. M. Charap and M. J. Duff, Phys. Lett. **B69** (1977) 445.

[25] H. La, CTP-TAMU-52/92.

[26] M. Cvetič, UPR-560-T.

[27] A. Dabholkar, G. Gibbons, J. A. Harvey and F. Ruiz Ruiz, Nucl. Phys. **B340** (1990) 33.

[28] A. Font, L. Ibáñez, D. Lust and F. Quevedo, Phys. Lett. **B249** (1990) 35.

[29] A. Sen, TIFR-TH-92-41.

[30] J. Schwarz, CALT-68-1815.

[31] A. Sen, TIFR-TH-93-03.

Gravity's measurements in the 10÷100 m range of distance

S. Focardi

Department of physics Bologna University and INFN Bologna

1   Introduction

It is known that the gravitation law describes one of the fundamental interactions in physics. Electroweak and nuclear force, together with the gravitational one, complete the picture of the basic interactions.

One of the most ambitious projects, in physics, is the demonstration that all these forces are different manifestations of a single force. In other words, at sufficient high energies all these forces would be unified, and their actual apparent difference would be a consequence of the low energies.

In the last years, some prospectives have arised for a possible unification of the electroweak and nuclear forces.   More difficulties exist for the unification of the gravitational force with the other ones. The main trouble is that gravitation can be described within the framework of the general relativity theory which gives essentially a geometric interpretation for this force; whilst the other interactions are treated in terms of virtual particle exchange. Gravitation requires definite information on position and momentum in contrast to quantization rules.   The contradiction may be solved by means of three different possibilities: make partial changes in quantum mechanics, modify general relativity or correct both the theories. An attempt to unify gravitation with the other forces is pursued from quantum gravity theories.   A very interesting prediction of the quantum gravity theories based on local supersymmetry is the existence of a spin-1 partner of the graviton (graviphoton) . In some cases, a second graviton's partner, having spin-0 (graviscalar), is also expected. Both particles would acquire masses different from zero because of the symmetry breaking. Forces mediated by the graviphoton and by the graviscalar are, respectively, repulsive and attractive.

2   The gravitation

As it is well known Newton formulated the attracting force between pointlike massive particles as

$$F = G\frac{m_1 m_2}{r^2}.$$

145

V. de Sabbata and H. Tso-Hsiu (eds.), Cosmology and Particle Physics, 145–158.

The same physical phenomenon can be described in terms of potential energy

$$U = -G\frac{m_1 m_2}{r}.$$

The masses m_1 and m_2 of the two interacting particles indicate their gravitational effect (gravitational masses). The mass appears also in another fundamental law of the physics

$$\mathbf{F} = m\mathbf{a}.$$

In this case, the mass is qualified as inertial mass.

Newton was the first who posed himself the matter of the independence of the ratio $\frac{m_i}{m_g}$ from the material composition. This hypothesis of independence is a postulate of the general relativity theory and is known as the weak equivalence principle.

Two important limits to the ratio R defined as

$$R = \frac{\left(\frac{m_i}{m_g}\right)_1 - \left(\frac{m_i}{m_g}\right)_2}{\left(\frac{m_i}{m_g}\right)_1}.$$

have been set experimentally.

The first one was determinated by Eötwös in 1909

$$R_E < 3.10^{-9}$$

and presented to the Beneke Foundation in Gottingen. The text of this comunication is now unknown. The detailed description of the experiment was published only in 1922[1].

The second limit was setted by Braginskii [2] in 1972

$$R_B < 0.9.10^{-12}.$$

An important difference between R_E and R_B is that R_E has been determined by using the Earth as gravitational source, whereas for R_B the Sun has been used.

In 1986 Fischbach et al [3] published a paper about a new analysis of the Eötwös experiment. They showed that the original data are compatible with a value for R_E different from zero. This result is in contrast with the weak equivalence principle.

Fischbach et al. write the gravitational energy in the form

$$U = -G\frac{m_1 m_2}{r}(1 + \alpha e^{-r/\lambda})$$

The coefficient α, which sets the amplitude of the new term relative to the old newtonian one, must depend from the composition of the materials. In fact the existence of the Braginskii's limit requires the introduction of a term vanishing at distances as large as the average sun-earth distance.

[1] Eötwös R.v, Pekar D., Fekete E., Annalen der Physik, 68, 11, 1992

[2] Braginskii V.B., Panov V.I., Soviet Physics JEPT 34, 463, 1972

[3] Fischbach E., Sudarsky D., Szafer A., Talmadge C., Aronson S.H., Physical Review Letters, 56, 1, 1986

3 Recent experiments on the gravitation

The Fischbach's paper contained three very exciting prospects to be subjected to a rigorous experimental check:
a) a possible violation of the weak equivalence principle
b) a possible modification of the Newton's law
c) an indirect support to the supersymmetric theories.
The experiments started after that paper can be separated in two different classes: experiments searching for violation of the weak equivalence principle and experiments searching for a correction of the Newton's law. The experiments belonging to the first class are more ambitious because they would prove also point b) and c). The experiments belonging to the second class would prove point a) only if the Fischbach's analysis were correct.
First class experiments will not be discussed here. An exhaustive discussion can be found in reference[4]. Most of them did not find any violation effect whilst few experiments showed evidence for anomalous effects. At the moment there is no hypothesis able to conciliate the contrasting results.
The experiments belonging to the second class can be further classified in two different categories. In the first one we can put experiments wich are model dependent, and in the second one model independent ones. The model dependent experiments were performed by measuring the gravitational field at different heights on towers or at differnt depths in mines, ice-pack or ocean. In all these cases, deviations from Newton's law can be unambiguously proved only if the gravitational field is known without uncertainty. Such a knowledge cannot be reached because the gravitational field depends on the masses distributions which can be calculated only by mapping the gravity acceleration at the ground. A review relative to such experiments can be found in reference [4]. Model independent experiments require, obviously, a fixed position for the measuring instrument. In such a cases the local gravitational field is constant and its knowledge is not required for the treatment of the experimental data.

4 The lake experiment

On the basis of the present experimental knowledge, the exixtence of a non-Newtonian term with a coefficient α of the order of 0.01, at distances between 10 and 1000 meters, cannot be excluded (see, e.g. Cook[5] and De Rujula [6]) .
In order to verify the Newton's law in the range of distances between 10 and 100 meters, an experiment was planned[7,8] which uses the changing in time of the gravitational field produced by the water-level's variations of a lake. The lake Brasimone (845 m above

---

[4]Fischbach E. and Talmadge C., Nature, 356, 207, 1992

[5]Cook A., Rep. Prog. Phys., 51, 707, 1988

[6]De Rujula A. Physics Letters B, 180, 213, 1986

[7]Achilli V., Baldi P., De Sabbata V., Focardi, S., Palmonari F., Pedrielli F., Proceedings of the XII Warsaw Symposium on elementary particle physics, Word Scientific ed., 589, 1990

[8]Achilli V., Baldi P., Focardi, S., Gasperini, P., Palmonari F., Sabadini R., Cahiers du Centre Europeen de Geodynamique et de Seismologie, 3, 81, 1990

sea level, located in Italy, about midway between Bologna and Florence) together with the Suviana lake constitutes a power storage used by ENEL (italian national electrical company) . During the night, water is pumped up from Suviana to Brasimone and the inverse process is used to produce electric power, at hours of maximun need. A research centre of ENEA (italian national institut for energy development) lies by the side of the Brasimone lake. 30 m below the ground level, a 100 m length tunnel departs from the ENEA centre and extends towards the centre of the lake. Such a happy combination of circumstances gave us the idea of putting a gravimeter at the end of the tunnel in order to measure gravity variations produced by the moving water masses. The aim of the experiment was to push the limit on α parameter toward 10^{-4}. To this end, the following measurements must be performed:

a) a survey of the shore morphology
b) a continuous record of the water level
c) a continuous record of the water and air temperature
d) a continuous record of the gravity acceleration
e) a measurement of the water density
f) a study of the relation between the water level in the lake and the subsidence
g) a control of the water-bearing stratum

Only a few of these measures will be discussed below.

5   The gravimeter

The gravimeter used in the experiment is a GWR superconducting gravimeter (number of our instrument T015). For the superconducting gravimeter the mechanical spring of previous instruments is replaced with a magnetic suspension. A niobium superconducting sphere floats in a magnetic field produced by two superconducting coils. The weak field gradient, adjusted by the ratio of the currents trapped in the coils, makes the sphere position extremely sensitive to the variations of the vertical component of gravity. The variation of the gravity accelaration g are recorded measuring the voltage variations induced on the terminals of a third coil whose current is changed by a feedback system wich keeps sphere position centered between two electrodes of a capacitive bridge[9]. The most important characteristic of a superconducting gravimeter is its sensitivity which can reach 10^{-2} μGal (1 Gal = 1 cm/s^2). During the day, g variations due to the tides produced by the moon and the sun are about 150 μGal. Fig 1 shows a diurnal variation of g, as measured by the gravimeter. The largest g variation, at the end of the tunnel, due to maximum water level variations is about 300 μGal.

The tidal variations of g depend on the place and are not completely known *a priori*. For such a reason, tides cannot be used for calibration of the gravimeter. The calibration method adopted by us consists in the mooving up and down around the gravimeter of a steinless steel ring whose mass and geometrical dimensions were accurately determinated. Since the gravitational effect produced by moving the ring between the two positions corresponding to maximum gravitational field can be calculated with

---

[9]Goodkind J.M., Cahiers du Centre Europeen de Geodynamique et de Seismologie, 3, 81, 1990

a relative precision of the order of 10^{-4}, the calibration coefficient can be determined, in principle, at such a precision level. The actual value, because of the limitated number of measurements,

$$C = (65.56 \pm 0.1)\mu Gal/volt$$

is not so precise. We plan to improve of a factor three the c value, in a next future. The c value was measured in a laboratory far from the lake and is associated with a particular set of superconducting currents. Fig. 2 shows the output voltage of the gravimeter versus the ring deplacement. From the graph it is possible to determinate the maxima positions to be used for successive calibrations. In the same laboratory, in the ENEA centre, far from the lake, tides were measured over a six months period, from july to december 1992. This data will be used in order to determine the amplitudes and the phases of the different tidal components. Data below the lake were also taken during two distint periods, from mars to july 1992 and from january 1993 to now. As shown in fig 3, diurnal variations of g, under the lake depend from the tides and also from the level variations of the lake. That variations are represented, for the same day, in Fig. 4.

6 The lake

Very precise measurements were performed in order to obtain an accurate knowledge of the lake's shores. A system of 21 datum point was set around the lake and the relative distances were measured by two different methods, the classic geodesic one and the GPS (global position system) which uses artificial earth satellites. Differences between the two methods are of the order of 2 mm on distances of 1 km. An aerophotogrammetry was successively carried out when the water level in the lake was near the minimum of its periodic variations. That data will be used to calculate newtonian gravitational effects in the position occupied by the gravimeter, under the lake.

Samples of water were drawed from the lake, in different days, in various positions and at different depths. Their density was measured and no differences between them were observved. The density distribution for all samples is shown in fig. 5. From these samples the water density at 21 ^{0}C was determinated as

$$\delta = (988.146 \pm 0.016)Kg/m^3)$$

Water temperature was continuously recorded by two thermometers placed respectively 1 m and 5 m below water level hanging from a buoy floating not far from the vertical over the gravimeter. In any case differences in temperature larger than a degree, between different points in the lake, were seldom observed.

Air temperature, relative humidity and atmospheric pressure were also continuously recorded, in order to deduce air density.

Water level was also continously taken with a very simple system which makes use of the existence of a well connected with the lake. A buoy in the well is connected by a wire to a digitizer wich permits to know the well level with a precision of 0.1 mm. Three termometers hanging in the well, at different depths permits to take in account thermal differences between lake and well.

7 The subsidence

The experiment is model independent, as affirmed in section 3. Therefore, this is not completely true because the sinking and the lifting produced at the end of the tunnel by the load changes dues to the deplaced water. In order to evaluate that effect, a clinometer, able to measure vertical relative deplacements, was setted in the tunnel. It consists in a water pipe whose two extremes, at the ends of the tunnel, are joned with level capacitive sensors. The clinometer can appreciate 1 μ relative deplacements over the 100 m tunnel length.

In fig. 6 and 7 the sinking and the lifting are respectively reported versus the level of the lake. Both the figures refer to changes produced by 10 cm water level variations. It is clearly shown that the phenomenon is not reversible and that an hysteresis occour wich is very probably attribuable to the clays.

We hope that this differential behaviour can permit the determination of the absolute vertical displacement in the gravimeter position.

8 Data analysis

At this moment, we can not still present the final conclusions of the experiment. To do this, we need to know the amplitudes and the phases of the different tidal components. The data taken at this end—from july to december 1992—are now ready to be processed. When the tide will be determined, it will be possible to subract them from the data taken below the lake. That process will permit to determine the effect wich corresponds to a given variation of the level.

The theoretical value expected from the Newton's gravitational law, for every 10 cm of water level variation, is shown in fig. 8 versus the level of the lake. This is the curve wich will be used for the comparason with the experimental data.

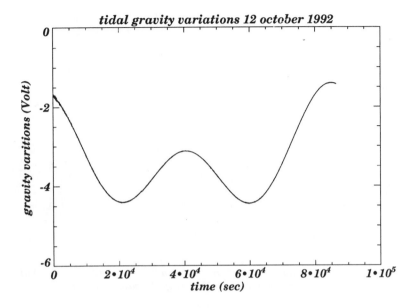

fig 1   A diurnal variation of g

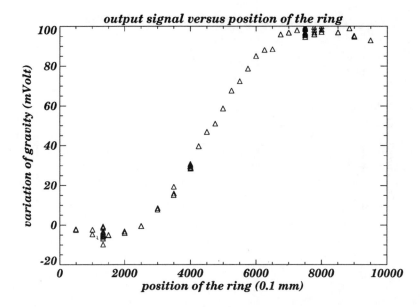

fig 2   The calibration curve of the gravimeter

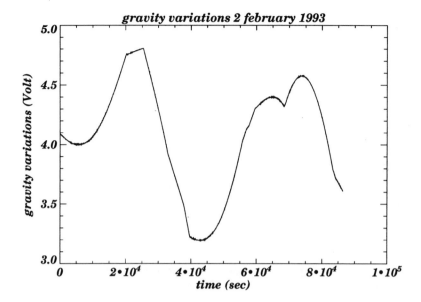

fig 3   A diurnal gravity variation under the lake

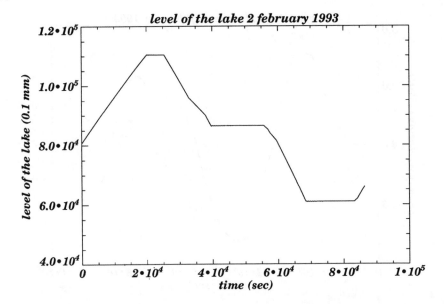

fig 4   A diurnal variation of the lake level

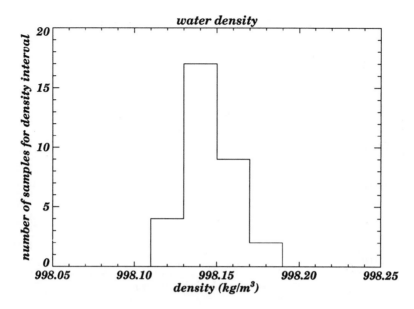

fig 5   The water density

156

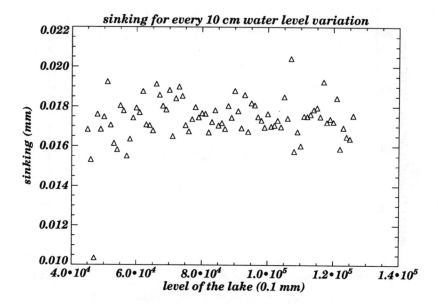

fig 6   The sinking for every 10 cm lake level variation

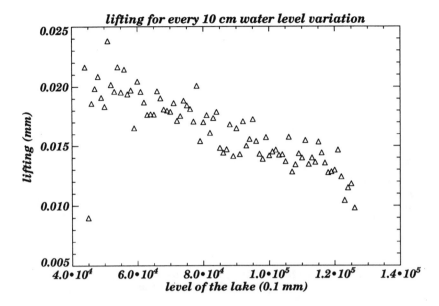

fig 7   The lifting for every 10 cm lake level variation

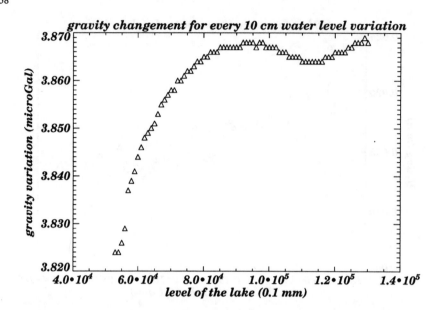

fig 8   The gravit signal expected for every 10 cm lake level variation

Long Distance Correlations of Matter Distribution in the Universe

H.E.Jørgensen,[1*] P.D.Naselsky,[1,2,4*] D.I.Novikov,[2,3*] I.D.Novikov,[1*,2,3]

1) University Observatory, Østervoldgade 3, Copenhagen, Denmark

2) NORDITA, Blegdamsvej 17, DK-2100 Copenhagen, Ø Denmark

3) Astro Space Center of P.N.Lebedev Physical Insitute

Profsoyuznaya 84/32, Moscow, 117810, Russia

4) Rostov State University, Stachki 194, Rostov-Don, Russia

*) Permanent address

Abstract

We discuss the specific anomalies in the spatial correlation function $\xi(r)$ at long distances and corresponding peculiarities in projected angular correlation function $w(\theta)$ of galaxy and galaxy cluster distributions in some cosmological models. We call these anomalies the Long Distance Correlations.

Such peculiarities could be the result of either the specific shape of the inflation potential in the early Universe or the consequences of sound waves in the ionized primordial plasma.

We investigate the dependence of the properties of the LDC from the parameters of the models and emphasize that the properties are very sensitive to the composition of dark matter.

The possibility to observe the LDC-effect is discussed. We propose a special method of a filtering of this effect in observational data.

Subject headings: cosmology: large scale structure of the Universe - galaxies: formation.

1. Introduction

The modern data on large scale distribution of galaxies, clusters of galaxies and QSO's[1-8] play a most significant role in determining the nature of dark matter in the Universe. It is very important that on large scales the power spectrum analysis

159

V. de Sabbata and H. Tso-Hsiu (eds.), Cosmology and Particle Physics, 159–182.

detects significant inhomogeneities in the Abell cluster correllations with a scale $r \geq 10^2 h^{-1} \text{Mpc}$[4,5,7].

This analysis of the cluster-cluster two-point correlation function indicates the existence of structure in the matter distribution on scales much larger than those given by the correlation length of galaxies. One of the most important characteristics of the large-scale distribution of matter is the existence of typical scales of the order of a few dozens or a hundred Mpc which are related to typical scales in the density spectrum after recombination. For example, in the baryonic dark matter model, with isocurvature perturbations there are typical natural scales of the size of the acoustic horizon at the moment of recombination (scaled to the present moment) $r_{acous} \simeq 100 \div 200 h^{-1} \text{Mpc}$. This fact was used by Peebles[9] for interpretation of the peculiar velocity field of galaxies. Gnedin and Ostriker[10] have investigated the problem of initial chemical composition of pregalactic matter in such models. Gouda et al.[11] and Makine and Sato[12] have discussed the expected small scale anisotropy of cosmic background radiation in the Peebles model.

Recently Jørgensen et al.[13] proposed the crucial test for the ionization history of the pregalactic plasma and the nature of dark matter to be an investigation of abnormal correlations on large scales in the distribution of galaxies and cluster of galaxies. These correlations are called Long Distance Correlations (LDC) as was done in Jørgensen et al.[13]. We use the standard definition for the dimensionless Hubble constant $h = (H_0/100 \text{km s}^{-1} \text{Mpc}^{-1})$ and Ω is the ratio of total to critical density at present time. These correlations are connected with the secondary adiabatic perturbations generated by primordial isocurvature perturbations before the epoch where the plasma becomes transparent.

At that moment the wavelengths of acoustic waves correspond to the comoving linear scales

$$\lambda_S \leq r \leq r_{rec} \frac{v_{acoust}}{c} = \tilde{R} \simeq 100 \div 200 h^{-1} Mpc$$

where λ_S is the Silk damping scale, r_{rec} is the event horizon at the redshifts z_{ls} of the last scattering of radiation ($z_{ls} \simeq 1100$ for the standard ionization history and $z_{ls} \simeq 20$ for the model with reionization); $\frac{v_{acoust}}{c} \approx 3^{-1/2}\sqrt{1 + 3\rho_b/4\rho_r}$ and ρ_b/ρ_r is the ratio of baryon and radiation densities at $z = z_{ls}$, c is the speed of light. After the plasma became transparent the primordial isocurvature and the second adiabatic perturbations grow due to gravitational instability and form galaxies and the large scale structure in the baryonic dark matter. Moreover, the present spectrum of spatial distribution of baryonic dark matter perturbations preserves the signature of the acoustic motions at the moment $z = z_{ls}$.

Peebles[14] showed that for the initial adiabatic perturbations and "standard" process of recombination the spatial correlation function $\xi(r)$ has a bump in the vicinity of $r = 2\tilde{R}$. The value of $\xi(r = 2\tilde{R})$ is connected directly with the acoustic modulations of the spectrum. In the model under consideration with the isocurvature initial perturbations the correlation function $\xi(r)$ has two peaks at $r \simeq \tilde{R}$ as well as at $r \simeq 2\tilde{R}$, see[13]. Note, that in the Peebles model (with reionization of the hydrogen plasma) the large velocity coherence scale is the result of the prominent peak in the power spectrum of density distribution and this peak must also exist in the correlation function $\xi(r)$ [9,13].

If the dark matter in the Universe is nonbaryonic (for example, cold dark matter (CDM)) then the spectrum of density perturbations after recombination has monotonic form[15]. For the standard CDM model with $h = 0.5$, $\Omega = 1$ the charateristic scale of correlation is less than $l \sim 20h^{-1}$Mpc. As discussed by Efstathiou et al.[16], it is difficult to account the correlations for the large value of $l \sim 50 \div 100h^{-1}$Mpc, if the large scale mass fluctuations grew out of primordial adiabatic perturbation with Harrison-Zeldovich spectrum $P_0(k) \propto k$. Recently Efstathiou et al.[16] have discussed the low-density model with the cosmological constant $\Omega_\lambda = 1 - \Omega$ and scale-invariant primordial spectrum and have demonstrated that this model fit the observational data mentioned above. Einasto et al.[7] have shown that Abell clusters correlation function has the spectral index $-2 \leq n \leq -1$ on intermediate scale $30 \div 50$Mpc, and $n = 1$ on very large scales $r \gg 150h^{-1}$Mpc. The transition from the spectral index $n = 1$ to a lower index occurs at the scale $\lambda \simeq 150 \pm 50h^{-1}$Mpc. All these facts tell us that the scales $\sim 100 \div 200$Mpc are specific for the observable matter distribution. Thus there is a problem to look for processes responsible for these peculiarities.

Note, that in modern inflationary cosmological models, there are a few possibilities to explain this additional power. At first, the primordial perturbation spectrum could be non scale invariant. The scale invariance could be broken, for example, if primordial perturbations are created by a power-law inflation or by complicated potentials of the inflation scalar field, see[17]. Second, additional structure of the spectrum can be easily explained within a model with two subsequent inflationary stages[18,19].

Recently Starobinsky[17] proposed a non-standard model of inflation which is based on a very specific form of the potential $V(\varphi)$ for the scalar field φ. Starobinsky[17] showed that the "smoothed" nonanalytical character of $V(\varphi)$ in the vicinity of $\varphi = \varphi_{cr}$, where the first derivative $\frac{dV}{d\varphi}(\varphi = \varphi_{cr})$ has a singular point, produces oscillations of the primordial spectrum with characteristical scale $R = R(\varphi_{cr}) \simeq 50 \div 75h^{-1}$Mpc (scaled to the modern epoch).

In this paper we discuss both the expected specific peak-like anomalies in the spatial correlation function $\xi(r)$ for Starobinsky's model of initial CDM-spectrum and the LDCs in baryonic dark matter models. If peak-like LDCs are discovered, it would be a direct indication of peculiarities of the scalar field potential $V(\varphi)$ in the nonstandard CDM-model or the manifestation of isocurvature perturbations in a baryonic Universe. The specific behavior of each of these peculiarities allows us to distinguish them from each other and from other possible peculiarities.

Thus the purpose of the paper is the following. We shall describe the long distance correlations which arise in some cosmological models at scales much longer than the typical correlations of galaxies and clusters of galaxies distributions. LDCs reflect the physical processes in the early Universe. They can be used for the determination of some fundamental parameters of the Universe and determination of the nature of dark matter.

2. LDC-effect in the spatial correlation function for small perturbations in the Universe

This section is devoted to the manifestation of primordial spectrum of perturbations and to the effect of processes in primordial plasma in barionic models on the spatial correlation function $\xi(r)$. We require that the initial fluctuation density field is smoothed on the scale $r = R_*$ [20]. For a Gaussian filter of scale R_* the smoothed density perturbation field $\delta(\vec{r}, R_*)$ is given by

$$\delta(\vec{r}, R_*) = \frac{1}{(2\pi R_*^2)^{3/2}} \int d^3x \delta(\vec{x}) \exp\left(-\frac{(\vec{r} - \vec{x})^2}{2R_*^2}\right), \tag{1}$$

where $\delta(\vec{x})$ is the initial density perturbation.

It is straightforward to show that the correlation function of primordial density fluctuation (PDF)

$$\xi(r) = < \delta(\vec{x_1}, R_*)\delta(\vec{x_2}, R_*) >; \qquad |\vec{x_1} - \vec{x_2}| = r$$

is the covariance between the fields at different points (Peebles 1980)

$$\xi(r) = \frac{b^2}{2\pi^2 r} \int_0^\infty dk k P_0(k) T(k) \sin(kr) \exp(-k^2 R_*^2), \tag{2}$$

where $P_0(k)$ is the primordial spectrum of density perturbations, $T(k)$ is the transfer function, and b is the bias-parameter. Note, that we consider the two-point correlation function of density perturbations in the present Universe on scales where perturbations are small (the linear regime).

We shall discuss the following models in which there are LDC-effects: 1) the baryonic cosmological model with isocurvature initial perturbations ($\Lambda = 0$) and 2) adiabatic perturbations in cold dark matter (CDM) and more complicated mixed models (for example, CDM+HDM etc.).

In the framework of the isocurvature baryonic dark matter model the transfer function can be written in the following form [21]:

$$T(k) \approx \frac{k^4 R^4}{(1 + k^2 R^2)^2}[1 + e^{-\frac{kr_d}{2}} \sin(kR)]^2 \tag{3}$$

where $R = 32.8h^{-1}$Mpc, $r_d = 2.9h^{-1}$Mpc for $\Omega_b = 0.3$, $R = 52.0h^{-1}$Mpc, $r_d = 4.6h^{-1}$Mpc for $\Omega_b = 0.2$ and $R = 81.4h^{-1}$Mpc, $r_d = 14.5h^{-1}$Mpc for $\Omega_b = 0.1$, h is the Hubble parameter in the unit $H_0 = 100 \mathrm{km}\ \mathrm{s}^{-1}\mathrm{Mpc}^{-1}$. Various numerical models (see [14], [22-24]) give the values of the parameters r_d and R with dispersion $\sim 30\%$. With the same accuracy we can use the expression (3) and corresponding values for r_d and R. The expression in the brackets of eq.(3) describes the generation of the secondary density perturbations by the primary isocurvature perturbations. On small scales $(kr_d \gg 1)$ the secondary perturbations are erased by Silk damping. On large scales $(kr_d \ll 1$, but $kR > 1)$ they evolve as sound waves up to the recombination epoch.

For nonbaryonic cold dark matter models (CDM) we discuss the following form of the transfer function [15]:

$$T(k) = \{1 + [ak + (bk)^{3/2} + (ck)^2]^\nu\}^{-2/\nu} \tag{4}$$

where $a = (6.4/\Gamma)h^{-1}\mathrm{Mpc}$; $b = (3.0/\Gamma)h^{-1}\mathrm{Mpc}$; $c = (1.7/\Gamma)h^{-1}\mathrm{Mpc}$; $\Gamma = \Omega h$ and $\nu = 1.13$ [16].

Efstathiou et al.[25] have shown that the galaxy clustering in the APM and QDOT surveys is well approximated by equation (4) with $\Gamma = \Omega h = 0.2$ for the low density spatially flat CDM model with cosmological constant $\Omega_\Lambda = 1 - \Omega$. The "standard" CDM model is described by equation (4) with $\Gamma = \Omega h = 0.5$ $(\Omega = 1;\ h = 0.5)$. For more complicated mixed models, for example, baryonic dark matter+CDM, CDM+HDM etc. we use the analytical approximation of the transfer function $T(k)$ given by Holzman[26]. It is easy to show that for $P_0(k) \propto k^n$, $r \gg R_*$, and $T(k) \equiv 1$ the correlation function $\xi(r)$ has the following form: $\xi(r) \propto r^{-\frac{n+3}{2}}$ and $\xi(r)$ does not depend on the smoothing scale R_*.

Now let us discuss the characteristic of the spectrum of primordial density fluctuations in nonstandard model of inflation.

According to Starobinsky[17], the nonanalytical inflation potential $V(\varphi)$ is connected with the following form of the CDM density perturbation spectrum

$$P_0(k) = A\phi_0(k)D(k) \tag{5}$$

where A is an amplitude; $\phi_0(k) \propto k^n$ is the "standard" part of spectrum for the simplest model of inflation, and $D(k)$ is the modulation function [17]

$$D(y) = 1 - 3(\varepsilon - 1)\frac{1}{y}\left[\left(1 - \frac{1}{y^2}\right)\sin 2y + \frac{2}{y}\cos 2y\right] + \tag{6}$$

$$+\frac{9}{2}(\varepsilon - 1)^2\frac{1}{y^2}\left(1 + \frac{1}{y^2}\right)\left[1 + \frac{1}{y^2} + \left(1 - \frac{1}{y^2}\right)\cos 2y - \frac{2}{y}\sin 2y\right],$$

where $y = kR$. The parameter $\varepsilon = \frac{A_-}{A_+}$ in (6) is the ratio of $\frac{dV}{d\varphi}(\varphi \to \varphi_{cr})$ before (A_-) and after (A_+) the critical point $\varphi = \varphi_{cr}$ (see Fig.1). Note, that $\xi(r) \equiv \xi_0(r)$ for $\varepsilon = 1$, see eq.(9). The behavior of $D(y = kR)$ in the models with $\varepsilon = 0.5; 1.0; 2.0; 3.0$

is represented on Fig.2. One can see from Fig.2 and eq.(6) that for $y \geq 4 \div 5$ this function the $D(y)$ has a very simple analytical form

$$D(y) \approx 1 - 3(\varepsilon - 1)\frac{\sin 2y}{y}; \quad y \geq 4 \div 5. \tag{7}$$

This approximate formula is very important for the PDF correlation function calculation in the most interesting region $y \gg 1$, where the LDC effect takes place.

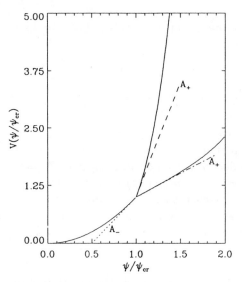

Fig.1. *Qualitative behavior of the potential $V(\varphi)$ according to Starobinsky*[17].

Now we consider the behavior of $\xi(r)$ at $y \gg 1$. We write down the equation (2) in the following form

$$\xi(r) = \xi_0(r) - \xi_1(r); \tag{8}$$

where

$$\xi_0(r) = \frac{Ab^2}{2\pi^2 r} \int_0^\infty dk\, k \Phi_0(k) T(k) \sin kr; \tag{9}$$

$$\xi_1(r) = \frac{3A(\varepsilon - 1)b^2}{2\pi^2 r R} \int_0^\infty dk\, \Phi_0(k) T(k) \sin(2kR) \sin kr; \tag{10}$$

$$\Phi_0(k) = \phi_0(k) \exp(-k^2 R_*^2)$$

In eq.(8) $\xi_0(r)$ is the standard correlation function for the monotonic spectrum $\phi_0(k)$. The second term in eq.(8) describes the LDC-effect. It may be shown that the function $\xi_1(r)$ satisfies the following relation

$$\frac{1}{r}\frac{d}{dr}(r\xi_1(r)) = -\frac{3(\varepsilon - 1)}{2R}\left[\frac{r - 2R}{r}\xi_0(r - 2R) - \frac{r + 2R}{r}\xi_0(r + 2R)\right]. \tag{11}$$

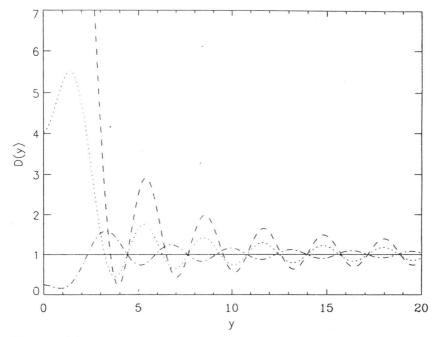

Fig.2. $D(y)$ *for different values of* ε. *The dotted, dash-dotted and dashed lines correspond to* $\varepsilon = 0.5$, $\varepsilon = 2$ *and* $\varepsilon = 3$ *respectively.* $D(y) \equiv 1$ *for* $\varepsilon = 1$.

In the region $\left| \frac{2R-r}{2R} \right| < \frac{r_c}{2R}$, (where r_c is the correlation scale of $\xi_0(r)$, approximately being equal to R_*) the second term in eq.(11) $\xi_0(r + 2R)$ is negligible in comparison to $\xi_0(r - 2R)$. Note that $\xi_0(r)$ is an even function, and the definition (2) may be formally used for $r - 2R < 0$. Taking these properties into account and writing $\xi_0(r)$ in eq.(11) as a Taylor series in the considered region we obtain

$$\xi(r \simeq 2R) \approx -\xi_1(r \simeq 2R) = -\frac{3(\varepsilon - 1)}{4\langle k^2 \rangle R^2} \xi_0(0) \left\{ 1 - \frac{\langle k^2 \rangle x^2}{2} \left(1 - \frac{x^2}{4r_c^2} \right) \right\} \qquad (12)$$

where

$$\xi_0(0) = \frac{b^2}{2\pi^2} \int_0^\infty dk k^2 \Phi_0(k) T(k);$$

$$\langle k^2 \rangle = \int_0^\infty dk k^2 \Phi_0(k) T(k) / \int_0^\infty dk \Phi_0(k) T(k);$$

$$x = r - 2R \ll r_c.$$

If at $r \simeq 2R \gg r_c$ the correlation function $\xi_0(2R)$ is negligible in comparison to $\xi_1(2R)$, then the asymptotical value of $\xi(r)$ is negative (for $\varepsilon > 1$) and positive (for $\varepsilon < 1$), and has a resonance described by (12). The amplitude of the peak is $\xi(r = 2R) = -\frac{3(\varepsilon-1)}{4\langle k^2 \rangle R^2} \xi_0(0)$, and the width of the peak is of order $\langle k^2 \rangle^{-1/2}$.

Now let us discuss the characteristic of the correlation function in a baryonic Universe with primordial isocurvature perturbations.

For the subsequent estimates we rewrite (3) in the following form:

$$T(k) = T_0(k)[1 + \mu(k)\cos(2k\tilde{R}) + \nu(k)\sin(k\tilde{R})], \tag{13}$$

where

$$T_0(k) = \frac{k^4 \tilde{R}^4}{(1 + k^2 \tilde{R}^2)^2}(1 + \frac{1}{2}e^{-kr_d}),$$

$$\mu(k) = -(2e^{kr_d} + 1)^{-1},$$

$$\nu(k) = 2e^{-\frac{kr_d}{2}}/(1 + \frac{1}{2}e^{-kr_d}).$$

Now the correlation function can be written as follows:

$$\tilde{\xi}(r) = \tilde{\xi}_0(r) + \tilde{\xi}_1(r) + \tilde{\xi}_2(r) \tag{14}$$

where

$$\tilde{\xi}_0(r) = \frac{b^2}{2\pi^2}\langle\langle\frac{\sin(kr)}{kr}\rangle\rangle, \tag{15}$$

$$\tilde{\xi}_1(r) = \frac{b^2}{2\pi^2}\langle\langle\mu(k)\frac{\sin(kr)}{kr}\cos(2k\tilde{R})\rangle\rangle, \tag{16}$$

$$\tilde{\xi}_2(r) = \frac{b^2}{2\pi^2}\langle\langle\nu(k)\frac{\sin(kr)}{kr}\sin(k\tilde{R})\rangle\rangle, \tag{17}$$

and $\langle\langle...\rangle\rangle$ means the following operation:

$$f(r) \equiv \langle\langle f(k,r)\rangle\rangle = \int_0^\infty dk k^2 P_0(k) T_0(k) f(k,r) \exp(-k^2 \tilde{R}_*^2).$$

For the case $(\tilde{\xi}(r))$ from eq.(14) the relation between $\tilde{R}_*$ and r_d is important. In the case $r_d \ll \tilde{R}_*$ we can formally put $r_d \to 0$. Doing so the expression (16) can be written in the form

$$\tilde{\xi}_1(r) = -\frac{b^2}{6\pi^2}\langle\frac{\sin(kr)\cos(2k\tilde{R})}{kr}\rangle =$$

$$= \frac{1}{6}\left[\frac{2\tilde{R}+r}{r}\tilde{\xi}_0(2\tilde{R}+r) - \frac{2\tilde{R}-r}{r}\tilde{\xi}_0(2\tilde{R}-r)\right]. \tag{18}$$

We shall now demonstrate the existence of resonances in the correlation function around $r \simeq \tilde{R}$ and $r \simeq 2\tilde{R}$. Firstly we study the "second resonance point" $r \simeq 2\tilde{R}$ evaluating $\tilde{\xi}(r)$ in the two intervals

$$\frac{\tilde{r}_c^2}{4\tilde{R}^2} < \left|\frac{2\tilde{R}-r}{2\tilde{R}}\right| < \frac{\tilde{r}_c}{2\tilde{R}}, \tag{19}$$

where $\tilde{r}_c$ is the correlation scale of $\tilde{\xi}_0(r)$ being equal to $\tilde{R}_*$ by order of magnitude. It is easy to show from eqs. (15), (16) and (17) that $\tilde{\xi}_0(r)$ and $\tilde{\xi}_2(r)$ are smaller than $\tilde{\xi}_1(r)$ in the interval given by (19). Furthermore, in eq.(18) the first term is small compared to the second. Taking these properties into account and writing $\tilde{\xi}_o(r)$ as a Taylor's series we obtain for the considered region:

$$\tilde{\xi}_1(y) = -\frac{\tilde{r}_c}{12\tilde{R}}\sigma y(1 - \frac{y^2}{2} + \beta\frac{y^4}{4!}), \tag{20}$$

where

$$y = \frac{r - 2\tilde{R}}{\tilde{r}_c}, \qquad \sigma \equiv \tilde{\xi}_0(0),$$

$$\beta = \frac{\tilde{\xi}_0(0)\tilde{\xi}_0^{(IV)}(0)}{[\tilde{\xi}_o''(0)]^2}, \qquad \tilde{\xi}_0''(0) = \frac{d^2\tilde{\xi}_0}{dr^2}|_{r=0}, \qquad \tilde{\xi}_0^{IV}(0) = \frac{d^4\tilde{\xi}_0}{dr^4}|_{r=0}.$$

In the interval given by (19) the maximum of $\tilde{\xi}(r)$ is at $y_{max} = -\sqrt{2/3}$ and the minimum at $y_{min} = \sqrt{2/3}$. At these points

$$\tilde{\xi}(y_{max}) \approx \frac{1}{3\sqrt{3}}\frac{\tilde{r}_c}{\tilde{R}}\sigma; \qquad \tilde{\xi}(y_{min}) = -\tilde{\xi}(y_{max}). \tag{21}$$

The characterisctic widths of the extrema are of order of $\tilde{r}_c$. When $|y| \gg 1$ the function $\tilde{\xi}_1(r) \ll \tilde{\xi}_0(r)$. In the very vicinity of $y = 0$ $[|\frac{2\tilde{R}-r}{2\tilde{R}}| < \frac{r_c^2}{4\tilde{R}^2}]$ our approximation is not correct but all functions $\tilde{\xi}_0$, $\tilde{\xi}_1$ and $\tilde{\xi}_2$ are small in this case and we may not care about the exact expressions and still use the approximation (20).

Thus the term with $\cos(2k\tilde{R})$ in the transfer function leads to the origin of a LDC in the spatial correlation function caused by the secondary acoustic waves.

We shall now consider the behavior of $\tilde{\xi}_2(r)$ at $r \simeq \tilde{R}$. We write down the expression (17) in the following form

$$\tilde{\xi}_2(r) = \tilde{\xi}_2^-(r) - \tilde{\xi}_2^+(r), \tag{22}$$

where

$$\tilde{\xi}_2^-(r) = \frac{b^2}{3\pi^2}\langle\langle\frac{\cos k(\tilde{R} - r)}{kr}\rangle\rangle,$$

$$\tilde{\xi}_2^+(r) = \frac{b^2}{3\pi^2}\langle\langle\frac{\cos k(\tilde{R} + r)}{kr}\rangle\rangle,$$

and $r_d < \tilde{R}_*$.

In the region

$$\frac{l_c^2}{\tilde{R}^2} < |\frac{\tilde{R} - r}{\tilde{R}}| < \frac{l_c}{\tilde{R}},$$

with $l_\xi^2 = \langle\langle k^{-1}\rangle\rangle/\langle\langle k\rangle\rangle$ the term $\tilde{\xi}_2^+$ is negligible compared to $\tilde{\xi}_2^-$. For this region we have $\tilde{\xi}_2 > (\tilde{\xi}_0, \tilde{\xi}_1)$ and in analogy to the previous case we write

$$\tilde{\xi}(r) \approx \tilde{\xi}_2^-(r) \approx A_2 \left[1 - \frac{(\tilde{R} - r)^2}{2l_c^2} + \beta_1 \frac{(\tilde{R} - r)^4}{4! l_c^4} \right], \tag{23}$$

where

$$A_2 = \frac{b^2}{3\pi^2 \tilde{R}} \langle\langle k^{-1}\rangle\rangle, \qquad \beta_1 = \frac{\langle\langle k^{-1}\rangle\rangle\langle\langle k^3\rangle\rangle}{\langle\langle k\rangle\rangle^2}$$

We can rewrite A_2 in the form

$$\sigma \approx 2\overline{k} \tilde{R} A_2, \tag{24}$$

where

$$\overline{k} = \frac{\langle\langle k^0\rangle\rangle}{\langle\langle k^{-1}\rangle\rangle}.$$

In the case of the power-function $\varphi(k) \propto k^{-\gamma}$ for initial isocurvature perturbations a good estimate for $\overline{k}$ is

$$\overline{k} \approx (\kappa l_c)^{-1}$$

where κ is a numerical coefficient of the order of unity. Now we have

$$\tilde{\xi}(r) \approx \tilde{\xi}_2^-(r) \approx \frac{2\kappa}{3} \frac{l_c}{\tilde{R}} \sigma \left[1 - \frac{x^2}{2} + \beta_1 \frac{x^4}{4!} \right]; \tag{25}$$

where $x = \frac{\tilde{R} - r}{l_c}$ and $|x| < 1$. The structure of eq.(25) is similar to that of eq.(20). A LDC resonance as discussed above occurs again but this time at $r \simeq \tilde{R}$ being due to the term with $\sin(k\tilde{R})$ in the transfer function.

The amplitude of the LDC in this case is of order $\kappa \frac{l_c}{\tilde{R}} \sigma \gg |\tilde{\xi}_0(\tilde{R})|$, and the width of the extremum is of order l_c.

Let us now consider the case $r_d \gg \tilde{R}_*$ (in this case the exponential cut off is not important at all). Qualitatively our conclusions discussed above are still valid. But in this case instead of $\tilde{R}_*$ we have to use the value of the Silk-damping r_d for the estimates.

The case $r_d \gg \tilde{R}_*$ could be important, in particular, for cosmological models with secondary ionization of hydrogen (see for example [14]) when $\tilde{R}$ and r_d increase considerably.

3. LDC in the Space Galaxy and Cluster Galaxy Distribution

In this section we shall compare our general qualitative analysis with some numerical examples. In our computation we used the explicit formulae (2)-(5) for $\Phi_0(k) \propto k^n \exp(-k^2 R_*)$ $(n = -1, 0, 1)$. We will see that the numerical examples confirm our estimates. As an illustration, we compare the appearance of the LDC effect with the experimental data on galaxy-galaxy ($\xi_{gg}(r)$) and cluster-cluster ($\xi_{cc}(r)$) distribution of Loveday et al. [3] and Efstatiou et al.[16].

It is very important that in the "bias" theory of galaxy and cluster formation the $\xi_{gg}(r)$ and $\xi_{cc}(r)$ on scales $\geq 10h^{-1}$Mpc are approximately constant multiplied by the correlation function $\xi(r)$ [20]. For the bias parameter $b = 1.7$ $\xi_{gg}(r) \simeq 2.89\xi(r)$ and $\xi_{cc}(r) \simeq 13.8\xi(r)$ [16].

Fig.3, 4a and 5a show the correlation functions for low-density CDM models in a spatially flat Universe with a cosmological constant ($\Gamma = 0.2$; $h = 1$) and Harrison-Zeldovich spectrum of primordial density perturbations $n = 1$ (Fig.3); $n = 0$ (Fig.4a) and $n = -1$ (Fig.5a). In Fig.4b and 5b we show the corresponding correlation function $\xi(r)$ in the vicinity of the local extremum at $r = 2R \simeq 10^2 h^{-1}$Mpc. One can see that the level of correlation depends on the parameter ε, and $\xi(r = 2R) < 0$, for $\varepsilon > 1$, and $\xi(r = 2R) > 0$, for $\varepsilon < 1$.

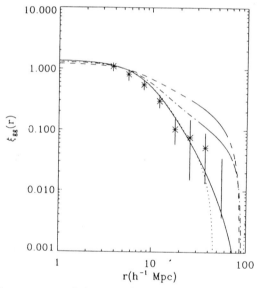

Fig.3. *Comparison of the correlation function $\xi_{gg}(r)$ for $n = 1$, $\Gamma = 0.2$, and $R = 50h^{-1}Mpc$, normalized to 1.0 at $r = 5h^{-1}Mpc$, with the observational data from Loveday et al[3]. The region $r < 5h^{-1}Mpc$ corresponds to non-linear perturbations. Symbols as in Fig.2.*

Fig.6a shows the galaxy-galaxy correlation function for different mixed models of dark matter considered by Holzman[26] but with the peculiarities in the primordial spectrum. One can see that in all models of dark matter the correlation function $\xi_{gg}(r)$ has peculiarities in the vicinity of $r = 2R \simeq 10^2 h^{-1}$Mpc. In Fig.6b we have plotted the cluster-cluster correlation function for cosmological models which are shown in Fig.6a.

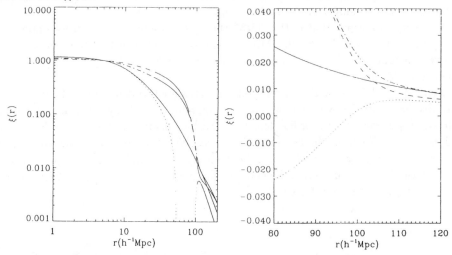

Fig.4. **(a)** *The correlation function $\xi_{gg}(r)$ for $n = 0$, $\Gamma = 0.2$ and $R = 50h^{-1}Mpc$, normalized to 1.0 at $r = 5h^{-1}Mpc$. The dotted, solid, dash-dotted and dashed lines correspond to $\varepsilon = 0.5$, $\varepsilon = 1$, $\varepsilon = 2$ and $\varepsilon = 3$ respectively.*
(b) *The correlation function $\xi_{gg}(r)$ for $n = 0$, $\Gamma = 0.2$ and $R = 50h^{-1}Mpc$, normalized to 1.0 at $r = 5h^{-1}Mpc$, and $\varepsilon = 0.5$, $\varepsilon = 1$, $\varepsilon = 2$ and $\varepsilon = 3$ in the vicinity of the point $r = 2R$. Symbols as in Fig.4(a).*

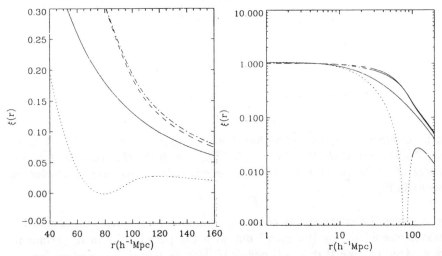

Fig.5. **(a)** *The same as on Fig.4(a) but for $n = -1$,* **(b)** *The part of the Fig.5(a) but in a linear scales.*

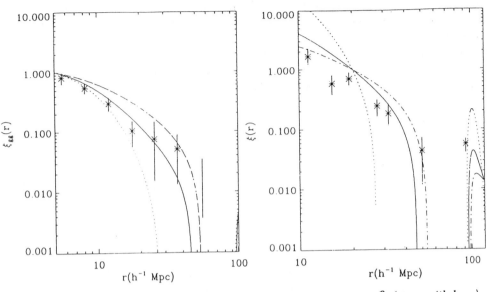

Fig.6. (a) *Comparison of observational data from Loveday et al*[3] *(stars with bars) with the correlation function* $\xi_{gg}(r)$ *for* $\Gamma = 0.2$; $n = 1$, $\varepsilon = 0.5$, $R = 50h^{-1}Mpc$ - *solid line, and three different models from* [26]: H_1 - *dash-dotted line;* H_2 *dashed line;* H_3 - *dotted line. Symbols:*

H_1, H_2, H_3 - *Holtzman's models:*

H_1 - $\Omega_{tot} = 0.2$; $\Omega_b/\Omega_{tot} = 0.5$; $\Omega_{CDM}/\Omega_{tot} = 0.5$; $\Omega_\Lambda = 0.0$; $\Omega_\nu = 0.0$; $N_{m\nu}^b = 0$.

H_2 - $\Omega_{tot} = 0.2$; $\Omega_b/\Omega_{tot} = 0.5$; $\Omega_{CDM}/\Omega_{tot} = 0.5$; $\Omega_\Lambda = 0.8$; $\Omega_\nu = 0.0$; $N_{m\nu}^b = 0$.

H_3 - $\Omega_{tot} = 1$; $\Omega_b/\Omega_{tot} = 0.01$; $\Omega_{CDM}/\Omega_{tot} = 0.69$; $\Omega_\Lambda = 0.0$; $\Omega_\nu/\Omega_{tot} = 0.3$; $N_{m\nu}^b = 1$.

$h = 1$ *for all these models;* Ω_{tot} - *total density excluding vacuum one;* Ω_b *is the density of baryon fraction;* Ω_{CDM} *is the density of CDM fraction;* Ω_Λ *is the vacuum density;* Ω_ν *is the density of fraction of massive neutrino;* $N_{m\nu}^b$ *is the number of sorts of neutrinos, which have non-zero mass.*

(b) *The cluster-cluster correlation function, normalized to 1.0 at* $r = 20h^{-1}Mpc$, *for the same models as on the Fig.6(a). The type of lines are the same as on Fig.6(a). Stars and bars are the observational data from Efstathiou et al*[16].

The behavior of $\tilde{\xi}(r)$ in the open baryonic model of the Universe with $\Omega_b = 0.3$, $h = 1$ is shown on Fig.7 for the case of the power spectrum $P_0(k) \propto k^{-\gamma}$, $R_* = 0.7h^{-1}Mpc$ on the Fig.7a, and $\tilde{R}_* = 3.5h^{-1}Mpc$ on Fig.7b.

One can see on the Fig.7a the local maximum at $r \approx \tilde{R} \approx 33Mpc$, as it should be according to the analytical estimates. The amplitude of the LDC-effect is greater for the smaller γ. The comparison Fig.7a with Fig.7b, which corresponds to $R_* = 3.5Mpc$, demonstrates that the LDC-effect is smaller for bigger R_*. Fig.8 and Fig.9

demonstrate the behaviours of $\tilde{\xi}(r)$ for the case $\Omega_b = 0.2$ and $\Omega_b = 0.1$ correspondingly. They also demonstrate the effect LDC at $r \approx \tilde{R}$ and its dependence on the parameters of the models. On the Fig.8c we plotted the observational data on $\xi(r)$ from the work Lovedey et al., 1992. The comparison theory and observations allow us to conclude that there is no contradiction between them, but for the detection of the LDC-effect one need the higher accuracy of the observational data.

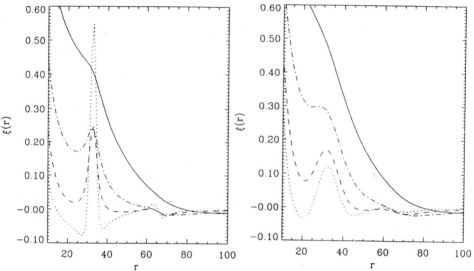

Fig.7. (a)*Correlation function $\xi(r)$ for the model with $\Omega = 0.3$, $h = 1$, $P_0(k) \propto k^{-\gamma}$, $\tilde{R}_* = 0.7h^{-1}Mpc$. Solid line corresponds $\gamma = 3$, dahsed line for $\gamma = 2$, dashed-dotted line for $\gamma = 1.2$, doted line for $\gamma = 0.5$. The correlation functions are normalized to 1 at $r = 5h^{-1}Mpc$.* (b) *The same as Fig.7a, but for $\tilde{R}_* = 3.5h^{-1}Mpc$.*

As we mentioned in the introduction the LDC in baryonic cosmological models with adiabatic initial perturbations has been discussed by Peebles [14]. We want to emphasize the principal difference of the properties of the LDC in the model with isocurvature initial perturbations from that in Peebles's discussion. In Peebles's case the LDC appears at $r \simeq 2\tilde{R}$, in our case the LDCs occur at both $r \simeq \tilde{R}$ and $r \simeq 2\tilde{R}$.

The reason for this difference is the following. For the adiabatic initial perturbations the transfer function is $T \propto (\sin k\tilde{R})^2 = \frac{1-\cos 2k\tilde{R}}{2}$ and it gives "a resonance" at $r \simeq 2\tilde{R}$. In the case of the isocurvature initial perturbations the acoustic waves are a secondary effect. They arise from the primordial isocurvature perturbations and the transfer function $T_2 \propto (1 + e^{-r_d k/2}\sin(k\tilde{R}))^2$. Rewriting this multiplier the terms with $\sin(k\tilde{R})$ and $\cos(2k\tilde{R})$ give rise to two "resonances" at $r \simeq \tilde{R}$ and $r \simeq 2\tilde{R}$. The new resonance at $r \simeq \tilde{R}$ is of principal importance since the correlation function $\xi(r)$ is considerably greater at $r \simeq \tilde{R}$ than at $r \simeq 2\tilde{R}$. It means that the possibility to observe this effect at $r \simeq \tilde{R}$ is much greater than at $r \simeq 2\tilde{R}$.

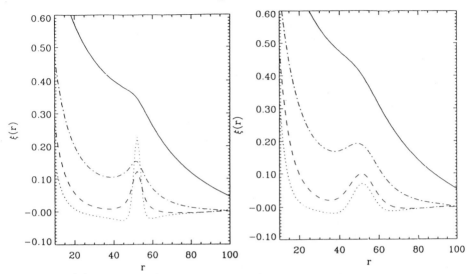

Fig.8. **(a)** *The same as Fig.1a, but for* $\Omega_b = 0.2$. **(b)** *The same as* Fig.1b, *but for* $\Omega_b = 0.2$.

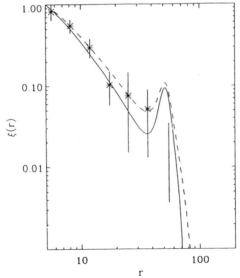

Fig.8. **(c)** *Comparison with the observations. Points with bars are the observational data from Loveday et al*[3]. *The type of line is the same as on* Fig.7(a).

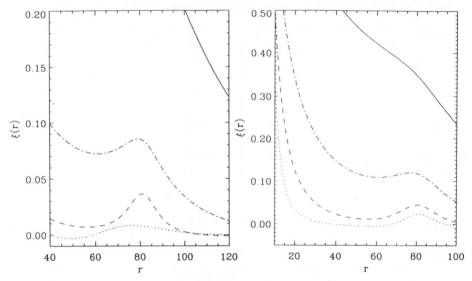

Fig.9. (a) *The same as* Fig.7a, *but for* $\Omega_b = 0.1$, (b) *The same as* Fig.7b, *but for* $\Omega_b = 0.1$.

4. LDC in the projected angular correlation function $w(\theta)$ for the isocurvature baryonic dark matter model.

In this section we discuss the manifestation of the LDC in the projected angular correlation function $w(\theta)$ for the baryonic dark matter model. This function was discussed by Peebles [27] and was used, for examples, by Maddox [1]. The function $w(\theta)$ can be written in the form [28]

$$w(\theta) = \frac{b^2}{\pi^2 D \chi^2} \int_0^\infty dk k P_o(k) T(k) W(kD\theta), \qquad (26)$$

where $\chi \equiv \int_o^\infty ds s^2 \tilde{\varphi}(s) = 18.6$ for the APM survey parameters (see [1]), and

$$\tilde{\varphi}(s) = \int_{b_l}^{b_u} \Phi dm = \Gamma[10^{-0.4(b_l-b_u)} s^2, 1-\alpha] - \Gamma[s^2, 1-\alpha],$$

$\Phi \propto \left(\frac{\mathcal{L}}{\mathcal{L}_*}\right)^{-\alpha} exp\left[-\frac{\mathcal{L}}{\mathcal{L}_*}\right]$ with $\alpha = 1,2$ is the Schechter luminosity function; $\mathcal{L}_*$ is the luminosity corresponding to the B_J absolute magnitude $M_* = -19.8$ for $h = 1$, $b_u = 20.5$, $b_l = 17.5$ are the blue magnitude limits of the APM catalogue; $s = r/D$;

$D = 10^{0.2(b_l - M_*) - 5} h^{-1} \sim 288 h^{-1} \mathrm{Mpc}$ is a characteristic scale length of the survey and $\Gamma(x, a)$ is the incomplete gamma-function. Following Kashlinsky [28], we get

$$W(y) = \frac{\pi}{2} \int_0^\infty ds s^4 \tilde{\varphi}^2(s) J_0(sy), \qquad (27)$$

where $J_0(x)$ is the Bessel-function.

On Figures 10-12 we plotted $w(\theta)$ for $\Omega_b = 0.1 - 0.3$, $h = 1$ $P_0(k) \propto k^{-\gamma}$. The plots are normalized in a such way that $w(\theta)$ is equal 0.07 at $\theta = 1°$. In all cases the LDC-effect at $\theta \approx \theta_R \simeq \frac{R}{D}$ is clearly seen. Its amplitude is higher for smaller γ and greater Ω. On Fig.11 one can see the existence of the secondary "resonance" at $\theta \approx 2\theta_R$ for $\gamma = 0.5$ but the amplitude of it is very small.

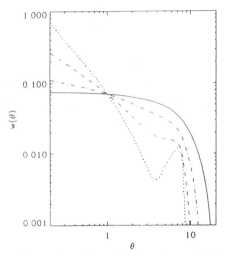

Fig.10. *Projected correlation function $w(\theta)$ for the model $\Omega = 0.3$, $h = 1$, $P_0(k) \propto k^{-\gamma}$. Solid line corresponds $\gamma = 3$, dahsed line for $\gamma = 2$, dashed-dotted line for $\gamma = 1.2$, doted line for $\gamma = 0.5$. The plots are normalized in a such way that $w(\theta)$ is equal 0.07 at $\theta = 1°$.*

The plots in Figures 10-11 were computed for the case $R_* \ll r_d$. If R_* is of order of r_d then the shape of plots changes, but the positions of the "resonances" are, of course, the same.

On Fig.12 we give the comparison of the theoretical curve for the model with $\Omega_b = 0.2$, $h = 1$, $\gamma = 1.2$ with the observational data from APM survey[1] scaled to the Lick catalogue ($D = 209 h^{-1} \mathrm{Mpc}$).

One can see that for the definite conclusion about the existence or absence of the LDC-effect we need higher accuracy of the observations. Probably the existing observations do not exclude the LDC-effect in baryonic dark matter.

176

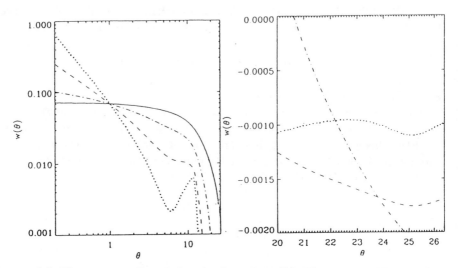

Fig.11. **(a)** *The same as Fig.10, but for* $\Omega_b = 0.2$, **(b)** *The same as* Fig.11a, *but for values of* θ *in the vicinity of the "second resonance" at* $\theta \approx 2\theta_R$.

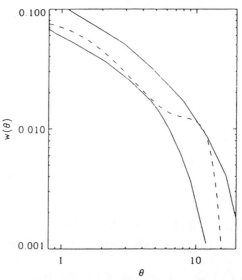

Fig.12. *The comparison of the* $w(\theta)$ *(dashed line) for the model* $\Omega_b = 0.2$, $h = 1$, $\gamma = 1.2$, $\theta_* = 0.5°$ *with the observational data from APM survey* [1] *scaled to the Lick catalogue* $(D = 209h^{-1} Mpc)$. *Solid lines correspond to the upper and lower boundaries of the distriction of points represented the observations.*

5. Method of the filtering of the structural properties of the LDC.

The LDC effects are thus a specific signature of the acoustic modes in baryonic Universe and oscillations of the primordial spectrum in CDM-models. So, what could be the way to filter the correlations in the galaxy distributions which 1)cancels the regular component of the correlations and 2)emphasizes/preserves the effect of LDC?

We propose the following method which is the analogy to the method proposed by Naselsky and Novikov [29] for the correlations in the measurments of the anisotropy of the microwave background radiation. It is based on the filtering of the second derivatives of the correlation function. The importance of the determination of the $\frac{d^2\xi}{dr^2}$ was pointed out in[14].

Let us start with the space correlations.

We introduce an auxiliary function of two variables

$$\Phi(r, d) = \xi(r + d) + \xi(r - d) - 2\xi(r). \tag{28}$$

The scale of d should be less than the correlation scale r_c, $d < r_c$. One can see that . in the case $d \ll r_c$ we have

$$\Phi(r, d) = \frac{d^2\xi(r)}{dr^2} d^2. \tag{29}$$

Using eqs.(20) and (25) we obtain for the baryonic dark matter model $r \sim \tilde{R}$ and $r \sim 2\tilde{R}$:

$$\Phi(r \sim \tilde{R}, d) = -\frac{2}{3}\kappa\frac{d^2}{\tilde{R}l_c}\sigma\left(1 - \beta_1\frac{x_1^2}{2}\right); \qquad x_1 = \frac{\tilde{R} - r}{l_c};$$

$$\Phi(r \sim 2\tilde{R}, d) = \frac{d^2}{8\tilde{R}\tilde{r}_c}\sigma x_2\left(1 - \frac{5}{18}\beta x_2^2\right); \qquad x_1 = \frac{2\tilde{R} - r}{\tilde{r}_c}.$$

The parabolic shape of $\Phi(r, d)$ is indicative of peaks in $\tilde{\xi}(r)$ at the resonances at $r \simeq \tilde{R}$ and $r \simeq 2\tilde{R}$. The widths of these peaks are determined by β_1 and β correspondingly.

On Fig.13 we plot the function $\Phi(r, d)$ for the case $d = 3h^{-1}\text{Mpc}$, $\Omega_b = 0.2$; $h = 1$. As it is seen from Fig.13 the function Φ emphasizes in a very spectacular way the structure of the LDC in the baryonic dark matter model. For Starobinsky's model the behaviour of the function $\Phi(r, R_d)$ in the vicinity of the extremum is plotted on Fig.14. It is not so difficult to investigate the behavior of the function $\Phi(r, R_d)$ in the vicinity of $r \simeq 2R$, where we get

$$\Phi(r, R_d) \approx \frac{3(\varepsilon - 1)}{4}\xi_0(0)\frac{R_d^2}{R^2}\left\{1 - \frac{3}{2}\frac{x^2}{r_c^2}\right\}; \qquad x = r - 2R. \tag{30}$$

From eq.(30) and Fig.14 one can see that the second derivative of the correlation function has a "resonance" near $r = 2R$.

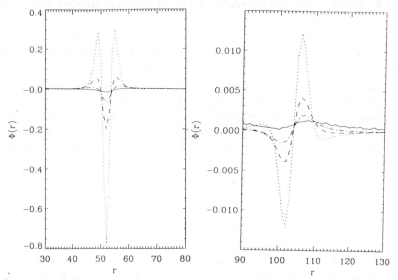

Fig.13. (a) *Function* $\Phi(r,d)$ *for the model* $\Omega_b = 0.2$, $h = 1$ *and* $d = 3h^{-1}Mpc$ *at* $r \approx \tilde{R}$ *for different values of* γ *Symbols are the same as on* Fig.7a.
(b) *The same as Fig.13a, but for* $r \approx 2R$.

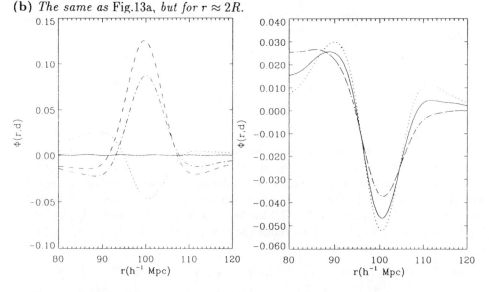

Fig.14. (a) $\Phi(r,d)$*for the galaxy-galaxy correlation function plotted on the* Fig.6(a), $d = 3h^{-1}Mpc$. *Symbols for different models are the same as in* Fig.6(a).
(b) *The behavior* $\Phi(r,d)$ *in the vicinity of the point* $r = 2R$ *for the model* $n = 1$, $\Gamma = 0.2$, $R = 50h^{-1}Mpc$, $d = 3h^{-1}Mpc$ *and different values of* ε: $\varepsilon = 0.5$ - *dotted line;* $\varepsilon = 1$ - *solid line;* $\varepsilon = 2$ - *dashed-dotted line;* $\varepsilon = 3$ - *dashed line.*

The same method is applicabable to the description of the LDC in the projected correlation function $w(\theta)$. In this case we shall use the function

$$f(\theta) = \frac{\theta^4}{\theta_R^2}\left[\left(\frac{\theta}{\theta_c}\right)^{2-\gamma}w(\theta)\right]''; \quad \theta_c = \left[-\frac{w(\theta)}{w''(\theta)}\right]^{1/2}\Big|_{\theta=0}; \quad w''(\theta) = \frac{d^2w}{d\theta^2} \quad (31)$$

where a prime means the derivative with respect θ. This form (31) is useful, because outside the regions $|\theta - \theta_R| < \theta_c$ and $|\theta - \theta_{2R}| < \theta_c$ the values of $f(\theta)$ are small and inside of these regions the resonances manifestate themselves clearly.

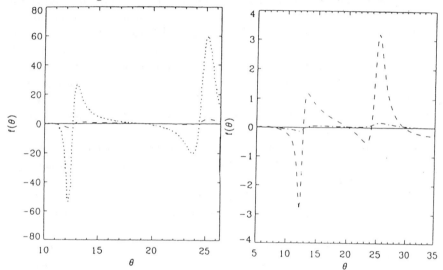

Fig.15. (a) *The function $f(\theta)$ for the model $\Omega_b = 0.2$, $h = 1$. One can see the "resonance" at $\theta_R = 12.4°$ and $2\theta_R = 24.8°$ for the different values of γ. Symbols the same as in Fig.7a.*
(b) *The same as Fig.15(a) but the scale on the ordinate is different to show details.*

In the vicinities of the "resonances" $r \sim \tilde{R}$ of $w(\theta)$ the function $f(\theta)$ has the following approximations:

$$f(\theta) \approx \frac{\theta^{6-\gamma}}{\theta_R^2\theta_c^{2-\gamma}}w''(\theta) \quad (32)$$

On Fig.15 the plots of $f(\theta)$ are represented for the model with $\Omega_b = 0.2$. One can see very clearly the structure of both "resonances" at $\theta = \theta_R$ and $\theta = 2\theta_R$.

6. Conclusions

We have shown that the LDC-effect in a baryonic universe caused by the secondary adiabatic modes of perturbations leads to the specific behaviours of $\xi(r)$ and $w(\theta)$ at the scales $r \approx \tilde{R}$ and $r \approx 2\tilde{R}$; and at the corresponding angles θ_R and $2\theta_R$, determined by the acoustic horizon $\tilde{R}$ at the moment of the last scattering of the radiation. The numerical examples (see all figures) show that the amplitude of the LDC is higher for the smaller values of the exponent γ in the power spectrum $P_0(k) \propto k^{-\gamma}$. Our discussion concentrated on the baryonic models with $\Omega_b = 0.1 - 0.3$ and $h = 1$. The results could be generalized easily on the models with Λ term and $\Omega_{tot} = \Omega_b + \Omega_\Lambda$ with the same values Ω_b. According to Holtzman [26], Sugiyama and Gouda [30], Loveday et al [3] in the case of CDM models the generalization of the dynamics of the evolution of the perturbations on the models with $\Omega_{tot} = 1$ and $\Lambda \neq 0$ has no principal difficulties. The ratio of the growth factors in both models from the moment of the recombination of the plasma up to our epoch can be 2-3. For our models these conclusions are also correct.

We are not discussing here the problem of comparison of the theory with the observational data on $\frac{\Delta T}{T}$. Some remarks see [9,10]. General consideration, which includes the possibility of nonstandard thermal history of the Universe allow us to conclude that there are not contradictions between the theory and observations on $\frac{\Delta T}{T}$ at various angular scales.

Note also that the nonstandard thermal history of the Universe discussed in the models proposed by Peebles [9] and Gnedin and Ostriker [10] leads to very broad variations of the main parameters r_d and $\tilde{R}$ of our model.

We want to emphasize that both CDM models with $\Omega_{CDM} = 0.2$ [25] and open baryonic models with isocurvature perturbations fit the projected correlation function of the APM survey.

But the LDC-effect exists not only in the baryonic models. We have shown that the existense of the singular point ($\varphi = \varphi_{cr}$) of the inflation potential $V(\varphi)$ leads to an extremum of galaxy-galaxy and cluster-cluster correlation functions on a scale $r = 2R$ in CDM and mixed nonbaryonic dark matter models.

For $\varepsilon < 1$ the qualitative behavior of $\xi(r)$ in the vicinity of the extremum looks like the LDC-effect for baryonic dark model with the isocurvature initial perturbations. There is an important difference between the LDC-effect in CDM models with the spectrum (5) and the analogical effect in the baryonic Universe. In CDM models we see a single local maximum (or minimum) for $r = 2R$, while in the baryonic model there are two local extrema on scales $r \simeq \tilde{R}$ and $r \simeq 2\tilde{R}$, which are connected with the acoustic horizon of perturbations at the moment of plasma recombination. This is a most important point.

At the same time the LDC-effect is most pronounced at $r = \tilde{R}$. If $\tilde{R} \sim 2R$ and $\varepsilon < 1$ the character of the distribution of correlations in the vicinity of an extremum is qualitatively similar to those in the CDM models for a baryonic universe. In this case, the effect of the presence of the secondary extremum for baryonic dark matter will be a crucial criterium for determination of the character of dark matter in the universe. In addition, the character of dark matter can be determined by studying the angular anisotropy of the microwave background radiation in the range of $\alpha \simeq 20' \div 1°$ [29]. This issue will be considered in a separate paper.

7. Acknowledgements

This paper was supported in part by the Danish Natural Science Research Council through grant 11-9640-1. We acknowledge E.Kotok, who was a coauthor of our work in this field and helped us in the preparation of this paper.

8. References

1. Maddox S.J., G.Efstathiou, W.Y.Sutherland & J.Loveday, 1990. MNRAS **242**, 43P.

2. Vogeley, M.S., C.Park, M.J.Geller & J.P.Huchra, 1992. Ap.J.,**391**, L5.

3. Loveday, J., G.Efstathiou, B.A.Peterson & S.J.Maddox, 1992. Ap.J. **400**, L43.

4. Mo, H.Y., Z.G.Deng, X.Y.Xia, P.Schiller & G. Börner, 1992. A.A. **257**, 1.

5. Peacock J.A. & M.J.West, 1992. MNRAS, **259**, 494.

6. Palumbo, G.G.,*et al.* 1993. Ap.J., **405**, 413.

7. Einasto, J., M.Gramman, E.Saar & E.Tago, 1993. MNRAS, **260**, 765.

8. Boyle, B.Y. & H.Y.Mo, 19,93. MNRAS, **260** 925.

9. Peebles, P.J.E., 1987. Nature, **327**, 210.

10. Gnedin, N.Yu. & J.P.Ostriker, 1992. Ap.J. **400**, 1.

11. Gouda. N., M.Sasaki & Y.Suto, 1989. Ap.J., **341**, 557.

12. Makino, N., & Y.Sato, 1993. Ap.J., **405**,1.

13. Jørgensen,H., E.Kotok, P.Naselsky & I.Novikov, 1993. NORDITA preprint N93/3A, MNRAS (in press).

14. Peebles, P.Y.E., 1981, Astrophys.J., **248**, 885

15. Bond Y.R. & G.Efstathiou, 1984. Ap.J., **285**, L45.

16. Efstathiou, G., G.B. Dalton, W.J.Sutherland, S.L.Maddox, 1992. MNRAS, **257**, 125.

17. Starobinsky, A.A., 1992. JETF Lett, **55(9)**, 489.

18. Kofman, L.A., A.D.Linde & A.A.Starobinsky, 1985. Phys.Lett., **B157**, 361.

19. Gottlöber, S., & J.P.Mücket, 1993. A.A. in press.

20. Bardeen J.M., J.R.Bond, N.Kaiser & A.S.Szalay, 1986. Ap.J. **304**, 15.

21. Shandarin S.F., Doroshkevich, A.G., and Zel'dovich, Ya.B., 1983, Sov. Phys. Usp., **26(1)**, 46.

22. Press, W.H., and Vishniac, E.T., 1980, Ap.J., **236**, 323.

23. Wilson, M.L. and Silk, J., 1981, Ap.J., **243**, 14.

24. Jones, B.Y.T., and Wyse, R.F.G., 1983, Mon.Not.R.astr.Soc., **205**, 983.

25. Efstathiou, G., W.J.Sutherland, & S.L.Maddox, 1990. Nature, **348**, 705.

26. Holtzman, J.A., 1989. Ap.J.Suppl., **71**, 1.

27. Peebles, P.Y.E., 1980, *The Large - Scale Structure of the Universe* (Princeton; Princeton University Press).

28. Kashlinsky, A., 1992 Astrophys.J.Lett, **386**, L37.

29. Naselsky P.D. & I.D.Novikov, 1992. NORDITA preprint N92/72A, 1993. Ap.J. **413** No 1.

30. Sugiyama, N. and Gouda, N., 1990, Astrophys.J., **365** 432.

CRITICAL PHENOMENA IN BLACK HOLES AND THE EMERGENCE OF A TWO DIMENSIONAL QUANTUM DESCRIPTION

C. O. LOUSTO*

Universität Konstanz,

Fakultät für Physik,

Postfach 5560,

D-78434 Konstanz, Germany.

ABSTRACT. We study the occurrence of critical phenomena in black holes, derive the critical exponents and show that they fulfill the scaling laws. Correlation functions critical exponents and Renormalization Group considerations assign an effective dimension, $d = 2$, to the system. The two-dimensional Gaussian approximation to critical systems is shown to reproduce all the black hole's critical exponents. Higher order corrections (which are always relevant) are discussed. Identifying the two-dimensional surface with the event horizon and noting that generalization of scaling leads to conformal invariance and then to string theory, we arrive to 't Hooft's string interpretation of black holes.

1. Critical Phenomena in Black Holes

The scaling of critical phenomena[1,2,3,4] applies to a great variety of thermodynamical systems. Those ranging from the internal structure of elementary particles to ferroelectricity and turbulent fluid flow, passing through superconductivity and superfluidity. The scaling is found to hold (within experimental error) in almost every case. The renormalization group approach [2,5] use the scaling hypothesis and provides a sound mathematical foundation to the concept of universality. On the other hand black hole dynamics is governed by analogues of the ordinary four laws of thermodynamics[6,7,8]. This two facts lead us to conjecture that black holes also obey the scaling laws or fourth law of thermodynamics[9]:

Let us suppose that a rotating charged black hole is held in equilibrium at some temperature T, with a surrounding heat bath. If we consider a small, reversible transfer of energy between the hole and its environment; this absorption will be isotropic, and will occur in such a way that the angular momentum J and charge Q remain unchanged, on the average. The full thermal capacity (not per unit mass) corresponding to this energy transfer can be computed by eliminating M between the equations for the temperature and the area of the black hole, and differentiate keeping J and Q constant,

$$C_{J,Q} = T\frac{\partial S}{\partial T}\bigg|_{J,Q} = \frac{MTS^3}{\pi J^2 + \frac{\pi}{4}Q^4 - T^2S^3} \ . \tag{1}$$

---

* Permanent Address: IAFE, Cas. Corr. 67, Suc. 28, 1428 Buenos Aires, ARGENTINA.

V. de Sabbata and H. Tso-Hsiu (eds.), Cosmology and Particle Physics, 183–192.

This heat capacity goes from negative values for a Schwarzschild black hole, $C_{Sch} = -M/T$, to positive values for a nearly extreme Kerr - Newman black hole, $C_{EKN} \sim \sqrt{M^4 - J^2 - M^2Q^2} \to 0^+$. Thus, $C_{J,Q}$ has changed sign at some value of J and Q in between. In fact, the heat capacity passes from negative to positive values through an infinite discontinuity. This feature has lead Davies[10] to classify the phenomenon at the critical values of J and Q as a second order phase transition. The values J_c and Q_c at which the transition occurs are obtained by making to vanish the denominator on the right hand side of eq (1). We can then define the following parametrization,

$$J_c^2 = \frac{j}{8\pi}M^4 \quad \text{and} \quad Q_c^2 = \frac{q}{8\pi}M^2 .$$

Eliminating S and T in eq (1) by use of the expressions for the temperature and entropy of a black hole[9], the infinite discontinuity in $C_{J,Q}$ takes place at[10]

$$j_{JQ}^2 + 6j_{JQ} + 4q_{JQ} = 3 . \tag{2}$$

For an uncharged, i.e., Kerr, hole, $q_{JQ} = 0$. Thus, $j_{JQ} = 2\sqrt{3} - 3$. Then we have

$$\Omega_c = \frac{\sqrt{2\sqrt{3} - 3}}{4\sqrt{3} - 3}T_c \cong 0.233T_c . \tag{3}$$

While for a non rotating, i.e., Reissner - Nordstøm, hole, $j_{JQ} = 0$. Thus, $q_{JQ} = 3/4$. And the critical value of the electric potential is given by

$$\Phi_c = \frac{1}{\sqrt{3}} , \tag{4}$$

independent of the other parameters of the black hole such as its mass or charge.

It can also be shown[11] that the four isothermal compressibilities are divergent as their corresponding heat capacities. For example,

$$K_{T,Q}^{-1} = J\frac{\partial\Omega}{\partial J}\bigg|_{T,Q} \sim \frac{\pi(2\Phi Q - M)(1 - 4\pi TM)}{S^2[1 - 12\pi TM + 4\pi^2 T^2(6M^2 + Q^2)]} , \tag{5}$$

diverges as $C_{J,Q}$. Also $K_{T,J}^{-1} = C_{J,Q}(\partial\Phi/\partial Q|_{S,J})/C_{J,\Phi}$ diverges as $C_{J,Q}$ on the singular segment given by eq (2).

By use of the expressions for the temperature and entropy of black holes[9], the heat capacity C_{JQ} can be expressed as [11]

$$C_{JQ} = \frac{4\pi TSM}{1 - 8\pi TM - 4\pi ST^2} \sim \frac{1}{T - T_c} , \tag{6}$$

where the critical temperature is given by $T_c^{JQ} = \{2\pi M[3 + \sqrt{3 - q_{JQ}}]\}^{-1}$, and q_{JQ} is given by the critical curve Eq. (2).

We can obtain the first two critical exponents directly by inspection of eq (6):

$$\alpha = 1 \quad , \quad \varphi = 1 \ . \tag{7}$$

Analogously, from eq (5) (that diverges as $C_{J,Q}$), we obtain

$$\gamma = 1 \quad , \quad 1 - \delta^{-1} = 1 \Rightarrow \delta^{-1} \to 0 \ . \tag{8}$$

To obtain the corresponding critical exponents we choose a path either along a critical isotherm or at constant angular momentum $J = J_c$ or constant charge $Q = Q_c$. However, in this case the black hole equations of state just reproduce the critical curves (such as eqs (3)-(4), and others deduced from them). In this case, we can formally assign a zero power corresponding to critical exponents:

$$\beta \to 0 \quad , \quad \delta^{-1} \to 0 \ ,$$

$$1 - \alpha = 0 \quad , \quad \psi \to 0 \ . \tag{9}$$

One can easily check that the set of critical values given by eqs (7) - (9) satisfy the scaling laws (with $\beta\delta = 1$):

$$\alpha + 2\beta + \gamma = 2 \quad , \quad \alpha + \beta(\delta + 1) = 2 \ ,$$

$$\gamma(\delta + 1) = (2 - \alpha)(\delta - 1) \quad , \quad \gamma = \beta(\delta - 1) \ , \tag{10}$$

$$(2 - \alpha)(\delta\psi - 1) + 1 = (1 - \alpha)\delta \quad , \quad \varphi + 2\psi - \delta^{-1} = 1 \ .$$

Other five heat capacities can be computed, of which $C_{\Omega,Q}$ and $C_{J,\Phi}$ exhibit also a singular behavior. The remaining $C_{\Phi,Q} = C_{J,\Omega}$ and $C_{\Omega,\Phi}$ being regular functions in the allowed set of values of the parameters[11]. Heat capacities and isothermal compressibilities at fixed (Ω, Q) and (J, Φ) give the same critical exponents as in the previous case where we held (J, Q) constant. This result can in fact be understood as a realization of the *Universality hypothesis*: For a continuous phase transition the static critical exponents depend only on the following three properties:

a) the dimensionality of the system, d.

b) the internal symmetry dimensionality of the order parameters, D.

c) whether the forces are of short or long range.

The critical curves for the three cases studied are all of them different, but the critical exponents, according to the above mentioned hypothesis, are the same within each class as specified after eq (10). We also observe that the equality between the primed $(T \to T_c^-)$ and unprimed $(T \to T_c^+)$ critical exponents is trivially verified in each one of the three transitions studied.

The lack of qualitative change in the properties of the black hole can be understood as in analogy to what happens in the case of a liquid-vapor system; where near criticality no qualitative distinction can be made between phases. Note that in this case there is not such thing as a latent heat [12] (since M remains continuous through the transition), as it happens in magnetic critical transitions. Besides, it can be seen that the critical transitions occur when we cold down the black hole with respect to the corresponding Schwarzschild temperature, $T_S = 1/(8\pi M)$, by increasing its charge or angular momentum at fixed total mass. Further, we have seen how black holes fulfil the scaling laws and universality hypothesis, both characteristics of critical phase transitions.

It worth noting[12] that although this phase transition does not affect the internal state of the system it is physically important as it indicates the transition from a region $(C_{JQ} < 0)$ where only a microcanonical ensemble is appropriate (stable equilibrium if the system is isolated from the outside world) to a region $(C_{JQ} > 0)$ where a canonical ensemble can be also used (stable equilibrium with an infinite heat bath).

2. Correlation functions, the Gaussian model and the Renormalization Group

Not only relations among critical exponents corresponding to thermodynamic functions can be obtained, but also relations concerning correlation functions exponents.

The static two-point connected correlations can be defined as

$$G_c^{(2)}(|\vec{r}|) = < \phi(0) \cdot \phi(\vec{r}) > - | < \phi > |^2 \ , \tag{11}$$

where ϕ is the order parameter of the system in question and may have, in principle, D different internal components. For example, in the Ising model, the order parameter has only one component, in a Heisenberg system, three, and in the He^3 superfluid transition as many as eighteen[3].

Away but not far from the critical region, one can write

$$G_c^{(2)}(r) \sim \frac{\exp\{-r/\xi\}}{r^{d-2+\eta}} \ , \quad r \text{ large} \ . \tag{12}$$

Here d is the dimensionality of the system, η is a further critical exponent and ξ is the correlation length. As one approaches the critical curve ξ diverges as

$$\xi \sim |T - T_c|^{-\nu} \ . \tag{13}$$

Here ν is another critical exponent.

Kadanoff[13] studied the scaling properties of the correlation functions and found a new scaling law relating the critical exponents

$$(2 - \eta)\nu = \gamma \tag{14}$$

With an additional assumption about the scaling behavior of the correlation function [14] one obtains the hyperscaling law

$$\nu d = 2 - \alpha \ . \tag{15}$$

Note that only here the dependence with the dimensionality of the system appears. By use of the Renormalization Group equations, one can show[15] that hyperscaling laws hold for $d \leq 4$ but break down for $d > 4$.

Now, let us consider the black hole in equilibrium with a radiation bath. By use of the Quantum Field Theory technics in the curved spacetime of the black hole one can obtain the correlation function of fluctuations of fields in this curved background. In equilibrium, the field will be in the Hartle-Hawking vacuum state. The correlation function of a scalar field in the Schwarzschild background, for large distances r is given by[16] (ω being the frequency of the mode considered)

$$G_\omega(r) \sim \frac{\omega}{2\pi \left[\exp\left(\frac{2\pi}{k_H}\right) - 1\right]} \ . \tag{16}$$

And thus, independent of the distance r. We expect that in equilibrium, gravitational correlations behave in a similar qualitative way, even considering charged and rotating black holes.

From Eq. (12) we thus conclude that

$$d - 2 + \eta = 0 \ . \tag{17}$$

The correlation length can be computed from [3]

$$\xi^2 = -\frac{1}{2\pi} \left(\frac{\partial^2 G(\omega)}{\partial \omega^2}\right)_{\omega=0} \ . \tag{18}$$

By use of Eq. (16) and since $K_T^{-1} \sim |T - T_c|^{-\gamma}$ we find that

$$\xi^2 \sim |T - T_c|^{-1} \quad \Rightarrow \quad \nu = \frac{1}{2} \ , \tag{19}$$

where we have used the definition of ν, Eq. (13), and that in our case $\gamma = 1$.

We see that our two new critical exponents (Eqs (17) and (19)) take values that fulfil the scaling relation (14) only if the dimension of the system is $d = 2$. In this case, the hyperscaling relation (15) is also satisfied, as expected for $d < 4$, and we have

$$\eta = 0 \ . \tag{20}$$

This is, in fact, the first hint that our system behaves as an effective two-dimensional one.

Additional insight can be gained by comparison with the Gaussian Model. This model can be described in its continuous version by the partition function[4]

$$\mathcal{Z}_G(J) = \mathcal{N} \int \mathcal{D}\phi \exp\left\{-\int d^d x \left[\frac{1}{2}c^2 |\vec{\nabla}\phi|^2 + \frac{1}{2}\mu\phi^2 - J\phi\right]\right\} \ . \tag{21}$$

The hamiltonian appearing in the exponential can be seen as a truncation of orders ϕ^4 or higher in a Ginzburg-Landau model. The Gaussian model was originally studied[17] for a discrete spin variable. It has the advantage of being exactly soluble and it presents a critical point with critical exponents (for a one-component field ϕ) given by [15]

$$\alpha = 2 - d/2 \ , \quad \beta = (d-2)/4 \ , \quad \gamma = 1 \ ,$$

$$\delta = \frac{d+2}{d-2} \quad , \quad \eta = 0 \quad , \quad \nu = \frac{1}{2} \quad . \tag{22}$$

It is worth to remark here that all this critical exponents can be made to take exactly the same values as for the black hole case, i.e. Eqs. (7)-(9), for $d = 2$. Thus, $d = 2$, appears here as the effective dimensionality of black holes near critical conditions.

The Gaussian model is not fully satisfactory because it has no "ordered" phase. The integral in Eq. (21) diverges for $T < T_c$ and thus one must include higher order terms (e.g. ϕ^4) to stabilize this integral. It is interesting to note here that black holes themselves pass through the critical curve from a region of canonical instability to a region of canonical stability as one lowers their temperature (see fig. 1 of ref.[9]).

One might think that the resulting effective dimension of the system, $d = 2$, relays only on comparison with the Gaussian model and that other possibilities are still open. To explore this possibility we can recall some results of the Renormalization Group Theory. Let us suppose that our effective Hamiltonian contains higher order terms than the Gaussian model. Then we can write

$$H_{eff}(\phi) = \frac{1}{2}c^2|\nabla\phi|^2 + \frac{1}{2}\mu\phi^2 + \frac{\lambda}{4!}\phi^4 + b\phi^2|\nabla\phi|^2 + ... \tag{23}$$

The scaling properties of the the additional operators, with n powers of ϕ and p derivatives, can be studied in terms of the sign of

$$\Delta = n - p - \frac{1}{2}d(n - 2) \quad . \tag{24}$$

If Δ is positive / negative the operator is relevant / irrelevant [4].

Thus, if the dimensionality of the system where larger or equal to four, the Renormalization Group analysis tell us [5,4] that the operators we have added to the gaussian hamiltonian are "irrelevant" in the sense that they do not contribute to modify the critical exponents, which will be those of the Gaussian model or the mean field (Landau) theory. Thus, no matching with the black hole results can be made for $d \geq 4$ models. There is still the possibility of having $d = 3$, as is the case of most realistic system, e.g. those studied in the Laboratory. In $d = 3$, ϕ^4 becomes a relevant operator. One can make a perturbation theory based on the gaussian part of the hamiltonian and obtain a set of critical exponents [4] that fit very well with experiments but are not those of black holes. Thus, we are left with $d = 2$ (since for $d < 2$ no critical phenomena takes place). The problem here is that *all* operators of the form ϕ^{2n} and $|\nabla\phi|^2\phi^{2n}$ are relevant and thus will modify the critical exponents. For instance, if we take the ϕ^4 term in Eq. (23), we will have [3] to a good approximation (practically the same values as in the Ising model) the following critical exponents in $d = 2$,

$$\alpha = 0 \quad , \quad \beta = 1/8 \quad , \quad \gamma = 7/4 \quad ,$$
$$\delta = 15 \quad , \quad \eta = 1/4 \quad , \quad \nu = 1 \quad . \tag{25}$$

Since all operators are relevant, we expect this theory to be renormalizable. In fact, we know that field theory (as well as gravity) in two dimensions is asymptotically free in the ultraviolet allowing us to can be built up a finite quantum fields theory.

We can now conclude that the first order approximation to Quantum effects in black holes correspond to the Gaussian approximation. Let us recall that the path integral formulation of the Hawking [18] radiation and black hole gravitational thermodynamics relies on the stationary phase approximation to obtain a convergent Gaussian integral. The next order approximation should include back reaction and self-interaction effects as well as higher order quantum corrections. In fact, whatever would be the final form of the quantum theory of gravity, we can assume that the Kerr metric should be a classical solution to the vacuum field equations. The critical exponents of this black hole solution will then be those given in Eqs. (7)-(9). By applying the universality hypothesis this exponents will be the same for the full family of black hole solutions to the theory. We can thus conjecture that the critical phenomena in black hole will survive to higher order corrections; that the scaling laws will continue to hold, but the critical exponents that will fulfil this laws, when quantum higher order corrections are taken into account, will be different from those given by Eqs. (7)-(9), and in particular, to the next quantum order to Hawking radiation approximation they will take Eq. (25) values.

3. The Black Hole horizon as a Quantum Critical System

Now that we have established that the dimensionality of the system is $d = 2$, it remains to identity this two-dimensional surface. A natural choice is the horizon of the black hole, which is a two-dimensional surface in Minkoswki space [†].

One observer far away from the black hole sees all the matter of the collapsing body that will form it to accumulate on the two-dimensional horizon (for more details on this membrane paradigm see Ref. 19.

By analogy to the models for spin systems able to suffer critical transitions we can think of the event horizon as having only a finite (and eventually discrete) number of degrees of freedom at every (lattice) cell of Planckian dimensions. We know that if there is a continuous internal symmetry in the order parameter, no long range order, or broken symmetry, will occur in two space dimensions. If the symmetry is discrete it is possible (e.g. Ising model).

It is interesting to compare our approach to black hole quantization with that of 't Hooft [20] since several points in common can be drawn. In this approach to the problem of black hole quantization it is postulated the existence of an S-matrix to describe the evaporation process. This hypothesis seems to be supported by new evidence revealing that the stimulated emission in the Hawking radiation might play an important role in solving the loss of quantum coherence paradox[21,22,23]. The horizon shift produced by light particles going out or coming into the horizon is an essential ingredient in the construction of the S-matrix. Its elements are given by

$$< p^{out}(\Omega)|p^{in}(\Omega) >= \mathcal{N} \exp \left\{ i \int \int p^{out}(\Omega) f(\Omega - \Omega') p^{in}(\Omega') d^2\Omega d^2\Omega' \right\} \quad (26)$$

[†] Finite size effects are not expected to affect the scaling properties derived in the thermodynamic limit [3].

where $p^{in}(\Omega)$ and $p^{out}(\Omega)$ are the momentum distribution at angle $\Omega = (\vartheta, \varphi)$ of the in- and out-going particles respectively. The shift function f, is the Green function defined on the event horizon [24,25] satisfying

$$\nabla_\perp^2 - af = b\delta^2(\Omega - \Omega') \ , \tag{27}$$

$$a = 2r_+ K(r_+) \ , \quad b = 32\pi pr_+^2 g_{uv}(r_+) \ .$$

This expression can be integrated by using the properties of Legendre polynomials

$$f(\Omega - \Omega') = -\frac{\pi b}{\sqrt{2}} \frac{P_{1/2+i\sqrt{3}/2a}[-\cos(\Omega - \Omega')]}{\cosh\left(\frac{\sqrt{3}}{2}\pi a\right)} \ . \tag{28}$$

The Legendre functions $P_{1/2+i\lambda}(z)$ are called conical functions and are defined positives for $z < 1$.

A functional integral representation can be given for the S-matrix[20]

$$< p^{out}(\Omega)|p^{in}(\Omega) >= \mathcal{N} \int \mathcal{D}x(\Omega) \exp\left\{\int d^2\Omega \left[-\frac{i}{2\pi}\left((\partial_\Omega x)^2 + ax^2\right) + ixp^{ext}\right]\right\} \ . \tag{29}$$

It is apparent the similarity between this expression and the partition function of our model, Eq. (21), if we identify there the two-dimensional surface with the event horizon and the scalar order parameter with the "membrane coordinates", x.

Near criticality the "mass term", μ in Eq. (21), vanishes like $\mu \sim |T - T_c|$. Thus, the model becomes conformaly invariant. In this case we can write the functional integral formulation in a covariant "Stringy" way

$$\mathcal{Z}_G(J) = \mathcal{N} \int \mathcal{D}\phi^\mu(\sigma)\mathcal{D}g^{ab}(\sigma) \exp\left\{-\int d^2\sigma \left[-i\sqrt{g}g^{ab}\partial_a\phi^\mu\partial_b\phi^\mu + i\phi^\mu J^\mu\right]\right\} \ . \tag{30}$$

where σ stands for the two horizon coordinates (in Euclidean space), the order parameter has now $[\mu]$ internal dimensions and g^{ab} is the metric on the horizon surface.

4. Discussion

Summarizing, we have started by showing the scaling of black holes near criticality. This property of critical systems can be embodied in the conformal invariance theory[26,27]. Then we are lead to string theory which is a realization of a conformal field theory on the two-dimensional world-street[28]. Eq. (30) corresponds to the bosonic string case. The fermionic degrees of freedoms can be eventually incorporated in it, this corresponding to the addition of a fermionic order parameter in the effective hamiltonian Eq. (23).

Actually, for a continuum model, the lower critical dimension is precisely two[4]. This means that to have critical phenomena we should consider $d = 2 + \epsilon$, where ϵ is of the order of the Planck lengh[29]. Otherwise, if we consider a discrete model, the lower critical dimension is one. We can thus keep $d = 2$ and deal with a discrete order parameter on the surface of black hole transformed now in a lattice. This corresponds to consider a quantization in the charge and spin of the particles lying on the sites of the two-dimensional lattice.

Of particular interest here is the result that the continuum limit of the two-dimensional tricritical Ising model near the critical point is supersymmetric[30] (in ref. [9] we remarked the existence of tricritical points in extreme Kerr-Newman black holes).

This elements lead us to a two-dimensional field model for black holes that could provide new insights to its quantum properties. In particular, the quantization of the black hole mass and the origin of its gravitational entropy are presently under study by the author.

Acknowledgements

This work was partially supported by the Directorate General for Science, Research and Development of the Commission of the European Communities and by the Alexander von Humboldt Foundation.

References

1. H. E. Stanley, *Introduction to Phase Transitions and Critical Phenomena*, Oxford Univ. Press, Oxford (1971).

2. K. G. Wilson, *Rev. Mod. Physics.*, **55**, 583 (1983).

3. D. J. Amit, *Field Theory, the Renormalization Group, and Critical Phenomena*, World Sci., Singapore, 2nd Ed. (1992).

4. J. J. Binney, N. J. Dowrik, A. J. Fisher and M. E. J. Newman, *The theory of Critical Phenomena*, Clarendon Press, Oxford (1992).

5. K. G. Wilson and J. Kogut, *Phys. Rep.*, **C12**, 75 (1974).

6. J. M. Bardeen, B. Carter and S. W. Hawking, *Commun. Math. Phys.*, **31**, 161 (1973).

7. J. D. Bekenstein, *Phys. Rev.*, **D7**, 949 (1973); ibid, **D7**, 2333 (1973); ibid, **D9**, 3292 (1974).

8. R. M. Wald, in *Black Hole Physics*, V. De Sabbata and Z. Zhang Eds, Nato Asi series, Vol. 364 (1992), p 55-97.

9. C. O. Lousto, *The fourth Law of Black Hole thermodynamics*, Nucl. Phys. B, (1993).

10. P. C. W. Davies, *Proc. R. Soc. Lond.*, **A353**, 499 (1977).

11. D. Tranah and P. T. Landsberg, *Collective Phenomena*, **3**, 81 (1980).

12. P. Hut, *Mon. N. R. Astr. S.*, **180**, 379 (1977).

13. L. P. Kadanoff, Physics, **2**, 263, 1966

14. A. Hankey and H. E. Stanley, *Phys. Rev.*, **B6**, 3515 (1972).

15. M. E. Fisher, in *Critical Phenomena*, F. J. W. Hahne Ed, Springer Verlag, Berlin (1982), p 1-139.

16. D. W. Sciama, P. Candelas and D. Deutsch, Advances in Physics, **30**, 327 (1981).

17. T. H. Berlin and M. Kac, Phys. Rev., **86**, 821 (1952).

18. J. B. Hartle and S. W. Hawking, *Phys. Rev. D* **13**, 191 (1976).

19. K. S. Thorne et al., *Black holes: The membrane Paradigm*, Yale Univ. Press, New Heaven (1986).

20. G. 't Hooft, *Nucl. Phys. B* **355**, 138 (1990).

21. J. D. Bekenstein, preprint UCSB-TH-93-02 (1993).

22. M. Schiffer, preprint CERN-TH 6811/93 (1993).

23. C. O. Lousto and R. Müller, preprint Konstanz (1993).

24. T. Dray and G.'t Hooft, *Nucl. Phys. B* **253**, 173 (1985).

25. C. O. Lousto and N. Sánchez, *Phys. Lett. B* **220**, 55 (1989).

26. J. L. Cardy, in *Phase Transitions and Critical Phenomena*, Eds. C. Domb and J. L. Lebowitz, Vol. 11, Academic Press, London, p 55 (1987).

27. C. Itzykson, H.Saleur and J.B. Zuber, *Conformal invariance and Applications to Statistical Mechanics*, World Sci., Singapore (1988).

28. M. B. Green, J. H. Schwarz and E. Witten, *Supestring Theory*, Vol. 1, Cambridge Univ. Press., Cambridge (1987).

29. G. 't Hooft, Nucl. Phys., **B 256**, 727 (1985).

30. Z. Qiu, Nucl. Phys., **B 270**, 205 (1986).

QUANTIZATION OF THE METRIC CREATED BY ULTRARELATIVISTIC PARTICLES

C. O. LOUSTO*

Universität Konstanz,

Fakultät für Physik,

Postfach 5560,

D-78434 Konstanz, Germany.

and

N. SÁNCHEZ

Observatoire de Paris, Section Meudon,

*Demirm**,*

92195 Meudon Principal Cedex,

France.

ABSTRACT. We study the quantization of the curved spacetime created by ultrarelativistic particles at Planckian energies. We consider a minisuperspace model based on the classical shock wave metric generated by these particles, and for which the Wheeler - De Witt equation is solved exactly. The wave function of the geometry is a Bessel function whose argument is the classical action. This allows us to describe not only the semiclassical regime ($S \to \infty$), but also the strong quantum regime ($S \to 0$). We analyze the interaction with a scalar field ϕ and apply the third quantization formalism to it. The quantum gravity effects make the system to evolve from a highly curved semiclassical geometry (a gravitational wave metric) into a strongly quantum state represented by a weakly curved geometry (essentially flat spacetime).

1. Introduction

The spacetime geometry created by ultrarelativistic sources, that is, the gravitational shock wave spacetimes, have arisen great interest recently[1,2,3,4,5,6,7,8,9,10,11,12]. These backgrounds are relevant to describe the particle scattering at the Planck energy scale. Quantum scattering of particle fields and strings by this class of metrics have been studied in the approximation in which the geometry is treated classically, i.e., as a background field. A further step in this direction is to quantize the geometry itself. As it is known, so far there is not a quantum theory of gravitation to fully carry out this program. Although

---

* Permanent Address: IAFE, Cas. Corr. 67, Suc. 28, 1428 Buenos Aires, ARGENTINA.

** UA 336 Laboratoire associé au CNRS, Observatoire de Meudon et École Normale Supérieure.

V. de Sabbata and H. Tso-Hsiu (eds.), Cosmology and Particle Physics, 193–199.

a conventional quantum field theory of gravitation lacks to be renormalizable, information about the quantization of the spacetime geometry can be obtained by solving the Wheeler - De Witt equation[13]:

$$\left[G_{ijkl} \frac{\delta}{\delta g_{ij}} \frac{\delta}{\delta_{kl}} + \sqrt{^{(3)}g} \; ^{(3)}R \right] \Psi(^{(3)}\mathcal{G}) = 0 \; . \tag{1}$$

In the canonical description of General Relativity, the space-time metric has the $3 + 1$ decomposition

$$d^2 s = (N^2 - N_i N^i) dt^2 - 2N_i dx^i dt + g_{ij} dx^i dx^j \; , \tag{2}$$

that is

$$g_{\mu\nu} = \begin{pmatrix} -N^{-2} & N^{-2} N^i \\ N^{-2} N^i & g^{ij} - N^{-2} N_i N_j \end{pmatrix} \; . \tag{3}$$

In the region between two space like hypersurfaces $t = t_i$ and $t = t_f$, the Einstein equations determine the sequence of three-geometries $^{(3)}\mathcal{G}$ on the space-like surfaces of constant t, the dynamical important object being the 3-geometry $^{(3)}\mathcal{G}$. The dynamics of the gravitational field is entirely described by the so called Hamiltonian constraint $\mathcal{H} = 0$. In the quantum theory, this becomes an equation for the state vector $\mathcal{H}\Psi = 0$, which takes the form of the functional differential equation (1). There are also the others constraints of the classical theory, but at the quantum level, they express merely the gauge invariance on Ψ. Classically, one can know both $^{(3)}\mathcal{G}(t_0)$ and $\frac{\partial}{\partial t} {}^{(3)}\mathcal{G}(t_0)$ at some time parameter t_0 and determine the 4-geometry $^{(4)}\mathcal{G}$, but quantum mechanically, one can only know $^{(3)}\mathcal{G}(t_0)$ or $\frac{\partial}{\partial t} {}^{(3)}\mathcal{G}(t_0)$ and therefore, one has a certain probability for $^{(3)}\mathcal{G}(t)$. The manifold of all possible $^{(3)}\mathcal{G}$ - the so called superspace - in which each point is a metric $g_{ij}(\vec{x})$, has the metric

$$G_{ijkl} = \sqrt{g}(g_{ik}g_{jl} + g_{il}g_{jk} - g_{ij}g_{kl}) \; , \tag{4}$$

with signature $(-, +, +, +, +, +)$. In order to solve Eq. (1), one considers spacetime symmetries and restricts the degrees of freedom to be quantized (minisuperspace models).

2. Pure Gravity Model

Let us quantize now the gravitational shock wave geometries. The minisuperspace metric to quantize is

$$ds^2 = -(N^2 - N_u N^u) dv^2 + 2N_u du dv + F(\rho, u) du^2 + dx^2 + dy^2 \tag{5}$$

where $u = z - t$, $v = z + t$ are null variables and $\rho = \sqrt{x^2 + y^2}$. The classical shock wave metrics have the form of Eq. (5) with[10]

$$F(u, \rho) = f(\rho)\delta(u) \; , \tag{6}$$

and the Lagrange multipliers taken $N_u = 1$, $N = 0$. This expression of the metric is generic for any ultrarelativistic source; its particularities entering only in the form of the function $f(\rho)$. This expression can be extended to D - dimensions and to shock waves superimposed to curved backgrounds[4,7]. Eq. (5) can also represent a sourceless plane gravitational wave. In such case

$$F(x, y, u) = (x^2 - y^2)D(u) ,$$

$D(u)$ being an arbitrary function of u.

The metric on superspace has the components:

$$G_{1111} = \frac{1}{2}F^{3/2} ,$$

$$G_{1122} = G_{1133} = -\frac{1}{2}F^{1/2} ,$$

$$G_{2222} = G_{3333} = -G_{2233} = \frac{1}{2}F^{-1/2} ,$$

the others components vanish.

The three - curvature $^{(3)}R$ is given by

$$^{(3)}R = -g^{11}\nabla_\perp^2 g_{11} + \frac{1}{2}(g^{11})^{-2}(\vec{\nabla}_\perp g_{11})^2 = \frac{1}{F}[-\nabla_\perp^2 F + \frac{1}{2F}(\vec{\nabla}_\perp F)^2]$$

where the subscript $\perp$ refers to the (2, 3) spatial transverse coordinates.

Thus, the Eq. (1) in our case reads,

$$\left[-\frac{1}{2}F^{3/2}\frac{\delta^2}{\delta F^2} + F^{1/2}\left(F^{-1}\nabla_\perp^2 F - \frac{F^{-2}}{2}\left(\vec{\nabla}_\perp F\right)^2\right)\right]\Psi(F) = 0 . \tag{7}$$

Here, for the sake of convenience, we have taken a 3+1 decomposition with respect to the hypersurface $v = const$, that will play the role of time. Notice also that the supermomentum constraint is already satisfied classically, since for the metric (5), one has $R_{vu} = 0$. Thus, we do not need to consider the quantization of this equation[14].

Let us consider now a minisuperspace model by freezing out all the transversal degrees of freedom and quantizing the longitudinal one, i.e. we write,

$$F(u, \rho) = f(\rho)D(u) , \tag{8}$$

where $f(\rho)$ is the profile function characterizing the classical shock wave metric[10]. The function $D(u)$ will represent the degree of freedom to quantize (classically, $D(u) \equiv \delta(u)$).

By discretizing the variable $u \rightarrow u_n$ and making the change of variable $s^2(u_n) = D(u_n)$ we can transform the functional differential equation (7) into the following ordinary differential equation,

$$\left[\frac{d^2}{ds^2} + (2p - 1)s^{-1}\frac{d}{ds} + 4C(\rho)\right]\Psi(s_n) = 0 , \tag{9}$$

where p accounts for the arbitrariness in the operator ordering, and

$$C(\rho) \doteq -2 \left(\nabla_\perp^2 f - \frac{f^{-1}}{2} \left(\vec{\nabla}_\perp f \right)^2 \right) = 2f \,^{(3)}R(\rho) . \tag{10}$$

The whole wave function will read

$$\Psi(D) = \prod_n \Psi(D(u_n)) .$$

For $p = 1$ (which corresponds to the Laplacian ordering prescription), Eq. (9) can be brought into a Bessel equation of index $\nu = 0$. We choose the solution that remains finite when $D \to 0$. Thus, our solution reads,

$$\Psi(F_n) = J_0 \left(2\sqrt{C(\rho)D(u_n)} \right) = J_0 \left(2\sqrt{2 \,^{(3)}RF(\rho, u_n)} \right) . \tag{11}$$

We can obtain the semiclassical limit by considering large arguments of the Bessel function in equation (11). Thus,

$$\Psi(F) \simeq \left(\pi \sqrt{2 \,^{(3)}RF(\rho, u_n)} \right)^{-1/2} \cos \left[2\sqrt{2 \,^{(3)}RF(\rho, u_n)} - \pi/4 \right] . \tag{12}$$

On the other hand, the semiclassical regime can also be directly studied from the Hamilton - Jacobi equation for our system,

$$-\frac{1}{2}F^{3/2} \left(\frac{dS}{dF} \right)^2 + F^{1/2} \,^{(3)}R(\rho) = 0 ,$$

where S is the Hamilton - Jacobi principal function that by direct integration reads,

$$S = \pm 2\sqrt{2 \,^{(3)}RF} . \tag{13}$$

We see that the argument of the wave function (11) is the *action* of the classical Hamilton - Jacobi equation. Thus, for large S, the Bessel function has an oscillatory behavior and the wave function Eq. (11), is given by

$$\Psi \simeq \frac{A}{S} \exp\{iS/\hbar\} + \frac{B}{S} \exp\{-iS/\hbar\} . \tag{14}$$

that is, Eq. (12) gives the right semiclassical limit.

Semiclassically, the wave function Ψ is picked around the classical metric geometry. It appears natural to interpret the oscillatory behavior Eq. (14) as describing Ψ in the classical allowed regime. The classically forbidden regime, instead, appears with a *real* exponential behavior (Euclidean signature region).

For small S, $(S \ll \hbar)$, the behavior is non - oscillatory. In particular, the Bessel function reaches a finite value in the limit $S \to 0$, i.e. $J_0(0) = 1$. This can be interpreted as a genuine quantum behavior, as opposite to the large S sector which describes the semiclassical regime. Notice that $S \to 0$, i.e. $D(u) = 0$, is flat spacetime and that this appears in the strong quantum regime. For $F > 0$ the action is real and Eq. (14) describes Ψ in the classically allowed regime. For each S, i.e. each F, Ψ describes a classical configuration. All configurations are classically allowed.

There, $\mid \Psi \mid^2$ (which to some extent can be seen as proportional to a probability), is a maximum. We will come back to this interpretation by the end of the paper.

For $^{(3)}R > 0$ the action is real and Eq. (12) describes the classically allowed regime. If $^{(3)}R < 0$ then $\Psi(S \to \infty) \sim \exp\{S\}/S$. This is an exponential growing of the semiclassical wave function. It is the opposite of what happens in the case of tunnel effect, and like the situation of the falling of a particle into a well potential, indicating the presence of an instability. That is, the classically forbidden regime does not corresponds to tunnel effect, but to an unstable (exponentially growing) behavior.

3. Third Quantization of the Gravity plus Matter model

We will include a scalar field in the analysis of the wave function. The Wheeler - De Witt equation in this case reads

$$\left\{ -\frac{\delta^2}{\delta F^2} - \frac{2}{F}\,^{(3)}R(\rho) + F^{-2}\frac{\delta^2}{\delta\phi^2} + (\phi_{,u})^2 + \frac{1}{F}[(\phi_{,\rho})^2 + m^2\phi^2] \right\} \Psi(F,\phi) = 0 . \tag{15}$$

here it is understood that the discretization of the variable u is already made in an analogous way to that of the previous section.

In this section we study the possibility of third quantizing the wave function of the ultra-relativistic particles. A motivation for the third quantization is to overcome the problem of negative probabilities just as was the case for the Klein - Gordon equation (see for example Ref. 15).

Let us then consider the Eq. (15) for an ultrarelativistic particle and an homogeneous massless scalar field ϕ, that up to the operator ordering ambiguity reads,

$$\left[\partial_t^2 - \partial_\phi^2 + 72\,^{(3)}R \exp\{-6t\} \right] \Psi(t,\phi) = 0 , \tag{16}$$

where we have made the change of variables (at each point u_n)

$$t_n = -\frac{1}{6}\ln(F_n) , \tag{17}$$

which allows the variable t_n to run from $-\infty$ to $+\infty$ and to give to it a time - like interpretation.

The general solution to Eq. (16) can be written as

$$\Psi_j(t,\phi) = \prod_n Z_{2ij}\left(\sqrt{8 \; ^{(3)}R} e^{3t_n}\right) \exp\{ij\phi\} \; , \tag{18}$$

where Z_ν is a Bessel function of first or second kind.

It is interesting to note here that the case of pure gravity is recovered for $j = 0$. As j is related to the energy of the matter field ϕ, it seems that the presence of matter fields excite the gravitational modes and allow for particle production as we shall immediately see.

Following the work of Ref. 16, we can define *in* states proportional to the Bessel function $J_{2ij}\left(\sqrt{8 \; ^{(3)}R} e^{3t_n}\right)$ for $t \to -\infty$ (which are natural positive frequency modes in the *in* region), and *out* states proportional to the Hankel function $H_{2ij}^{(2)}\left(\sqrt{8 \; ^{(3)}R} e^{3t_n}\right)$ for $t \to +\infty$. Then, one can compute the Bogoliubov transformation coefficients between these two basis, and since the Hankel function is a linear combination of positive and negative subindex (frequency) Bessel functions one obtains particle production of the outgoing modes with respect to the *in* vacuum. It appears that the spectrum of produced particles has a Planckian distribution at temperature, $T = 18\sqrt{8\pi G/3}$, of the order of the Planck temperature.

Thus, the interpretation of our results can be made in terms of the creation of ultrarelativistic particles carrying with them its own geometry. The interaction with the matter fields deplete the gravitational energy bringing the ingoing semiclassical state, for instance picked around a shock wave metric, (that is, a strongly curved geometry) into a state in the quantum regime represented by a weakly curved geometry (essentially flat spacetime). The evolution of the system is from the classical into the quantum regime. The initial semiclassical configuration is a highly curved geometry, the final configuration is a weakly curved one.

Acknowledgements

The authors thank the Ettore Majorana Center at Erice and IAFE at Buenos Aires, where part of this work was done, for kind hospitality and working facilities. This work was partially supported by the Directorate General for Science, Research and Development of the Commission of the European Communities and by the Alexander von Humboldt Foundation.

References

1. G. 't Hooft, *Phys. Lett.* B **198**, 61 (1987)

2. D. Amati, M. Ciafaloni and G. Veneziano, *Int. J. Mod. Phys.* A **3**, 1615 (1988)

3. D. Amati and C.Klimčik, *Phys. Lett.* B **210**, 92 (1988)

4. C. O. Lousto and N. Sánchez, *Phys. Lett.* B **220**, 55 (1989)

5. C. O. Lousto and N. Sánchez, *Phys. Lett.* B **232**, 462 (1989)

6. H. J. de Vega and N. Sánchez, *Nucl. Phys.* B **317**, 706 and 731 (1989)

7. C. O. Lousto and N. Sánchez, *Int. J. Mod. Phys.* A **5**, 915 (1990)

8. H. J. de Vega and N. Sánchez, *Phys. Lett.* B **244**, 215 (1990)

9. G. T. Horowitz and A. R. Steif, *Phys. Rev. Lett.* **64**, 260 (1990)

10. C. O. Lousto and N. Sánchez, *Nucl. Phys.* B **355**, 231 (1991)

11. C. O. Lousto and N. Sánchez, *Phys. Rev.* D **46**, 4520 (1992)

12. C. O. Lousto and N. Sánchez, *Nucl. Phys.* B **383**, 377 (1992)

13. B. S. De Witt, *Phys. Rev.* **160**, 1113 (1967) ; J. A. Wheeler, in *Batelles Rencontres*, ed. C. De Witt and J. A. Wheeler, Benjamin, N.Y. (1968)

14. C. O. Lousto and F. D. Mazzitelli, *Int. J. Mod. Phys.* A **6**, 1017 (1991)

15. M. Mc Guigan, *Phys. Rev.* D **38**, 3031 (1988) ; M. Mc Guigan, *Phys. Rev.* D **39**, 2229 (1989)

16. A. Hosoya and M. Morikawa, *Phys. Rev.* D **39**, 1123 (1989)

GRAVITATIONAL WAVE SEARCH WITH RESONANT DETECTORS

G.Pizzella

Physics Department, University of Rome "Tor Vergata"
Institute for Nuclear Physics, Frascati

ABSTRACT

The gravitational wave basic properties and their principal sources are schematically reviewed. The performances of the resonant cryogenic antennas are illustrated and their sensitivity is compared with those of the interferometer antennas. Finally the main results so far obtained by the Rome group are briefly presented.

KEYWORDS: General Relativity, Gravitational Waves

1. Introduction

Professor Joe Weber at the University of Maryland started in the early sixties the experimental search for gravitational wave. He worked alone for about ten years and finally he showed experimental evidence for coincidences between two g.w. antennas located at large distance between them, one in Maryland and the other one at the Argonne Laboratory. It appeared difficult to interpret Weber's results in terms of g.w. and various other groups tried, in the period 1970-1975, to repeat the experiment with similar techniques, which consisted in using ton heavy aluminium bars at room temperature, equipped with piezoelectric ceramics for the detection of the bar small vibrations.

Although they did not confirm Weber's observations, other groups entered in the field, developing new techniques but all based on those invented by Weber, in spite that, if the g.w. sources were just those predicted by the General Relativity, no hope to detect them existed before several tens of years. In particular, the groups of Stanford, Louisiana State and Rome Universities started in 1971 to construct cryogenic antennas, and the groups of MIT, Max-Planck in Munich and Glasgow University started to develop laser interferometers. One of the most important final goals for all groups was to reach a sensitivity such to be able to detect gravitational collapses occurring in the Virgo Cluster. Although many progress have been made, this goal appears to be still very far because of the extreme weakness of the signal due to the

V. de Sabbata and H. Tso-Hsiu (eds.), Cosmology and Particle Physics, 201–212.
© 1994 *Kluwer Academic Publishers. Printed in the Netherlands.*

gravitational force as compared with the noise due to the other forces in Nature.

Only in May 1991 two large cryogenic antennas entered in simultaneous and continuos operation with high sensitivity: the Explorer antenna of the Rome group and the antenna of the Louisiana State University. The search for coincidences between these two antennas has just started, marking a new phase in this fascinating field of fundamental physics. It is expected that three other groups will soon joint with their cryogenic antennas for coincidence analysis: the group of Stanford, the group of Perth in Australia and the group of Legnaro in Italy.

An important milestone was also the decision by the scientific community all over the world to start the construction of very long arm interferometers, which with respect to the resonant antennas, have similar sensitivity in the frequency range of 1 kHz but have the advantage of a larger bandwidth reaching lower frequencies. At present, there are two laser interferometers in construction or in advanced stage of planning: the MIT-Caltech and the Pisa-Orsay collaborations while other groups (Glasgow, Max-Planck, Tokyo, Perth) are also planning to joint in the new adventure.

2. Theoretical aspects

We recall, very briefly, a few theoretical properties of the gravitational radiation. This radiation is expressed by the perturbation h_{ik} of the metric tensor g_{ik}, $g_{ik}=g_{ik}^0+h_{ik}$, where g_{ik}^0 is the minkowskian metric tensor and h_{ik} is a tensor whose component absolute values are much smaller than unity. For a plane wave propagating in the x direction the only non zero components are

$$h_{yz} = h_{zy} = h_x$$
$$h_{yy} = -h_{zz} = h_+$$

These quantities are functions of the argument (t-x/c). Notice that the g.w. propagate with the speed of light c and they are transverse. It can be also shown that, because of the conservation laws, the g.w. can be associated to a particle with spin 2 (the mass is obviously zero): the graviton. Therefore the two polarisation states are changed one in the other by a spatial rotation of $\pi/4$.

The g.w. are not a unique feature of GR. In fact, it could be shown that any metric theory of gravity, which incorporates the Lorentz invariance in its field equations, foresees the propagation of g.w. The speed and the polarisation status will depend on the particular theory.

G.w. are generated by the movement of gravitational masses. The irradiated power has a null dipole term, because of the momentum conservation law. The first non-zero term is the quadrupole one. For this reason, and also because of the small value of the gravitational constant G, the irradiated power is so small that, at present, it is not possible to detect the g.w. that could be generated in an earth laboratory. It is necessary that the source be constituted by enormous masses moving with extreme accelerations. Astrophysical sources, like supernovae for instance, are, therefore, the best candidate supply of g.w.

We distinguish between continuous and impulsive sources. Example of continuous sources are the pulsars and the binary systems. In the first case it is necessary that the rotating collapsed star be non-spherical, in order to have a time varying mass quadrupole. Possible values for the metric tensor perturbation could be of the order of $h \sim 10^{-26} - 10^{-27}$, time-varying with twice the pulsar frequency. In the second case there should always be a generation of g.w., but with very small frequency, well outside the presently explored range. For the impulsive sources one of the most interesting one is the supernova. It is very difficult to guess the degree of non-sphericity during the collapse. If we assume that a fraction $M_{gw}c^2$ of all available energy $M_{star}c^2$ is entirely converted into g.w., it is found, at a distance R on the Earth

$$h = \frac{1}{R\,\omega_g} \sqrt{\frac{8\,G\,M_{gw}}{c\tau_g}} = 3.11 \; 10^{-20} \; \frac{1 \text{ kHz}}{\nu_g} \frac{10 \text{ Mpc}}{R} \sqrt{\frac{M_{gw}}{M_o} \frac{1 \text{ ms}}{\tau_g}}$$

where $\tau_g = 1$ ms is the typically assumed duration of the g.w. burst and n_g is the antenna resonance frequency. If the source is located at the centre of our Galaxy we obtain for $M_{gw} = 10^{-2}M_o$: $h \sim 3 \; 10^{-18}$. If the source is located in the Virgo Cluster we get $h \sim 3 \; 10^{-21}$.

3. The resonant antenna

This detector is, typically, a metallic cylindrical bar, usually of high quality aluminium alloy (for the Perth experiment the material is niobium, which has higher Q). For capturing as more energy as possible from the g.w. the mass of the bar needs to be as large as possible. The LSU-Rome-Stanford collaboration uses bars 3 m long, weighting about 2300 kg. When the bar is hit by a g.w. burst with amplitude h(t) and very short duration t_g impinging perpendicularly to its axis, it starts to vibrate at those resonance modes that are coupled to the g.w. The odd longitudinal modes are the most coupled ones. The first mode has frequency $v_0 = \pi v/L$, about 900 Hz for the above considered bars ($v=5400$ m/s is the sound velocity in aluminium). The vibration amplitude, at its ends, for a wave direction perpendicular to the bar axis, is

$$\xi(t) \approx \frac{2 L}{\pi^2} e^{-\frac{t}{\tau_V}} H(\omega_o) \sin(\omega_o t)$$

where $H(\omega_o)$ is the Fourier component of h(t) at the bar resonance and τ_V is the relaxation time related to the merit factor $\tau_V = Q/\omega_0$. A quick calculation for a supernova in the Galaxy gives the unthinkable very small value $\xi(t) \approx 3 \cdot 10^{-18}$ m. The maximum energy associated to this vibration is

$$E = \frac{1}{4} M \omega^2 \xi_o^2 = \frac{M v^4}{L^2} H^2(\omega_o)$$

from which, in the rough approximation $H(\omega_0) = h(t) \tau_g$, we derive

$$h = \frac{1}{\tau_g} \frac{L}{v^2} \sqrt{\frac{E}{M}}$$

For M=2300 kg, L=3 m, $\tau_g = 1$ ms, for a g.w. due to a SN ($M_{gw} = 0.01 M_o$, h=3 10^{-18}) coming from the Galactic Centre we obtain E=2.7 10^{-24} joule = 0.2 kelvin. If the same g.w. comes from the Virgo Cluster we get E=2 10^{-30} joule= 1.4 10^{-7} kelvin. The values in kelvin are usually given in order to compare this small signal with the noise of the apparatus that it is more convenient to express in kelvin units. The bar vibrations become electrical signals by means of proper electromechanical transducers. The transducers, which we shall not discus here, represent the most delicate part of the experiment and still need to be improved considerably.

The above formulas are derived from the classical cross-section theory. J.Weber and G.Preparata have derived cross-sections larger than the classical one by several order of magnitude.

There are several types of noise. Some noise can be reduced only by comparing the data taken with two independent antennas located very far one from each other. Some other noise can be also reduced by taking proper precautions and with a suitable data analysis. The data analysis allows to reduce considerably two of the most important noises: that coming from the thermal motions of the bar constituents (brownian noise) and that due to the electronic amplifier (electronic noise).

The electrical signal from the transducer is amplified and then processed by means of optimum filters in order to make the signal to noise ratio (SNR) as large as possible. It can be shown that if one looks for signals due to short bursts of g.w. as described above, then the minimum E_{min} vibration energy of the bar that can be detected with SNR=1 is, in a simplified treatment,

$$E_{min} = kT_{eff} \approx k\,T\,/\,\beta\,Q + 2\,k\,T_n$$

where T_n is the noise temperature of the electronics, T is the bar temperature, β is the fractional part of energy available to the amplifier and Q is the quality factor of the entire apparatus. T_{eff} is called the effective noise temperature and represents the sensitivity for short bursts. It is clear that T and T_n must be as small as possible and $\beta\,Q$ must be large. As far as T_n a FET amplifier can have, at best, $T_n \approx 0.1$ K. The SQUID amplifiers could go down to the quantum limit, that is $T_n \approx 6\ 10^{-8}$ K, but it is extremely difficult to properly match this very low noise amplifier to the transducer. By lowering the bar temperature to T=40 mK, with $\beta\,Q \approx 10^6$ and $T_n \approx 10^{-7}$ K one could get $T_{eff} \approx 2\ 10^{-7}$ K that should allow to observe the collapses in the Virgo Cluster.

5 Planned detectors sensitivities

Many years of work are needed before the ultimate sensitivity will be reached both for the resonant and non resonant detectors. In this section we show the sensitivities that, according to the various research groups, should be reached by the end of our century.

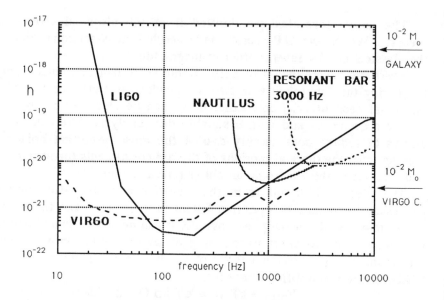

Fig. 1
Sensitivity for interferometers and for bars

In fig. 1 we show the minimum value of h with SNR=1 that should be observed, as function of frequency. Firstly we illustrate the bar sensitivity. The curve Nautilus is calculated assuming : L=3 m, M=2300 kg, T=0.05 K, Q= 5 10[7] and the SQUID amplifier operating at the quantum limit. The curve indicated with resonant bar is valid for a detector resonating at higher frequency: L=1 m, M=500 kg, T=0.05 K, Q= 5 10[7]. These curves have been calculated assuming a sinusoidal gravitational wave lasting just one cycle, with various values of the frequency. We notice that the resonant bar is sensitive also to waves with frequency very different from the bar resonant frequency, because, due to the wave short duration, the Fourier component at the resonance frequency is large.

The sensitivities both for LIGO and for VIRGO [1,2] show clearly that the interferometers are unquestionably to be preferred at low frequency. On the other end, at higher frequencies the bars seem to possess better performances.

6. Experimental results

We give here a brief review of the experimental results so far obtained by the Rome group in collaboration with other experimental groups.

a) we recall the result by the Rome and Maryland groups during the SN1987A. The two room temperature resonant bars of the two groups showed coincidences with the neutrino detectors Mont Blanc, Kamioka, Baksan and IMB [3,4,5,6]. The g.w. antennas recorded about a dozen of signals (with energy larger by a factor of 10^5 than that predicted with the classical cross-section) preceding the signals coming from the neutrino detectors by 1.2 ± 0.5 s. These signals occurred in a period of about two hours centred at the time 2:52 UT of 23 February 1987 when the Mont Blanc detectors recorded a burst of five neutrinos. The probability for this correlation to have occurred by accident was very small, less that 10^{-6}, but a physical interpretation was lacking unless one makes use of the much larger Weber and Preparata cross-section. This result, therefore, needs an additional support from new experimental data.

b) The gravitational wave antenna of the LSU group and Explorer of the Rome group, installed at CERN, have recorded data simultaneously from 24 June 1991 to 16 December 1991 for a period of 180 days [7]. The antennas were aligned parallel to each other, so that the same gravitational wave burst would have produced in the two antennas signals with the same amplitude. Each group performed its own filtering on the data of its own antenna, both filters aimed at the detection of short bursts of gravitational radiation . The filtering procedure gives the energy innovations, which we call also events. Both LSU and Explorer have thresholded these events such to select about 100 events per day. This corresponds to consider only events with energy of the order of 80 or 100 mK (a few times the effective noise temperature), corresponding to values $h \approx 2.5 \; 10^{-18}$.

While the data analysis is still underway, we can report some preliminary result [8], obtained by limiting the analysis to events with energy larger than 200 mK. There are 2440 LSU events and 10609 Explorer events above this threshold. One coincidence is found with energy 203 mK, while expecting on the average 1.55 accidental coincidences, as determined by shifting 200 times one event file with respect to the other one. Thus our search for gravitational wave bursts in the above period of time, able to produce in the antennas signals with energy larger than 200 mK, gives a negative result For smaller amplitudes the analysis is still in progress.

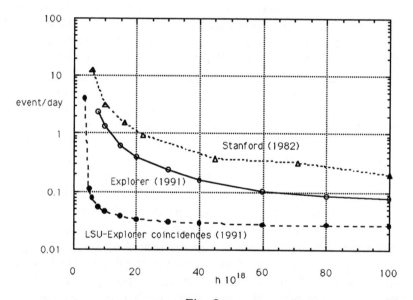

Fig. 2

Upper limits for gravitational wave bursts

From this null result we can calculate the upper limit for gravitational wave bursts. The upper limit is given in fig. 2 where, for comparison, we show also the upper limit obtained with the Stanford antenna alone in 1982 and with the Explorer antenna alone in 1991.

c) A resonant antenna can be used also for the detection of particles. We have used this capability for measuring the upper flux for nuclearites at sea level by means of the Explorer antenna.

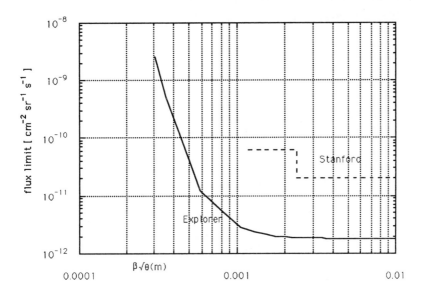

Fig. 3

Upper limit for nuclearites obtained with gravitational wave resonant detectors

The nuclearites are a new form of matter, consisting of aggregates of up, down and strange quarks in roughly the same proportion, which might exist and be stable. These particles may have masses ranging from a few GeV to the mass of a neutron star.

As shown in fig. 3 we find, at 90% confidence, there are less than 1.8 10^{-12} cm^{-2} sr^{-1} s^{-1} nuclearites with β range from 1 to 0.001.

d) A new program has just started, where we plan to study a possible correlation of our Explorer data with the gamma-bursts detected with the BATSE detector on the Compton Gamma Ray Observatory launched by NASA.

e) Finally the new ultra cryogenic antenna Nautilus, a 5056 Al bar, 3 m long and weighting 2300 kg, has been cooled to less than 0.1 K [9]. It has been the first time that so large a mass has been cooled to so small thermodynamic temperatures.

References

1. R.E. Vogt
   "The U.S. LIGO project"
   Internal report LIGO 91-7, Caltech
2. K.S.Thorne
   The LIGO/VIRGO Gravitational-Wave Detection System
   Proc. Fourth Rencontres on Particle Astrophysics,
   Ed. J. Tran Thanh Van,July 1992
3. E.Amaldi, P.Bonifazi, M.G.Castellano, E.Coccia, C.Cosmelli, S.Frasca, M.Gabellieri, I.Modena, G.V.Pallottino, G.Pizzella, P.Rapagnani, F.Ricci, G.Vannaroni
   "Data recorded by the Rome room temperature gravitational wave antenna, during the supernova SN 1987a in the Large Magellanic Cloud".
   Europhys. Lett. 3(12), 1325 (1987).
4. M.Aglietta, G.Badino, G.Bologna, C.Castagnoli, A.Castellina, W.Fulgione, P.Galeotti, O.Saavedra, G.Trinchero, S.Vernetto, E.Amaldi, C.Cosmelli, S.Frasca, G.V.Pallottino, G.Pizzella, P.Rapagnani, F.Ricci, M.Bassan, E.Coccia, I.Modena, P.Bonifazi, M.G.Castellano, V.L.Dadykin, A.S.Malguin, V.G.Ryassny,

O.G.Ryazhskaya, V.F.Yakushen, G.T.Zatsepin, D.Gretz, J.Weber, G.Wilmot
"Analysis of the data recorded by the Mont Blanc neutrino detector and by the Maryland and Rome gravitational-wave detectors during SN1987A".
Il Nuovo Cimento 12C, 75-103 (1989).

5. M.Aglietta, E.Amaldi, P.Astone, G.Badino, M.Bassan, G.Bologna, P.Bonifazi, C.Castagnoli, M.G.Castellano, A.Castellina, E.Coccia, C.Cosmelli, V.L.Dadykin, S.Frasca, F.Fulgione, P.Galeotti, D.Gretz, F.F.Khalchukov, I.V.Korolkova, P.V.Kortchaguin, V.A.Kudryatzev, A.S.Malguin, I.Modena, G.V.Pallottino, G.Pizzella, P.Rapagnani, F.Ricci, V.G.Ryassny, O.G.Ryazhskaya, O.Saavedra, G.Trinchero, S.Vernetto, M.Visco, J.Weber, G.Wilmot, V.F.Yakushev, G.T.Zatsepin,
"Coincidences among the Data Recorded by the Baksan, Kamioka and Mont Blanc Underground Neutrino Detectors, and by the Maryland and Rome Gravitational-Wave Detectors during Supernova 1987A ".
Il Nuovo Cimento, Vol. 14C, N. 2, Marzo-Aprile 1991.

6. M.Aglietta, P.Astone, G.Badino, M.Bassan, G.Bologna, P.Bonifazi, C.Castagnoli, M.G.Castellano, A.Castellina, E.Coccia, C.Cosmelli, S.Frasca, W.Fulgione, P.Galeotti, D.Gretz, E.Majorana, I.Modena, G.V.Pallottino, G.Pizzella, P.Rapagnani, F.Ricci, O.Saavedra, G.Trinchero, S.Vernetto, M.Visco, J.Weber, G.Wilmot
"Correlation between the Maryland and Rome Gravitational-Wave Detectors and the Mont Blanc, Kamioka and IMB Particle Detectors During SN1987A".
Il Nuovo Cimento, Vol. 106 B, N. 11, (1991).

7. P.Astone, M.Bassan, P.Bonifazi, P.Carelli, M.G.Castellano, G.Cavallari, E.Coccia, C.Cosmelli, V.Fafone, S.Frasca, E.Majorana, I.Modena, G.V.Pallottino, G.Pizzella, P.Rapagnani, F.Ricci, M.Visco,
"Long-term operation of the Rome Explorer cryogenic gravitational wave detector".
Physical Review D, 47, 2, January (1993).

8. P.Astone, M.Bassan, P.Bonifazi, E.Coccia, C.Cosmelli, S.Frasca, E.Majorana, I.Modena, G.V.Pallottino, G.Pizzella, P.Rapagnani,

F.Ricci, M.Visco, K.Geng, W.O.Hamilton, W.W.Johnson, E.Mauceli, S.Merkowitz, A.Morse, N.Zhu
"Result of a preliminary data analysis in coincidence between the LSU and Rome gravitational wave antennas".
Proceedings of the 10th Italian Conference on General Relativity and Gravitational Physics, Bardonecchia (Torino), September 1-5, 1992
9. P.Astone, M.Bassan, P.Bonifazi, M.G.Castellano, E.Coccia, C.Cosmelli, S.Frasca, E.Majorana, I.Modena, G.V.Pallottino, G.Pizzella, P.Rapagnani, F.Ricci, M.Visco
"First cooling below 0.1 K of the new gravitational wave antenna Nautilus of the Rome group".
Europhysics Letters, 16 (3), pp. 231-235, 1991.

STRING THEORY IN COSMOLOGY

N. SÁNCHEZ

Observatoire de Paris
Section de Meudon,Demirm
92195 Meudon Principal Cedex
FRANCE

ABSTRACT: In this lecture, I will describe those aspects of string theory which are relevant for Cosmology, with emphasis, for the purposes of this meeting, on the problem of connecting string theory to observational reality.I will talk on
- fundamental strings
- cosmic strings
- the possible connection among them

1. Fundamental Strings

There exists only three fundamental dimensional magnitudes (length, time and energy) and so three fundamental dimensional constants

$$(c, \hbar, G).$$

All other physical parameters being dimensionless, they must be calculable in an unified quantum theory of all interactions including gravity (" theory of everything"). The present interest in the theory of fundamental strings comes largely from the hope that it will provide such a theory. (There is no hope to construct a consistent quantum theory of gravity in the context of point particle field theory). As it is known, fundamental strings cannot exist in arbitrary space time and in particular they are consistent only in well determined (critical) dimensions $D_c > 4$. The unification of all interactions described by these theories takes place at the critical

V. de Sabbata and H. Tso-Hsiu (eds.), Cosmology and Particle Physics, 213–232.

dimensions $D = D_c > 4$ where the characteristic unification scale is the Planck scale:

$l_{Pl} = (2G\hbar/c^3)^{1/2} = 10^{-33}$ cm , $\qquad m_{Pl} = (\hbar c/2G)^{1/2} = 10^{-5}$ g. ,

corresponding to energies of order $E_{Pl} = 10^{19}$ GeV. Here $\mu = 1/\alpha'$, (α' is the characteristic string slope parameter with dimension of (length) ^{2} and $(\mu)^{1/2}$ characterizes the energy of string excitations. The problem of connecting string theory to reality is that of relating the theory in $D > 4$ dimensions where:

$$(G\mu)_D \approx 0 \quad (1) ,$$

to the real world where $D = 4$ and:

$$(G\mu)_{D=4} \quad \ll 1 ,$$

problem currently handled within the so-called compactification schemes (or alternatively by the "four dimensional models").(see for example refs.1 and 2). The effective low energy point particle field theory (containing GUTs and classical General Relativity) obtained by compactification is governed by physical couplings expressed in terms of μ via dimensionless numbers. (These numbers are vacuum expectation values of the different fields of the theory). The gauge (g) and gravitational couplings are (at the tree level) related by the Kaluza Klein (heterotic type) relation

$$g^2 = G\mu \quad in \quad D = 4 \tag{1}$$

The four and D dimensional couplings are related by

$$g_D^2 = g_4^2 \, V_{D-4} = (G\mu)_4 \, V_{D-4} \tag{2}$$

where V_{D-4} is the volume of the $(D - 4)$ dimensional compact manifold K_D of size l_{pl} (assuming the ground state of the metric to be $R^4 \times K_{D-4}$). It interesting here to plot the dimensionless parameter $G\mu$ against $M^2_{GUT}/$ as required by string unification constraints [3] on the renormalization group equations. Any perturbative compactification scheme leads to the non shaded triangle delimited in fig.1 (double logarithmic plot).

It is natural to require $M_{GUT} < (\mu)^{1/2}$ which excludes the upper shaded region. It is also natural to require that loop string corrections be under control, that is the ratio $(loops/trees) \ll 1$, which excludes the lower shaded region. The left shaded vertical region is also excluded in order to reproduce the value of the fine structure constant $g^2_{em} = 1/137$. This constraint is relevant here. The renormalization group equations:

$$M (dg^2/dM) = 2/3\pi (g^2)^2$$

yield
$$g^2 (M_0) = \frac{g^2(M)}{1 + (2/3\pi) \ g^2(M)\ln (M/M_0)} \qquad (3)$$

For $M_F < M < M_{GUT}$, the coupling constant $g(M)$ increases with M as it should be (QED is infrared stable), and with $M_0 = M_F = 200$ GeV, $M = M_{GUT} = 10^{16}$ GeV, we see:

$$1/g^2 (M_{GUT}) = 1/ g^2 (M_F) - 6.35,$$

i.e. :
$$g^2 (M_{GUT}) > g^2 (M_F) = 1/137$$

Since the string unification relation eq (1), one must have:

$$G\mu > \quad 1/137 \approx 10^{-2} \qquad (4)$$

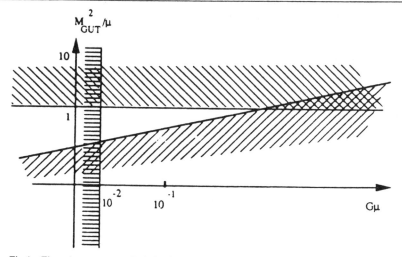

Fig.1. The elementary unshaded triangle restricting the values of $G\mu$ according to current compactification schemes (log-log plot as in ref.[3]).

Let us recall now that cosmic strings are macroscopic objects, finite energy solutions (vortices) of gauge theories coupled to Higgs fields. They have a mass per unit length μ:

$$G\mu = (<\Phi>/ m_{pl})^2 ,$$

and a radius $\delta \sim 1/M_{UNIF}$, where $<\Phi>$ is the vacuum expectation value of the Higgs field. For grand unified theories:

$$<\Phi>_{GUT} = 10^{-3} m_{Pl} .$$

and thus: $\qquad G\mu = 10^{-6}$ \hfill (5)

These thin strings are described by the Nambu equations.

On the other hand, cosmic strings should can be derived from fundamental strings. The thermodynamical model of strings (ideal gas of free strings at finite temperature) predicts the existence of a phase transition at the Hagedorn [4] critical temperature $\approx (\mu)^{-1/2}$ to a finite energy state in which all the energy is concentrated in a very long string [5] .When applied to the primordial universe,(at t about 10^{-32} s) this transition can be interpreted as the condensation of fundamental strings into macroscopic cosmic strings and this stringy phase could be the ancestor of the adiabatic era [6,7]. But if cosmic strings are created in this way , they must satisfy the constraint eq. (4).

Timing of millesecond pulsars as well as cosmic microwave radiation data on $\Delta T/T$ (see below Section 2) sets up an upper limit to the string parameter $G\mu$ such that $G\mu < 10^{-6}$-10^{-7}.

At present, we have two ways of connecting string theory to observational reality and thus of limiting the string parameter $G\mu$ [8]:

(i) radioastronomy observations (millisecond pulsar timing ; also microwave background data) which puts:

$$G\mu < 10^{-7} - 10^{-6} ;$$

(ii) elementary particle phenomenology (compactification schemes; also " four dimensional models") which require:

$$G\mu > 10^{-2} .$$

Cosmic strings derived from grand unified theories agree with (i). For cosmic strings derived from fundamental strings or for fundamental strings themselves there is contradiction between the possibilities (i) and (ii). One of these scenarii connecting string theories to reality must be revised (or the connection between fundamental and cosmic strings rejected). Meanwhile, FIRAS/ COBE data could help to solve the problem, select one scenario or reject both of them.

1.1 FUNDAMENTAL STRINGS IN COSMOLOGICAL BACKGROUNDS

We have first studied quantum string propagation in de Sitter space [9]. We have found the mass spectrum and vertex operator and found an string instability in de Sitter space. The lower mass states are the same as in flat space-time but heavy states deviate significantly from the linear Regge trajectories. We found that there exists a maximum (very large) value of order $1/g^2$ for the quantum number N and spin J of particles. There exists real mass solutions only for:

$$N < N_{max} = \pi / 2g^2 + O(g^{-2/3}) \quad , \quad g = 10^{-61}$$

Moreover, for states in the leading Regge trajectory, the mass monotonously increases with J up to the value

$$J_{max} = 1/ g^2 + O(1)$$

corresponding to the maximal mass $m^2_{max} = 0.76 + O (g^2)$. Beyond J_{max} the mass becomes complex. These complex solutions correspond to unstable states already present here at the tree (zero handle) level.

From the analysis of the mass spectrum , we find that the critical dimension for bosonic strings in de Sitter space-time is D=25 instead of 26 in Minkowski space-time). This result is confirmed by an independent calculation of the critical dimension from the path intergral Polyakov's formulation, using heat-kernel techniques: we find that the dilaton β- function in D-dimensional de Sitter space-time must be:

$$\beta^\phi = (D + 1 - 26) / (\alpha' 48\pi^2) + O(1) .$$

It is a general feature of de Sitter space-time to lower the critical dimensions in one unit. For fermionic strings we find D=9 instead of the flat value D=10.

We have found that for the first order amplitude η^i (σ, τ) , (i = 1,....D-1 refers to the spatial components), the oscillation frequency is:

$$\omega_n = [\, n^2 - (\, \alpha'mH)^2 \,]^{1/2},$$

instead of n , where H is the Hubble constant. For high modes n >> $\alpha'mH$, the frequencies $\omega_n \approx$ n are real. The string shrinks as the universe expands.This shrinking of the string cancels precisely the expanding exponential factor of the metric and the invariant spatial distance does not blow up. Quantum mechanically, these are states with real masses $(m^2H^2<1)$. This corresponds to an expansion time H^{-1} very much bigger than the string period $2\pi/n$, that is, many string oscillations take place in an expansion period H^{-1}(in only one oscillation the string does not see the expansion).

For low modes n < $\alpha'mH$, the frequencies become imaginary. This corresponds to an expansion time very short with respect to the oscillationtime $2\pi/n$ ("sudden" expansion, that is the string "does not have time" to oscillate in one time H^{-1}).These *unstable* modes are analyzed as follows. The n=0 mode describes just small deformations of the center of mass motion and it is therefore a physically irrelevant solution. When $\alpha'mH$ >1 relevant unstable modes appear. Then, the n=1 mode dominates η^i(σ, τ) for large τ. Hence, if $\alpha'mH$ > $(2)^{1/2}$, η^i diverges for large τ, that is fluctuations become larger than the zero order and the expansion breaks down. However, the presence of the above unstability is a true feature as it has been confirmed later by further analysis [10].

The physical meaning of this instability is that the string grows driven by the inflationary expansion of the universe. That is, the string modes couples with the universe expansion in such a way that the string inflates together with the universe itself. This happens for inflationary (ie accelerated expanding) backgrounds. In ref. [10] we have studied the string propagation in Friedman-Robertson-Walker (FRW) backgrounds (in radiation as well as matter dominated regimes) and interpreted the instability above discussed as Jeans-like *instabilities*. We have also determined under which conditions the universe expands, when distances are measured by stringy rods.

It is convenient to introduce the *proper amplitude* $\chi^i = C\eta^i$, where C is the expansion factor of the metric. Then, χ^i satisfies the equation:

$$\ddot{\chi}^i + [\, n^2 - \ddot{C}/C \,]\, \chi^i = 0$$

Here dot means τ-derivative. Obviously, any particular (non-zero) mode oscillates in time as long as $\ddot{C}/C$ remains < 1 and, in particular, when $\ddot{C}/C$ < O. A time- independent amplitude for χ is obviously equivalent to a fixed proper (invariant) size of the string. In this case, the behaviour of strings is stable and the amplitudes η shrink (like 1/C) .

It must be noticed that the time component , χ^0 or η^0 , is always well behaved and no possibility of instability arises for it. That is the string time is well defined in these backgrounds.

i) For non-accelerated expansions (e.g. for radiation or matter dominated FRW cosmologies) or for the high modes n >> α'Hm in de Sitter cosmology, string instabilities do not develop (the frequencies $\omega_n \simeq$ n are real). Strings behave very much like point particles : the centre of mass of the string follows a geodesic path , the harmonic-oscillator amplitudes η shrink as the univers expands. in such a way to keep the string's proper size constant. As expected , the distance between two strings increase with time, relative to its own size, just like the metric scale factor C.

ii) For inflationary metrics (e.g. de Sitter with large enough Hubble constant ,) the proper size of the strings grow (like the scale factor C) while the co-moving amplitude η remains fixed ("frozen") , i.e. $\eta \simeq \eta(\sigma)$.

Although the methods of references [9] and [10] allow to detect the onset of instabilities, they are not adequate for a quantitative description of the high instable (and non-linear) regime. In ref. [11] we have developped a new quantitave and systematic description of the high instable regime. We have been able to construct a solution to both the non-linear equations of motion and the constraints in the form of a systematic asymptotic expansion in the large C limit, and to classify the (spatially flat) Friedman-Robertson Walker (FRW) geometries according to their compatibility with stable and/or unstable string behaviour.

An interesting feature of our solution is that it implies an asymptotic proportionality between the world sheet time τ and the *conformal time* T of the background manifold. This is to be contrasted with the stable(point-like) regime which is characterized by a proportionality between τ and the *cosmic time*. Indeed, the conformal time (or τ) will be the small expansion parameter of the solution: the asymptotic regime (small τ limit) thus corresponds to the large C limit only if the background geometry is of the inflationary type. The non linear, high unstable regime is characterized by string configurations such that:

$$X'^0 \ll |\dot{X}^0| , \quad |\dot{X}^i| \ll |X'^i|$$

with:

$$X^0(\sigma, \tau) = C L(\sigma) , \quad L(\sigma) = (\delta_{ij} X'^i X'^j)^{1/2}$$

$$X^i(\sigma,\tau) = A^i(\sigma) + \tau^2 D^i(\sigma)/2 + \tau^{1+2\alpha} F^i(\sigma)$$

where A^i, D^i and F^i are functions determined completely by the constraints, and α is the time exponent of the scale factor of the metric: $C = \tau L^{-\alpha}$.

For power-law inflation: $1 < \alpha < \infty$, $X^0 = \tau L^{1-\alpha}$.

For de Sitter inflation: $\alpha = 1$, $X^0 = \ln(-\tau HL)$.

For Super-inflation: $0 < \alpha < 1$, $X^0 = \tau L^{1-\alpha}$ + const.

Asymptotically, for large radius $C \to \infty$, this solution describes string configurations with expanding proper amplitude .

These highly unstable strings contribute with a term of negative pressure to the energy-momentum tensor of the strings. The energy momentum tensor of these highly unstable strings (in a perfect fluid approximation) yields to the state equation $\rho = - P (D-1)$, ρ being the energy density and P the pressure (P< 0). This description corresponds to *large radius* $C \to \infty$ of the universe.

For *small radius of the universe*, highly unstable string configurations are characterized by the properties

$$|\dot{X}^0| \gg |X^{0'}| , \quad |X^{i'}| \ll |\dot{X}^i|$$

The solution for X^i admits an expansion in τ similar to that of the large radius regime. . The solution for X^0 is given by L/C, which corresponds to *small radius* $C \to 0$, and thus to small τ.

This solution describes, in this limit, string configurations with shrinking proper amplitude, for which CX'^i behaves asymptotically like C, while CX^i behaves like C^{-1}. Moreover, for an ideal gas of these string configurations, we found:

$$\rho = P(D-1),$$

with *positive pressure* which is just the equation of state for a gas of massless particles.

More recently [12], these solutions have been applied to the problem in which strings became a dominant source of gravity. In other words, we have searched for solutions of the Einstein plus string equations. We have shown, that an ideal gas of fundamental strings is not able to sustain, alone, a phase of isotropic inflation. Fundamental strings can sustain, instead, a phase of anisotropic inflation, in, which four dimensions inflate and simultaneously, the remaining extra (internal) dimensions contract. Thus, fundamental strings can sustain, simultaneously, inflation and dimensional reduction. In ref. [12] we derived the conditions to be met for the existence of such a solution to the Einstein and string equations and discussed the possibility of a successful resolution of the standard cosmological problems in the context of this model.

2. Cosmic Strings : Observational Tests

In January 1990) [13], the spectrum of the cosmic microwave background radiation (MBR) between 1cm and 500mm from regions near the North Galactic pole, observed by the FIRAS (Far Infrared Absolute Spectrophotometer) experiment on COBE satellite, has been reported. It is a "pure" spectrum of a black body with a temperature of 2.735 ± 0.06 K : the deviation from a black body is less than 1% of the peak intensity over the range 1 cm to 500mm ; this COBE spectrum thus disclaims the Berkeley-Nagoya excess reported by Matsumoto et al (1988)[14].

A stochastic background of gravitational waves can be detected by means of pulsar timing observations [15]. The stable rotation and sharp radio pulses of PSR 1937+21 make this millisecond pulsar a clock whose frequency stability may exceed that of the best atomic clocks.

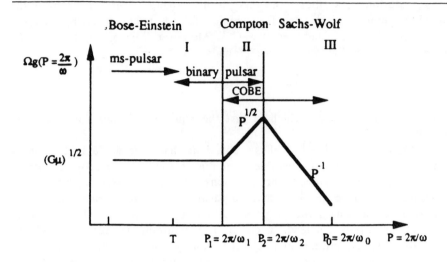

Fig.2. Schematic logarithmic spectrum $\Omega_g(P)$ as a function of the present period $P=2p/\omega$ of the gravitational waves generated by cosmic strings . $P_1 = 10^4$ yrs, $P_2 = 10^6$ yrs, $P_0 = 10^{10}$ yrs. The observational time spand is $T \approx 5$ yrs for millisecond pulsars; $T \approx 10\text{-}10^6$ yrs for the binary pulsar. The main effects of distortions in the cosmic microwave background due to electromagnetic and/or gravitational waves in each of the three regions are also indicated.

From the compatibility among these measurements, including the absence of distortion in the MBR spectrum recently reported by the COBE-FIRAS experiment, we obtained: (a) the coefficient f relating the electromagnetic and gravitational radiation rates released by SCS, (b) the chemical potential $\mu_{0\ scs}$ produced by SCS, (c) the string loop evolution parameters (α, β, γ), and (d) the string parameter $G\mu$. Here α measures the initial loop size relative to the horizon size, β accounts for the loop formation rate and γ for the total power energy emitted by each loop.

Up to now, the data yield a firm upper limit $\rho_g < 3.5 \ 10^{-36}$ g/cm^3 for the energy density of a cosmic background of gravitational radiation at the frequency of about 10^{-8} Hz [16]. This limit corresponds to approximately $2 \ 10^{-7}$ of the density required to close the Universe. On the other hand, accurate time of arrival measurements of pulses from the binary pulsar PSR 1913+16 over the last 14 years have led in particular to an accurate observed orbital decay rate Pb. The difference between predicted and observed orbital decay rates Pb of PSR 1913+16 [17] can be used to place the best available limit on the total energy density of a cosmic gravitational wave background $\Omega_{total} <$ $4 \ 10^{-2} \ h^{-2}$,(h ~ 0.7) at the ultra low frequencies of 10^{-9} to 10^{-13} Hz.

Cosmic superconducting strings (SCS) attract current considerable interest in astroparticle physics [18,19].If SCS exist,loops formed as a consequence of string interactions, decay during the expansion of the universe, by emitting gravitational and electromagnetic waves.The radiation emitted by loops at different epochs, from their formation at $t \sim 10^{-32}$ s till now , adds up into a stochastic gravitational wave background characterized by the (dimensionless energy density) spectrum $\Omega_g (\omega)$, known on a wide range of frequencies in three different regions[20] (as it is shown in Fig.2):
I: $\omega >> \omega_1 \sim 10^{-15}$ Hz , II: $\omega_2 \sim 10^{-13}$ Hz $< \omega < \omega_1$, and III: $\omega < \omega_2$. In ref[21] we have studied the constraints which can be placed on SCS through the results of five different measurements:

(i) $\quad \Omega_{gI} (\omega) = (\omega/\rho_{crit}) \ d\rho_{gI}/ \ d\omega \quad , \quad \rho_{crit} \sim 4 \ 10^{-29}$ g cm^{-3} ;

(ii) $\Omega_{gtotal} (\omega) = \int_{10^{-13} Hz}^{10^{-8} Hz} d\omega \ \Omega_g(\omega)/\omega$, through the binary pulsar residual R_b, (in region II).

(iii) MBR temperature fluctuation $\Delta T/T$ for angular separation of about 1°, (region III),

(iv) chemical potential μ_0 (region I) and

(v) Comptonization parameter y (region II), characterizing the MBR spectral distortions.

Recently, the measurable quantities (i)-(v) above referred have been expressed explicitly in terms of $G\mu$, the set (α, β , γ) and f covering the whole spectrum and SCS evolution [22] :

$$\Omega_{gl} = 1.8 \ 10^{-6} [G\mu/10^{-6}]^{1/2} (\alpha^{3/2}\beta) \gamma^{-1/2} (1+f)^{-3/2}$$

$$\Omega_{total} = 4.6 \ 10^{-4} [G\mu/10^{-6}]^{1/2} (\alpha^{3/2}\beta) \gamma^{-1/2} (1+f)^{-3/2}$$

$$\Delta T/T \leq 2.5 \ 10^{-6} [G\mu/10^{-6}] (\alpha\beta\gamma)^{1/2} (1+f)^{-3/4}$$

$$\mu_0 = 0.2 [G\mu/10^{-6}]^{1/2} (\alpha^{3/2}\beta) \gamma^{-1/2} f (1+f)^{-3/2}$$

$$y = 0.4 [G\mu/10^{-6}]^{1/2} (\alpha^{3/2}\beta) \gamma^{-1/2} f (1+f)^{-3/2}$$

At the present time, observations place the following constraints:

(i) millisecond pulsar timing measurements lead to:

$$\Omega_g (\omega) < 2 \ 10^{-7} \qquad \text{for} \quad (\omega/2\pi) \sim 0.2 \ \text{cy/yr} \qquad (6)$$

(Taylor,1989 [16]), but more generally:

$$\Omega_g (\omega) < 4 \ 10^{-7} (2\pi / T)^4 R_\mu^2 \qquad \text{for} \ \omega > 2\pi / T \qquad (7)$$

where T is the observation time-span in years and R_μ the residual in μs.

(ii) binary pulsar timing measurements lead to:

$$\Omega_{total} < 4 \ 10^{-2} \ h^{-2} \qquad \text{for} \quad 10^{-12} \ \text{Hz} < \omega < 10^{-9} \ \text{Hz} \qquad (8)$$

$$G\mu \;<\; 1.2 \; 10^{-8} \; (\; \alpha^{3/2} \; \beta \;)^{-2} \; \gamma$$

$$G\mu \;<\; 2.4 \; 10^{-5} \; (\alpha \; \beta \; \gamma \;)^{-1/2}$$

$$G\mu \;<\; 1.2 \;\; 10^{-8} \; (\; \alpha^{3/2} \beta \;)^{-2} \gamma$$

From the latest millisecond pulsar timing and from the preliminary COBE data , we obtain the following upper limits on $G\mu$:

	generic values $\alpha \sim \beta \sim 1$, $\gamma \sim 100$	latest numerical simulation values $\alpha \sim 2.8 \; 10^{-2}$, $\beta \sim 850, \gamma \sim 100$	kinky strings $\alpha \sim 10^{-2}$ $\beta \sim 750, \gamma \sim 100$
PSR 1937 + 21	$G\mu \;<\; 1.2 \; 10^{-6}$	$0.7 \;\; 10^{-7}$	$0.2 \;\; 10^{-7}$
COBE	$G\mu \;<\; 1.2 \;\; 10^{-6}$	$0.7 \; 10^{-7}$	$0.2 \;\; 10\text{-}7$

The first set of values for (α, β, γ) correspond to generic values [25], the second set correspond to latest numerical simulation results [26,27] and the third set to "kinky "string values[28] .
We see that the upper limits placed by COBE on $G\mu$ *converge* to those placed by the latest PSR 1937+21 timing measurements and that generic values of (α, β, γ) lead to $G\mu \sim 10^{-6}$ whereas numerical values yield $G\mu \sim 10^{-7}$.
We have placed new constraints on SCS and on the parameters governing their evolution by requiring that electromagnetic and gravitational energies released by loops, be compatible with the preliminary data on the MBR spectrum from COBE's - FIRAS experiment,

(Taylor and Weisberg,1989 [17]), and more generally:

$$\Omega_{total} < \quad (2H^2)^{-2} \ (R_b/T_b \ )^2 \tag{9}$$

where H= h x 100 km s^{-1} Mpc^{-1} is the Hubble constant , (h$_\sim$0.7).

(iii) angular temperature anisotropy measurements yield:

$$\Delta T / T \ < \ 6 \ 10^{-5} \qquad \text{for angular separations} \quad \theta \sim 1° \tag{10}$$

(Wilkinson 1986, Partridge 1988)[23] .

(iv) latest Rayleigh- Jeans MBR spectrum measurements yield:

$$\mu_0 \ < 10^{-2} \qquad \text{with} \quad T_{RJ} \simeq 2.74 \pm 0.02 \ K \tag{11}$$

(Smoot et al 1987)[24]. ·

The resuts from COBE[13] show a pure Black-Body spectrum at 2.735 $\pm$ 0.06 K within 1% the peak intensity over the range 1cm to 500mm.Using their conservative 1% error bands, Mather et al [13] set a 3σ-upper limit on the Comptonization y-parameter of 0.001, and then:
$$y \ < \ 10^{-3} \qquad \text{for} \quad T \sim 2.735 + 0.06 \ K \tag{12}$$

From a fit to a pure Bose-Einstein spectrum with a chemical potential μ_0 independent of frequency , Mather et al. gave also the 3σ-upper limit:
$$\mu_0 \ < \ 0.9 \ 10^{-2} \tag{13}$$
Notice that this preliminary value is about an order of magnitude greater than the value (12) on y .

Now, we can express the string parameter Gμ in terms of these measurable quantities with their present upper limits :

$$G\mu \approx 1.2 \ 10^{-8} \ (\Omega_{gI}/2 \ 10^{-7})^2 \ (\alpha^{3/2}\beta)^{-2} \ \gamma(1+f)^3 \tag{14}$$

$$G\mu \approx 0.4 \ 10^{-7} \ (\Omega_{total}/10^{-4})^2 \ (\alpha^{3/2}\beta)^{-2} \ \gamma(1+f)^3 \tag{15}$$

$$G\mu \approx 2.4 \ 10^{-5} \ ([\Delta T/T]/6 \ 10^{-5}) \ (\alpha\beta\gamma)^{-1/2} \ (1+f)^{3/4} \tag{16}$$

$$G\mu \approx 2 \ 10^{-9} \ (\mu_0/0.9 \ 10^{-2})^2 \ (\alpha^{3/2}\beta)^{-2} \ \gamma(1+f)^3 \ f^{-2} \tag{17}$$

$$G\mu \approx 6.2 \ 10^{-12} \ (y/10^{-3})^2 \ (\alpha^{3/2}\beta)^{-2} \ \gamma(1+f)^3 \ f^{-2} \tag{18}$$

Note that in eq.(14), the quoted upper limit 10^{-4} for Ω_{total} is not yet reached. From eqs. (14) and (16) and on the other hand, from eqs. (18) and (16) we get the following relations for the string evolution parameters and the ratio between electromagnetic and gravitational radiation rates :

$$\alpha^5 \ (\beta/\gamma)^3 \approx 6.8 \ 10^{-14} \ (1+f)^{9/2} \ f^{-4} \ (y/10^{-3})^4 \ ([\Delta T/T]/6 \ 10^{-5})^{-2}$$

$$\alpha^5 \ (\beta/\gamma)^3 \approx 0.25 \ 10^{-6} \ (1+f)^{9/2} \ (\Omega_{gI}/2 \ 10^{-7})^4 \ ([\Delta T/T]/6 \ 10^{-5})^{-2}$$

$$f \approx 2.3 \ 10^{-2} \ (y/10^{-3}) \ (\Omega_{gI}/2 \ 10^{-7})^{-1}$$

Finally, with the latest upper limits on Ω_{gI} (ω), $\Delta T/T$ at $1°$, y and the value $\sim 10^{-2}$ for f, from eqs. (14), (16), (18) we have for the string parameter :

with the already performed MBR angular $\Delta T/T$ measurements (Sachs-Wolf effect) for $\theta \sim 1°$, and with the latest millisecond pulsar timing measurements. These limits continue to descend but the breaking point is not yet reached .The absence of distortion lies within 1% of the peak intensity, but the FIRAS sensitivity for each spectral element is expected to be 1/1000 of the peak of the 2.735 K spectrum.In addition , new constraints are expected to be placed from limits on MBR anisotropies : from the anisotropy experiment[29] (Differential Microwave Radiometers) on COBE satellite, the sensitivity on $\Delta T/T$ in a field of view of about 7° is expected to be increased in an order of magnitude with respect to previous experiments. This will set up stringent limits on $G\mu$ for large straight moving SCS which produce $\Delta T/T$ fluctuations through lensing type effects.

3. Connection between Fundamental Strings and Cosmic Strings. Macroscopic Behaviour of Fundamental Strings

Fundamental strings exhibit macroscopic behaviour which appears quite similar to those of topological cosmic strings. These include :
(i) the reconnection or self-intersection processes,
(ii) the decay rate of fundamental strings into gravitational radiation,
(iii)the thermodynamical behaviour of strings at high temperatures.

It is important to know whether highly excited fundamental strings behave in a classical way.A quantity which is interesting and which can be obtained analytically is the decay rate of highly excited (open or closed) strings. One can compare then the gravity wave emission computed from the decay rate of a quantized closed bosonic string with the classical calculations for gravitational radiation already done for cosmic strings. Recently [30], the vertex operator for the absorption and emission of states from a closed string has been computed to obtain the one-loop scattering amplitude whose imaginary part gives the total decay rate Γ (total means summing over all the decay products). The computation is more easily performed for the states lying on the leading Regge trajectory corresponding to spin J massive particles with squared mass $m^2 = 4(J-2)$. Classically, these states correspond to classical trajectories which are rotating pairs of straight strings joined at the ends (with the ends moving at the speed of light); the quantum state (which is an occupation number eigenstate) is the superposition of such classical trajectories.

The decays in which one of the products is massless form the largest class. The amplitude Γ can be converted into a rate of radiation which gives the quantum state(which is an occupation number eigenstate) is the superposition of such classical trajectories.The decays in which one of the products is massless form the largest class. The amplitude Γ can be converted into a rate of radiation which gives:

$$P = \Gamma E \quad \sim 360 \, G\mu^2 \quad ,$$

E being the energy of the radiated massless particles.

Let us recall the gravitational radiation rate from cosmic strings, evaluated classically for some particular string trajectories, using the quadrupole formula:

$$P = \gamma \, \mu^2 \qquad \text{with} \quad \gamma \sim 100 \ .$$

We see that a highly excited fundamental string exhibits a classical behavior by emitting as it would be a *classical oscillating quadrupole*. It is very interesting that the purely quantum mechanical calculation of decaying fundamental strings into gravitons, agrees with the classical results of cosmic strings. For fundamental strings, the value of γ is larger than that obtained from cosmic strings, but this is due to the particular chosen trajectory-the leading Regge trajectory state corresponding classically to a rigid rotating rod for which the mass is proportional to the length.

The value of γ depends on the particular trajectory. It would be interesting to compute the rate Γ for other states (different from the leading Regge trajectory) and to compare with the values obtained for cosmic strings .

It must be noticed that besides the graviton, the rate Γ contains decays into the massless dilaton and antisymmetric tensor fields, and that , in general, the expression of Γ in terms of the string parameter μ is highly dependent on the compactification scheme.The results in D=4 and D=26 are rather different. (For instance, in D=26 Γ is inversely proportional to the string length, and decays into massive states are not suppressed as it is the case in D=4).

But in spite of these connections, important questions remain to be solved or to be revised and we are not still in a position to make contact with the precise results of observations, as has been done for cosmic strings:

a) we would like to know not only the parameter γ , but also the analogues of the α and β evolution parameters of cosmic string interactions , from a quantum fundamental string computation.

b) it would be interesting to compute the rate Γ of fundamental strings for other states (different from the leading Regge trajectory) and to compare with the values obtained for cosmic strings.

c) and more important, to derive the value of $(G\mu)$ in D=4 of fundamental strings, for which compactification and perturbative renormalization group techniques are in conflict with the radioastronomical constraints on cosmic strings.

4. Recent Developments

Interestingly, the complete integrability of the string propagation in D-dimensional De Sitter space time has been shown [30]. The string equations of motion -which correspond to a non-compact O(D,1) symmetric sigma model- plus the string constraints- are equivalent to a generalized Sinh-Gordon equation. In D=2, this is the Liouville equation, in D=3, this is the standard Sinh-Gordon equation and in D=4, this equation is related to the B_2 Toda model.We show that the presence of instability is a general exact feature of strings in the De Sitter space, as a direct consequence of the strong instability of the generalized Sinh Gordon Hamiltonian (which is unbounded from bellow), irrespective of any approximative scheme. We find all the classical solutions in D=2 and physically analyze them: they correspond to a string winding n times around de Sitter space (here the circle S^1) and evolving with it, the string inflates (or shrink) following the expansion (or contraction) rate.

More recently[31], exact and explicit string solutions propagating in 2+1 dimensional De Sitter spacetime were found. In all these solutions the strings generically tend to inflate or either to collapse.

The world sheet time τ interpolates between the cosmic time ($\tau \to \pm \infty$) and conformal ($\tau \to 0$) time. For $\tau \to 0$, the typical string instability is found, while for $\tau \to +\infty$, a new string behaviour appears. In that regime, the string expands (or contracts) but not with the same rate as the universe does.

REFERENCES

[1] M. Dine and N. Seiberg,Phys. Rev. Lett.55 ,366 (1985)
V. Klapunovsky,ibid.103
P. Candelas,G.T. Horowitz,A.Strominger and E. Witten,Nucl.Phys.285B,46 (1985)
D.J. Gross,J.A. Harvey,E. Martinec and R. Rohm, Nucl. Phys. 267B, 75 (1986)

[2]K.S. Narain, Phys. Lett. 169B, 41 (1986)
K.S. Narain,M.H.Sarmadi and E.Witten,Nucl. Phys. 279B, 369 (1987)

[3] R.Petronzio, G.Veneziano, Mod. Phys. Lett. A2, 707, (1987)

[4] R.Hagedorn, Nuovo Cimento Suppl., 3, (1965), 147
S.Fubini, G.Veneziano, Nuovo Cimento, 64A, (1969), 1640

[5] S.Frautschi, Phys.Rev., D 3, (1971) ,2821
R.D.Carlitz, Phys.Rev., D 5, (1972) ,3231
B.Sundborg, Nucl. Phys., B254, (1985), 583
M.J.Bowick, L.C.R.Wijewadhana, Phys. Rev. Lett., 54, (1985), 2485

[6] Y.Aharanov, F.Englert, J.Orloff, Phys. Lett., 199B , (1987) , 366

[7]F.Englert, J.Orloff, T.Piran, Phys. Lett.212,423 (1988)

[8] N. Sánchez and M. Signore, Phys. Lett. B214,14 (1988)

[9] H.J. de Vega and N.Sánchez, Phys. Lett. B197,320 (1987)

[10] N. Sánchez and G. Veneziano, Nucl. Phys.B333,253 (1990)

[11] M. Gasperini, N. Sánchez and G. Veneziano, IJMPA6, 3853 (1991).

[12] M. Gasperini, N. Sánchez and G. Veneziano, Nucl. Phys. B364, 365 (1991).

[13]J.C.Mather et al., Ap.J.354, L37 (1990).
J.C. Mather et al, COBE preprint N°90-03,1990.
J.C. Mather et al.COBE preprint 90-05, in the Proc. of "After the First Three Minutes" Workshop Oct. 15-17 1990, Univ. of Maryland, USA, Eds. S Holt, C. Bennett, V. Trimble, Ann. AAS (1991).

[14] T. Matsumoto et. al. Ap.J 329,567 (1988).

232

[15].B.Bertotti, B.J.Carr and M.J.Rees, Month. Notices. Roy. Astron. Soc. 203, 945, (1983).
L.A. Rawley, J.H.Taylor, M.M. Davis, D.W. Allan, Science 238, 761, (1987).

[16]J.H. Taylor,"Timing binary and millisecond pulsars",talk given at Observatoire de Paris-Meudon, (September 1988).

[17]J.H. Taylor and J.M. Weisberg, Ap.J. 345, 434, (1989).

[18] E. Witten, Nucl. Phys. B249, 557, (1985).
A. Vilenkin, TUTP 89-1-preprint, to appear in the Proc. of the 14th Texas Symposium on Relativistic Astrophysics.
J.P. Ostriker, C. Thompson, E.Witten, Phys. Lett. B180, 221, (1986).

[19]J.P.Ostriker and C. Thompson, Ap J. 323, L97, (1987).
B. Rudàk and M. Panek , Phys. Lett. B199 , 343, (1987).

[20] R. Brandenberger, A. Albrecht, N. Turok, Nucl. Phys. B277, 605, (1986).

[21] N.Sánchez and M. Signore, Phys. Lett. B214, 14, (1988).
M. Signore and N. Sánchez ,Mod. Phys. Lett. A4, 799, (1989).
N. Sánchez and M. Signore , Phys. Lett. B219 , 413 , (1989).

[22]N. Sánchez and M. Signore, Phys. Lett.241 B, 332 (1990)
M. Signore and N. Sánchez,I.J.M.P. A6,1591 (1991).
M. Signore and N. Sánchez, in the Proc. of "After the First Three Minutes" Workshop, Oct. 15-17 1990,Univ. of Maryland, USA, Eds. S Holt, C. Bennett, V. Trimble, Ann. AAS (1991).

[23] D.T.Wilkinson, Phil. Trans. R. Soc. London A320 , 595, (1986).
R.B. Partridge , Rep. Prog. Phys. 51 , 647 , (1988).

[24] G.F. Smoot et al , Ap. J. 317 , L45 , (1987).

[25]T. Vachaspati and A. Vilenkin , Phys. Rev. D31 , 3052 , (1985).

[26] A. Albrecht and N. Turok , Phys. Rev. D40, 973, (1989).
A. Albrecht and N. Turok ,Fermilab-Pub 89/140A ,PUPT-89-1133, (1989).
F.S. Accetta and L.M. Krauss, Phys. Lett. B233, 93, (1989).

[27] D.Bennett and F.R.Bouchet Princeton University Preprint PUPT-89-1126, (1989).

[28] B. Allen and R.R. Cadwell "Small scale structure on a cosmic string network",submitted to Phys.Rev. D

[29] Ch. L. Bennett and G.F. Smoot, Preprint LASP 88-14.
-G.F. Smoot et al, " COBE: The Differential Microwave Radiometers", talk given at " The 175th Meeting of the American Astronomical Society", Washington, DC (9-13 January 1990).

[30] H. J. de Vega and N. Sánchez, LPTHE Paris 92-31 and DEMIRM Meudon preprint to appear in Phys. Rev. D.

[31] H. J. de Vega, A. V. Mikhailov and N. Sánchez, LPTHE Paris 92-32 and DEMIRM Meudon preprint, to appear in Theor. Math. Phys. , special volume in the memory of M. C. Polivanov.

Basic Principles of Gravitational Wave Interferometers

Peter R. Saulson
Department of Physics
Syracuse University
Syracuse, New York 13244-1130
U. S. A.

Abstract

Modern interferometric gravitational wave detectors can trace their heritage to the Michelson-Morley experiment. In this lecture, I discuss the way such instruments elucidate the nature of space-time, and derive the fundamental limit to their sensitivity.

The existence of wave solutions to Einstein's field equations is, arguably, the most "relativistic" feature of the theory of gravitation known as the General Theory of Relativity. This is true in the sense that it is the instantaneous action-at-a-distance of Newtonian gravity that most offends our notions of causality as understood in the context of the Special Theory of Relativity. Newton's theory, taken literally, would predict that the gravitational field produced by a mass always has the familiar $1/r^2$ form, with r referring to its present position, no matter how fast it might move (or accelerate), and no matter how far away from it we consider its field.

General relativity fixes the problem posed by moving sources of a gravitational field. It proposes that the gravitational field (that is, the curvature of space-time) does not change instantly at arbitrary distances from a moving source. Instead, in a manner deeply analogous to electromagnetic waves, the "news" of the motion of a source of space-time curvature propagates at the speed of light.

In this lecture, I will briefly examine how gravitational waves of finite speed arise in general relativity. Then I will discuss in some detail the interaction of gravitational waves with systems of test bodies, and consider how such interactions may be observed. Finally, I will discuss the fundamental limit to measurement precision, photon shot noise.

1 Waves in General Relativity

One of the most fundamental concepts in the Special Theory of Relativity is that the *space-time interval* ds between any two neighboring points is given by the expression

$$ds^2 = -c^2 dt^2 + dx^2 + dy^2 + dz^2 \tag{1}$$

233

V. de Sabbata and H. Tso-Hsiu (eds.), Cosmology and Particle Physics, 233–255.
© 1994 *Kluwer Academic Publishers. Printed in the Netherlands.*

or

$$ds^2 = \eta_{\mu\nu} dx^\mu dx^\nu,$$ (2)

with the Minkowski metric $\eta_{\mu\nu}$ given, in Cartesian coordinates, by

$$\eta_{\mu\nu} = \begin{pmatrix} -1 & 0 & 0 & 0 \\ 0 & 1 & 0 & 0 \\ 0 & 0 & 1 & 0 \\ 0 & 0 & 0 & 1 \end{pmatrix}.$$ (3)

(Note that in Eq. 1 all of the superscripts indicate raising to the second power, while in Eq. 2 they range from 0 to 3 to represent t, x, y, and z, respectively. Eq. 2 also makes use of the famous "repeated index summation convention".)

The same physical concept is carried over into the General Theory of Relativity, with one key difference — space-time is no longer necessarily the "flat" space-time described by the Minkowski metric, but will in general be curved in order to represent what we call gravitation. The more general statement of the definition of the space-time interval is

$$ds^2 = g_{\mu\nu} dx^\mu dx^\nu,$$ (4)

where all of the information about space-time curvature is encoded in the metric $g_{\mu\nu}$. A lot of physics can be expressed in Eq. 4. Fortunately, we only need to work in one simple special case, that of a weak plane gravitational wave propagating through otherwise flat space-time. Then it makes sense to write the metric in the form

$$g_{\mu\nu} = \eta_{\mu\nu} + h_{\mu\nu},$$ (5)

where $h_{\mu\nu}$ represents the *metric perturbation* away from Minkowski space. As long as the gravitational wave is weak (a condition all too easily satisfied, as we will soon see), then $h_{\mu\nu}$ will be a small perturbation to the metric.

The key physics of the problem is thus carried in the form of $h_{\mu\nu}$. In the weak-field limit, the non-linear Einstein equations can be approximated in linear form, yielding, in appropriate coordinates, a wave equation[1]

$$\left(\nabla^2 - \frac{1}{c^2} \frac{\partial^2}{\partial t^2} \right) h_{\mu\nu} = 0.$$ (6)

Thus, the elements of $h_{\mu\nu}$ can take the form $h(\omega t - \mathbf{k} \cdot \mathbf{x})$, with $\omega = |\mathbf{k}|/c$, representing a plane wave propagating in the direction $\hat{k} \equiv \mathbf{k}/|\mathbf{k}|$ with the speed c. The appearance of the speed of light as the speed of gravitational waves is a result of the way space and time are brought together in relativity as space-time. As waves of space-time disturbances, c is the only "natural" speed in the problem.

There is a great deal of *gauge freedom* in the construction of explicit forms for $h_{\mu\nu}$, and this is the source of some confusion. But there is one particularly useful gauge choice in which both physics and mathematics become especially clear. This is the so-called *transverse traceless gauge*, or "TT gauge" for short. In this gauge, coordinates are marked out by the world lines of freely-falling test masses.

Consider the case of a wave propagating along the $\hat{z}$ axis. The statement that $h_{\mu\nu}$ be transverse and traceless means that it has the form

$$h_{\mu\nu} = \begin{pmatrix} 0 & 0 & 0 & 0 \\ 0 & a & b & 0 \\ 0 & b & -a & 0 \\ 0 & 0 & 0 & 0 \end{pmatrix}. \tag{7}$$

In other words, we can write this wave as a sum of two components, $h = a\hat{h}_+ + b\hat{h}_\times$, with

$$\hat{h}_+ = \begin{pmatrix} 0 & 0 & 0 & 0 \\ 0 & 1 & 0 & 0 \\ 0 & 0 & -1 & 0 \\ 0 & 0 & 0 & 0 \end{pmatrix}, \tag{8}$$

and

$$\hat{h}_\times = \begin{pmatrix} 0 & 0 & 0 & 0 \\ 0 & 0 & 1 & 0 \\ 0 & 1 & 0 & 0 \\ 0 & 0 & 0 & 0 \end{pmatrix}. \tag{9}$$

The "basis tensors" $\hat{h}_+$ and $\hat{h}_\times$ (pronounced "h plus" and "h cross") represent the two orthogonal polarizations for waves propagating along the $\hat{z}$ axis. This is true in spite of the fact that one goes into the other through a rotation of only 45°. The perturbation called h_+ momentarily lengthens distances along the $\hat{x}$ axis, simultaneously shrinking them along the $\hat{y}$ axis. The polarization $h_\times$ has its principal axes rotated 45°. Clearly, a 90° rotation takes one of these tensors into itself, up to a sign change. A particle physicist would say this is a reflection of the fact that the *graviton* is a spin-2 particle. I will have nothing else to say about gravitons — a classical picture is perfectly adequate for the situations we will be describing. The most general form for $h_{\mu\nu}$ can be generated by considering arbitrary spatial rotations of h_+ or $h_\times$, as parametrized by, say, the Euler angles.

It is important to understand how gravitational effects appear when we describe events in such a coordinate system. The TT gauge is one of many possible coordinate choices that embody the deepest principle of general relativity: gravitation is not a force, but a phenomenon of geodesic motion through curved space-time. This means that, from the relativistic point of view, a freely-falling mass (by definition an object that is subject to no influences of non-gravitational origin) does not accelerate when a gravitational wave passes. It just sits there. So, we can use a set of freely-falling masses to define a coordinate system in space-time.

This is a beautiful idea, but it immediately leads to two related questions. The first is, "If test masses don't respond to a gravitational wave, how can there be any observable physical manifestation of a gravitational wave?" This is tantamount to asking whether gravitational waves are "real" in any meaningful sense. We find the answer to this question by examining how changes in the metric can affect the measurable distance between our freely-falling test masses. The next two sections will explore this answer in some detail.

The second question has to do with making the translation between this relativistic language and the more conventional description of gravitation in the laboratory and in everyday experience. (Try going to the weight room at the gym and telling people there that gravity exerts no forces.) We will discuss below under what circumstances it can make sense to use the Newtonian concept of a gravitational force.

2 The Michelson-Morley Experiment

One way to determine the distance between widely separated objects is to determine the travel time of light from one to the other and back. This is a technique for mapping the structure of space-time that has a first-class relativity pedigree. The earliest actual use of light travel time to experimentally determine the structure of space-time is the measurement of Michelson and Morley.[2] This was the experiment that showed that light travelled at a constant speed in all inertial frames of reference. (That is not how Michelson and Morley described it, of course, since their work predated the theory of relativity. They would have described their experiment as an attempt to see if they could detect an apparent shift in the speed of light due to the Earth's motion through the *ether*.)

The experiment was carried out with a device that has come to be called a *Michelson interferometer*. In its simplest form, it consists of a source of light, a partially reflecting mirror (the *beam splitter*), and two mirrors located some distance away from the beam splitter in two orthogonal directions. Light transmitted through the beam splitter from the lamp travels toward one mirror, while reflected light travels to the other. The light beams are reflected from each of these mirrors back toward the beam splitter, where they are superposed. The measurement consists of inspecting the light emerging from the *output port* of the beam splitter.

Let's consider some of the features of the Michelson interferometer as a measuring device. Firstly, note that the function of this instrument is to compare the amount of time the light beams in the two arms take to complete their paths. Interferometry is well suited for this task. Recall that the light in the two arms comes from a common source. This means that the light enters each arm from the beam splitter in a well defined phase relationship with the light in the other arm. It is simplest to fix attention on one particular phase surface or *wave front* — this wave front enters each arm at the same instant, then propagates toward the far mirrors.

Assume, for the moment, that the two arms have precisely the same round trip length $2L$. If the ether theory were correct, then each beam travels through the ether at the speed of light, c. Since the laboratory is fixed on a moving, spinning Earth, then in most cases light will travel at different speeds in the two arms, as measured with respect to apparatus fixed in the moving laboratory. (It seemed a good working hypothesis in 1887 to assume that the Sun, as one of the "fixed stars", would be at rest with respect to the ether.) The yet-to-be-invented relativity theory would say instead that light travels at c in all directions as seen in the lab frame, no matter what that frame is. According to the ether theory, when the two wave fronts return to the beam splitter from their travels, they would have taken different amounts of time, and thus will have arrived slightly "out of sync". Or, returning to a picture of a continuous light wave, the two waves returning to the beam splitter will have suffered a phase shift with respect to each other. On the other hand, relativity says that if the two arms of the interferometer have precisely the same length, then the two beams will return precisely *in phase*.

The phenomenon of interference between the two returning beams enables one to make a surprisingly precise measurement of the relative phase, or difference in travel times, between the two beams. Let's consider how the state of the light at the output of the interferometer depends on the light travel time of the two beams. We'll follow here the analysis presented in Hermann Haus' excellent *Waves and Fields in Optoelectronics*.[3]

Call the electric field of the input light $E_0 e^{i(\omega t - kx)}$. We can represent the effect of the 50-50 beam splitter by an *amplitude reflection coefficient* $r = 1/\sqrt{2}$, and an *amplitude transmission coefficient* of $t = i/\sqrt{2}$. That is, the light that continues along the $\hat{x}$ axis after passing through the beam splitter has an electric field of $i\left(E_0/\sqrt{2}\right) e^{i(\omega t - k_x x)}$, while the light reflected into the $\hat{y}$ arm has amplitude $\left(E_0/\sqrt{2}\right) e^{i(\omega t - k_y y)}$. (We have written different wave numbers k_x and k_y to allow for the putative different speeds of light in the two arms, as proposed by the ether theory.)

Reflection at the far mirrors multiplies each wave by -1. When the light in each arm returns to the beam splitter, it again is partly transmitted, partly reflected, with the same coefficients as at the first encounter. The light exiting the interferometer through the output port has the field

$$
\begin{aligned}
E_{out} &= \tfrac{i}{2} E_0 e^{i(\omega t - 2k_x L_x)} + \tfrac{i}{2} E_0 e^{i(\omega t - 2k_y L_y)} \\
&= \tfrac{ie^{i\omega t}}{2} E_0 \left(e^{-i2k_x L_x} + e^{-i2k_y L_y} \right) \\
&= i e^{i(\omega t - k_x L_x - k_y L_y)} E_0 \cos(k_x L_x - k_y L_y).
\end{aligned} \tag{10}
$$

Light will also exit through the input port of the beam splitter, as if reflected back toward the lamp, with amplitude

$$
E_{refl} = -i e^{i(\omega t - k_x L_x - k_y L_y)} E_0 \sin(k_x L_x - k_y L_y). \tag{11}
$$

When all of the algebraic dust has settled, we can see that the amplitude of the light leaving the interferometer depends on the difference in the phase accumulated by the light travelling in the two arms, $k_x L_x - k_y L_y$. When that difference is zero, E_{refl} vanishes, and light of amplitude E_0 exits from the output port. If the difference were $\pi/2$, all of the light would instead be reflected from the input port, with no light at all leaving through the output port. Since the power in a beam of light is proportional to E^2, we see that

$$
\begin{aligned}
P_{out} &= P_{in} \cos^2(k_x L_x - k_y L_y) \\
&= \tfrac{P_{in}}{2}\left(1 + \cos 2(k_x L_x - k_y L_y)\right).
\end{aligned} \tag{12}
$$

In other words, a Michelson interferometer acts as a transducer from travel time difference to output optical power variation. By measuring the brightness of light at the output port, we can learn about the difference in travel time of the two beams, up to an integer number of intervals $\tau_\lambda \equiv \lambda/c$.

We can quickly verify that energy is conserved, by noting that the sum of the power exiting through the output port and the input port is

$$
\begin{aligned}
P_{out} + P_{refl} &= P_{in}\left(\cos^2(k_x L_x - k_y L_y) + \sin^2(k_x L_x - k_y L_y)\right) \\
&= P_{in}.
\end{aligned} \tag{13}
$$

We've taken a bit of historical liberty with this last part of the description of the Michelson interferometer. Today, we might set up the Michelson interferometer to operate in just the way described above, that is, as a transducer from travel time difference to brightness. But Michelson and Morley lacked a precise way to measure brightness. So they used a slight variation of the scheme.

To see how it worked, recall that the light travels through the interferometer in beams of finite width. When we calculated the amount of light exiting the interferometer, we implicitly assumed that the two beams were precisely aligned; otherwise, we would have had to describe how the relative phase varied with position across the beam.

Just that extra degree of freedom was what saved the day for Michelson and Morley. They used their interferometer as we described it, except for the fact that the beams were slightly misaligned. Then, a screen placed at an output port shows, not a spot of uniform brightness, but a set of alternating bright and dark bands, known as *fringes* from their resemblance to the fringed border at the end of a piece of woven cloth. Bright bands appear where constructive interference occurs, dark bands where the interference is destructive. The position of these fringes on the screen is what indicates the relative phase between the two returning beams. (You can think of the alternating bands as an interlaced set of bright and dark output ports.)

Michelson and Morley chose this deliberate misalignment because it was easier for them to see such fringes shift by a small amount from side to side than it was to estimate whether a spot had brightened or dimmed slightly. The situation is the opposite if one were using photoelectric light detectors to "interrogate the fringe". One bit of jargon has survived from Michelson's day to ours — even when we have well-aligned beams in the output port of an interferometer, we refer to the dependence of the uniform brightness superposition of the two beams on arm length as the "fringe pattern", or "fringe" for short.

You might wonder how it is possible to set up the interferometer with arms the same length to a precision of a small fraction of a wavelength of light. The answer is that it is hard (although possible), but also not necessary. The interferometer responds in nearly the same way if one arm is shorter than the other by a small integer number of wavelengths. The limit on the tolerable arm length difference is set by the fact that no source of light emits waves of precisely steady wavelength. The arm lengths must match substantially better than the "coherence length" of the light source. Michelson and Morley explain in their paper how they set the arm lengths to be sufficiently close to one another.

But if the prediction of the ether theory is a particular phase difference between the light in the two arms, didn't Michelson and Morley need to start with the interferometer arms set at precisely the same length in order to make a well-defined measurement? No. Regardless of the exact value of the length L of the arms, the ether theory predicts that for an interferometer moving through the ether at speed $v = \beta_\oplus c$ the travel-time difference should be modulated by the amount $2L\beta_\oplus^2/c$ as the interferometer is rotated by 90°. So the amplitude of the modulation is insensitive to small departures of the arm lengths from equality. That is why Michelson and Morley invested much effort in a way to rotate their interferometer gently and smoothly. Looking for a shift in the fringe positions is a natural and well-defined measurement, and does not depend crucially on the exact initial set-up of the interferometer. This is a particular example of a rule of great generality in experimental physics: It is easier to measure an effect you can modulate (or "chop") than to measure a constant effect.

The sensitivity Michelson and Morley were able to achieve is rather impressive. Previously, Michelson had used an interferometer in which light made only a single round trip across the apparatus, but good results were only obtained after additional mirrors were added to lengthen the light path in each arm to total optical length $L_{opt} = 11$ meters. With that value for L_{opt}, the predicted travel time difference between the two arms has a maximum value 4×10^{-16} sec. Michelson and Morley were able to read the shift in fringe positions to better than $\lambda/20$, equivalent to a time difference of 8×10^{-17} sec.

One of the greatest impediments to the success of this experiment was the sensitivity of the apparatus to externally-driven vibrations. This was true despite the fact that the mirrors were securely attached to a massive stone slab floating on mercury. The best results could only be obtained by working in the wee hours of the morning, when the noise of human activity is at a minimum. The switch to multi-pass arms of greater total length was of no help in making the ether's signal stand up above vibration noise, because multiple reflections increase this kind of noise as much as they do the signal.

3   A Schematic Detector of Gravitational Waves

It is indicative of the unity of physics that we find ourselves again turning a century later to the same sort of apparatus, in order to answer another important question about the nature of space-time. Inevitably, many of the same experimental physics issues will arise again. The Michelson-Morley experiment teaches many lessons we need to learn well, since in order to detect gravitational waves we'll have to achieve sensitivity greater by many orders of magnitude.

In this section, we'll discuss a schematic experiment to demonstrate how one could detect gravitational waves. Our *gedanken* version of an inteferometric gravitational wave detector looks very much like the Michelson interferometer of 1887. There is one key difference — the mirrors are not connected to a single rigid structure. Instead, each mirror rests on a freely-falling mass, so that it responds in a simple way to gravitational effects.

Since this is a thought experiment, let's consider the simplest possible version: a "50/50" beam-splitting mirror sits on one mass at the vertex of a symmetric "L", while at each end of the L sits another mass carrying a flat mirror. This works quite a bit like Michelson's original interferometer.

We want to imagine test masses separated by a very large distance. A rigid ruler isn't very practical for measuring the distance (if for no other reason than that a long ruler will not be well approximated as rigid, but will have longitudinal vibration modes of rather low frequency.) But we can determine the distance by measuring the round trip travel time of light beams sent over large distances, using a Michelson interferometer. So it is natural to consider sensing the effect of a gravitational wave on a set of test masses by installing a Michelson inteferometer to monitor their relative separations.

Before we proceed, it is worthwhile to have a brief discussion of the history of this idea. Interferometric detection of gravitational waves is one of those ideas of such beauty that it should not be surprising that it occurred independently to a number of people at roughly the same time. (It is also an experiment whose obvious difficulty makes it plausible that initial suggestions were not immediately followed up.) An early paper by Gertsenshtein and Pustovoit lays out the key idea, and gives a well-founded, if conservative, estimate of the possible sensitivity.[4] Unpublished work of Joseph Weber and his student Robert Forward in the 1960's formed the groundwork for the construction by Forward and his colleagues of the first operating gravitational wave interferometer.[5] The lineage of today's active interferometer community can be traced to independent work in the early 1970's by Rainer Weiss (who is also credited by Forward with crucial advice on the latter's work.) Inspired by Weber's claimed detections of gravitational waves, Weiss conceived of the large interferometers now starting construction, and performed a detailed analysis of their possible performance.[6]

4 Analysis of the schematic detector

The Special Theory of Relativity taught us that light's behavior is especially simple — it travels at a constant speed, c, in any inertial frame of reference. The mathematical expression of this idea is that a light beam connects sets of points separated by an interval of zero, or

$$ds^2 = 0. \tag{14}$$

This is the key bit of physics that will enable us to see how a Michelson interferometer can be used to detect a gravitational wave.

We simplify the mathematics, without loss of generality, by imagining that we've laid out our free-mass Michelson interferometer with its arms aligned along the $\hat{x}$ and $\hat{y}$ axes, with the beam-splitter at the origin. Then, since the paths taken by light in each arm will have only either dx or dy non-zero, we need only consider the 11 and 22 components of the metric (along with the trivial 00 component).

First, consider light in the arm along the $\hat{x}$-axis. The interval between two neighboring space-time events linked by the light beam is given by

$$\begin{aligned}
ds^2 = 0 &= g_{\mu\nu}dx^\mu dx^\nu \\
&= (\eta_{\mu\nu} + h_{\mu\nu})\, dx^\mu dx^\nu \\
&= -c^2 dt^2 + (1 + h_{11}(\omega t - \mathbf{k}\cdot\mathbf{x}))\, dx^2.
\end{aligned} \tag{15}$$

This says that the effect of the gravitational wave is to modulate the distance between two neighboring points of fixed coordinate separation dx (as marked, in this gauge, by freely-falling test particles) by a fractional amount $|h_{11}|$.

Here we're not using a ruler for measuring distances. Instead, we are taking advantage of the fact that what is measurable is a change in the time it takes light to travel through the apparatus. That will be revealed by a shift in the phase of the light wave with respect to another reference wave. This is almost the same as the effect Michelson set out to look for; but here, instead of a constant dependence of light's velocity on its direction of propagation with respect to the ether, we look for a transitory velocity modulation (i.e., a modulation of k_x with respect to k_y) with a more complicated direction dependence.

We can evaluate the light travel time from the beam splitter to the end of the $\hat{x}$ arm by integrating the square root of Eq. 15

$$\begin{aligned}
\int_0^{T_{out}} dt &= \frac{1}{c}\int_0^L \sqrt{1 + h_{11}}\, dx \\
&\approx \frac{1}{c}\int_0^L \left(1 + \tfrac{1}{2}h_{11}(\omega t - \mathbf{k}\cdot\mathbf{x})\right)\, dx,
\end{aligned} \tag{16}$$

(where, because we will only encounter situations in which $h \ll 1$, we've used the binomial expansion of the square root, and dropped the utterly negligible terms with more than one power of h.) We can write a similar equation for the return trip

$$\int_{T_{out}}^{T_{rt}} dt = -\frac{1}{c}\int_L^0 \left(1 + \frac{1}{2}h_{11}(\omega t - \mathbf{k}\cdot\mathbf{x})\right)\, dx. \tag{17}$$

The total round trip time is thus

$$\tau_{rt} = \frac{2L}{c} + \frac{1}{2c}\int_0^L h_{11}(\omega t - \mathbf{k}\cdot\mathbf{x})dx - \frac{1}{2c}\int_L^0 h_{11}(\omega t - \mathbf{k}\cdot\mathbf{x})dx. \qquad (18)$$

The integrals are to be evaluated by expressing the arguments as a function just of the position of a particular wavefront (the one that left the beam-splitter at $t = 0$) as it propagates through the apparatus. That is, we should make the substitution $t = x/c$ for the outbound leg, and $t = (2L - x)/c$ for the return leg. (Corrections to these relations due to the effect of the gravitational wave itself are negligible.)

A similar expression can be written for the light that travels through the $\hat{y}$ arm. The only differences are that it will depend on h_{22} instead of h_{11} and will involve a different substitution for t. The interferometer output will indicate the relative phase shift between the two beams that interfere upon returning to the beam-splitter. That phase shift is simply ω times the difference in travel-time perturbations in the two arms. We could write a general expression for the time travel difference as a function of the gravitational wave's arrival direction, polarization, and frequency, but it is probably more instructive to consider a few special cases. The simplest is for arrival along the $\hat{z}$ axis. Consider a sinusoidal wave in the $+$ polarization with frequency ω_{gw} and amplitude $h_{11} = -h_{22} = h$. If $\omega_{gw}\tau_{rt} \ll 1$, then we can treat the metric perturbation as approximately constant during the time any given wavefront is present in the apparatus. There will be equal and opposite perturbations to the light travel time in the two arms. The total travel time difference will therefore be

$$\Delta\tau(t) = \frac{2L}{c}h(t) = \tau_{rt0}h(t) \qquad (19)$$

(where we have defined $\tau_{rt0} \equiv 2L/c$.) Alternatively, we have

$$\Delta\phi(t) = \omega_{gw}\tau_{rt0}h(t). \qquad (20)$$

In other words, the phase shift between the light that traveled in the two arms is equal to a fraction h of the total phase a light beam accumulates as it traverses the apparatus. This immediately says that the longer the optical path in the apparatus, the larger the measurable effect, the phase shift, will be. This is the same scaling rule as in the Michelson-Morley experiment.

But this scaling law won't hold for arbitrarily long arms. Consider the case where the optical path is so long that $\omega_{gw}\tau_{rt} \ll 1$ is no longer valid. In contrast to the Michelson-Morley experiment, there is a maximum useful τ_{rt}, as can be seen by considering the case when $\omega_{gw}\tau_{rt} = 2\pi$. Then the light spends exactly one gravitational wave period in the apparatus; for every part of its path for which the light "sees" a positive value of h, there is an equal part for which it sees an equal but opposite value of h. Thus, in that case there would be no net modulation of the total travel time and thus zero measurable output at the beam-splitter. The travel time difference is proportional in this case to $\sin\omega_{gw}\tau_{rt}$.

A straightforward but involved manipulation of Euler angles gives the angular dependence of the sensitivity of the interferometer (in the low frequency limit). The idea is to determine the elements h_{11} and h_{22} in the $\hat{x} - \hat{y}$ plane, given a description of the wave in a coordinate system whose $\hat{z}'$ axis is aligned with the propagation vector of the wave. It is shown in several references (see for example Schutz and Tinto,[7] who followed Forward[5]) that a linearly polarized wave $h(t)$ arriving from a direction (Θ, Φ) gives an angular response function, or *beam pattern*, of

$$\Delta\phi = \omega_{gw}\tau_{rt}h(t)\left(\frac{1}{2}(1 + \cos^2\Theta)\cos 2\Phi\cos 2\Psi + \cos\Theta\sin 2\Phi\sin 2\Psi\right), \qquad (21)$$

where Ψ is the angle specifying the polarization of the wave. This expression reveals a rather non-directional response. Waves propagating along the $\hat{x}$ and $\hat{y}$ axes (with the proper polarization) show response down only by a factor of 2 from the maximum $\hat{z}$ axis response. The only true nulls are along the bisector between the $\hat{x}$ and $\hat{y}$ axes (and the three other directions in the detector plane separated by multiples of 90° from this direction); in these directions the interferometer arm length changes are always equal in the two arms, giving no response.

The broad angular response of a gravitational wave inteferometer is both a blessing and a curse for gravitational wave astronomy. An interferometer behaves nothing like a telescope that one points at an interesting portion of the sky. It is more like an ear placed on the ground; waves coming from nearly any direction are registered. On the one hand, this is an advantage for surveying the sky, especially if one fears there might be few signals strong enough to be detected. On the other hand, the task of understanding a source of gravitational waves as an astronomical object would get powerful assistance from being able to study the source with optical telescopes or other receivers of electromagnetic radiation. Identifying the source on the sky will require a rather accurate determination of the position of the source on the sky. That information will have to be extracted from determination of the differences in arrival times of signals at detectors at widely separated locations.

The statements made above concerning the frequency scaling and angular sensitivity need amendment at high frequencies. The reason is that once the light travel time becomes comparable to or larger than the gravitational wave period, the simplifying assumption that $h(\omega_{gw}t)$ is constant during a wave front's trip is no longer valid. Instead, the total phase shift accumulated by a wave front depends on a detailed consideration of the value of h_{11} or h_{22} as a function of time at the location of the optical wave front as it travels through the interferometer. The method we sketched out above in Eq.18 remains applicable, but the algebra becomes more involved. The results also depend in detail on whether, and if so how, the light makes multiple trips through each arm.

5   Effects of Gravitational Waves on Masses

In most respects, gravity in the laboratory can still be considered basically "a push or a pull",[8] even when we take into account General Relativity. Treating gravitational waves as a quasi-Newtonian phenomenon gives them a comfortable familiarity. It is almost a must when other "true" forces must be taken into account, such as the electrostatic forces that make a collection of atoms behave like an elastic solid. To take this into account, we need to use a slightly different way of defining coordinates than is given by the transverse traceless gauge. What corresponds most to our usual Newtonian intuition about experiments is the *proper reference frame* description.[9] Instead of coordinates being marked out by freely-falling masses, in this description of events we mark them by perfectly rigid rods arranged in an orthogonal framework.

Imagine that we have placed two freely-falling test particles of mass m a small distance apart along the $\hat{x}$ axis, at $x = \pm L$. If a gravitational wave is incident along the $\hat{z}$-axis in the +-polarization, how does the system respond? We saw above that if we used the travel time of a light beam to measure how the distance changed, then the distance change is equal to the distance times $h_{11}/2$ (for distances along the $\hat{x}$ axis.)

Measuring distances with rigid rulers ought to give us the same result. But if we define coordinates as points that are fixed along a rigid ruler, then we can only have the separation between masses changing if their coordinates change, as seen in such a frame. Thus, in such a description of physics, we need to describe a gravitational wave as something that can accelerate a freely-falling test mass. In other words, we need to describe a gravitational wave in terms of a force. It is not hard to show that we would get the same answers as before if we took as the *tidal force* due to a gravitational wave

$$F_{gw} = m\ddot{x} = \frac{1}{2}mL\frac{\partial^2 h_{11}}{\partial t^2}. \tag{22}$$

This is a perfectly acceptable description as long as separation of the test masses is small compared to the wavelength of the gravitational wave. (Recall that if we integrate the effect on an optical wavefront through a longer distance, the gravitational wave will have time to change sign, and reduce the net observable effect.) This is the only straightforward way to include other forces in the problem.

As an example, a resonant mass detector can be modelled as a pair of masses connected by a spring. We would analyze such a detector's interaction with a gravitational wave as we did above, with the single addition of the inclusion of a term for the spring force in the equation of motion. There are various ways to detect the relative motion of the ends of the bar. Weber's original scheme was to equip the middle of the bar (the "spring") with piezoelectric strain gauges.[10] Many of Weber's successors have instead placed an inertial sensor (i.e., an accelerometer) on one end of the bar.

It makes a difference to our description of gravitational effects whether we choose to use this proper reference frame picture, or whether instead we use the TT gauge and define points in our coordinate system by the positions of freely-falling test masses. In the proper reference frame, gravitation appears as a force that accelerate "free" test masses with respect to rigid rods. In the TT gauge, by contrast, there is no change in the coordinates of a freely-falling mass. The effect of gravitation shows up instead as a change in the metric describing the space between the non-moving test masses. But in spite of the differences in language, the two pictures give completely consistent descriptions of the results of experiments, as long as we are in the short distance regime where both are equally applicable. Whenever we need to consider systems comparable to or longer than a gravitational wavelength, then the TT gauge picture is the only sensible one.

6 Photon Shot Noise

Now we are ready to confront the question, "How well can an interferometer work?" The precision likely to be required is daunting. Metric perturbations h (and thus fractional changes in light travel time) of the order of 10^{-21} or smaller are typical of predictions of wave strength. It is convenient to express the wave's effect as an equivalent motion of the test masses. Recalling the "proper reference frame" description, we have

$$\Delta x = \frac{1}{2} h L, \tag{23}$$

or

$$\frac{\Delta x}{\lambda} = \frac{1}{2} h \frac{L}{\lambda}. \tag{24}$$

In an interferometer the size of Michelson and Morley's, with total length $L_{opt}/\lambda = 4 \times 10^7$ ($\lambda \approx 0.5 \mu$m), $h = 10^{-21}$ would cause a shift of only 10^{-14} of a fringe. Michelson and Morley only claimed to be able to detect shifts of $\lambda/20$ or so. Even if one could construct an interferometer with $L_{opt} = 1000$ kilometers, or $L_{opt}/\lambda = 2 \times 10^{12}$, $h = 10^{-21}$ only corresponds to an optical path length change of 10^{-9} of a fringe. Is it possible to confidently detect such a tiny change in path length? In the rest of this lecture, we'll see how to answer this question. We'll look at the basic physics issues involved, in the context of a simple interferometer in which the light only makes one round trip in an arm.

A key to finding the answer to this question is to remember that we can (and will) use the variant of the Michelson interferometer in which the recombining wave fronts are made strictly parallel. In this case the pattern of bright and dark "fringes" observed by Michelson and Morley widens out into a single spot over which the phase difference between the two beams is a constant.

This spot will brighten or darken uniformly as the phase difference is adjusted. The output power is given by

$$P_{out} = P_{in} \cos^2(k_x L_x - k_y L_y). \tag{25}$$

In other words, what we have is a device in which the path length difference between the two arms can be determined (up to an integer number of wavelengths) by a careful measurement of the optical power at the interferometer output.

This means we can recast the question "How small a gravitational wave amplitude can we detect?" as the question "How small a change in optical power can we detect?"

This latter question suggests that there may in fact be a fundamental limit to the measurement precision.

To see why, recall that light comes in finite sized chunks called photons. Measuring optical power is equivalent to determining the number of photons arriving during a measurement interval. Whenever we count a number of discrete independent events characterized by a mean occurrence $\bar{N}$ per counting interval, the set of outcomes is characterized by a probability distribution $p(N)$ called the *Poisson distribution*,

$$p(N) = \frac{\bar{N}^N e^{-\bar{N}}}{N!}. \tag{26}$$

(This is also colloquially referred to as "counting statistics".) When $\bar{N} \gg 1$, the Poisson distribution can be approximated by a Gaussian distribution with a variance $\sigma^2 = \bar{N}$. That is to say, the standard deviation σ is equal to $\sqrt{\bar{N}}$.

Consider an experiment in which we try to determine the rate of arrival of photons $\bar{n}$ (with units of sec^{-1}), by making a set of measurements each lasting τ seconds. The mean number of photons per measurement interval is $\bar{N} = \bar{n}\tau$. The fractional precision of a single measurement of the photon arrival rate (or, equivalently, of the power) is thus given by

$$\frac{\sigma_{\bar{N}}}{\bar{N}} = \frac{\sqrt{\bar{n}\tau}}{\bar{n}\tau} = \frac{1}{\sqrt{\bar{n}\tau}}. \tag{27}$$

This says that if we were to try to estimate $\bar{n}$ from measurements for which $\bar{n}\tau \sim 1$, then the fluctuations from instance to instance will be of order unity. But, if $\bar{n}\tau$ is very large, then the fractional fluctuations are small.

Let's carry through the calculation for the power fluctuations, and thence to the noise in measurements in h. Each photon carries an energy $\hbar\omega = 2\pi\hbar c/\lambda$. (I choose not to absorb the 2π into Planck's constant so that we can reserve the symbol h for the gravitational wave metric perturbation.) If there is a power P_{out} at the output of the interferometer then the photon flux at the output will be

$$\bar{n} = \frac{\lambda}{2\pi\hbar c} P_{out}. \tag{28}$$

Now we need to specify the operating point of the interferometer. (In order for this concept to be meaningful, the state of the interferometer must be nearly fixed, either because the noise is intrinsically small or because a servomechanism keeps it fixed. Although the latter will always be the case in practice, let's assume for simplicity's sake that the former is true). A sensible operating point would seem to be $P_{out} = P_{in}/2$. There, midway between maximum power and $P_{out} = 0$, the sensitivity dP_{out}/dL to arm length shifts is a maximum. At that point,

$$\frac{dP_{out}}{dL} = \frac{2\pi}{\lambda} P_{in}. \tag{29}$$

We can also consider this to be the sensitivity to the test mass position *difference* δL, since the interferometer is sensitive (with opposite signs) to shifts in the length of either arm.

Now consider the fluctuations in the mean output power $P_{out} = P_{in}/2$, averaged over an interval τ. The mean number of photons per measurement is $\overline{N} = (\lambda/4\pi\hbar c)P_{in}\tau$. Thus we expect a fractional photon number fluctuation of $\sigma_N/N = \sqrt{4\pi\hbar c/\lambda P_{in}\tau}$. Since we are using the output power as a monitor of test mass position difference, we would interpret such statistical power fluctuations as equivalent to position difference fluctuations of a magnitude given by the fractional photon number fluctuation divided by the fractional output power change per unit position difference, or

$$\sigma_{\delta L} = \frac{\sigma_N}{N} \Big/ \frac{1}{P}\frac{dP_{out}}{dL}$$
$$= \sqrt{\hbar c\lambda/4\pi P_{in}\tau}. \tag{30}$$

Recall that we can describe the effect of a gravitational wave of amplitude h as equivalent to a fractional length change in one arm of $\Delta L/L = h/2$, along with an equal and opposite change in the orthogonal arm. The net change in test mass position difference is $\delta L = Lh$, so if we interpret brightness fluctuations in terms of the equivalent gravitational wave noise σ_h, we have $\sigma_h = \sigma_{\delta L}/L$, or

$$\sigma_h = \frac{1}{L}\sqrt{\frac{\hbar c\lambda}{4\pi P_{in}\tau}}. \tag{31}$$

There is no preferred frequency scale to this noise; the arrival of each photon is independent of the arrival of each of the others. Note also that the error in h scales inversely with the square root of the integration time. These facts can be summarized by rewriting Eq. 31 as the statement that the *photon shot noise* in h is described by a white amplitude spectral density of magnitude

$$h_{shot}(f) = \frac{1}{L}\sqrt{\frac{\hbar c\lambda}{2\pi P_{in}}}. \tag{32}$$

(It is no accident that the amplitude spectral density looks almost identical to the rms fluctuation with τ set equal to 1 second. The correspondence would be exact if we were using two-sided spectral densities, but we pick up a factor of $\sqrt{2}$ from our choice here of one-sided spectra. All of the mathematical details are explained in Davenport and Root's discussion of the Schottky formula for the shot noise in a vacuum diode.[11])

To set the scale, we can rewrite the rms error in h as

$$\sigma_h = 3.7 \times 10^{-22} \left(\frac{1000 \text{ km}}{L}\right)\sqrt{\frac{\lambda}{0.545 \ \mu\text{m}}}\sqrt{\frac{1 \text{ watt}}{P_{in}}}\sqrt{\frac{10 \text{ msec}}{\tau}}, \tag{33}$$

or the spectral density of h as

$$h(f) = 5.2 \times 10^{-23} \text{ Hz}^{-1/2} \left(\frac{1000 \text{ km}}{L}\right)\sqrt{\frac{\lambda}{0.545 \ \mu\text{m}}}\sqrt{\frac{1 \text{ watt}}{P_{in}}}. \tag{34}$$

It is clear that there is little margin to spare — it is really necessary to reduce this fundamental limit, if we want to be able to confidently detect, and then study, burst signals with amplitudes of 10^{-21} or below. We've already pushed well beyond Michelson and Morley's $L = 11$ m. The arm length $L = 1000$ km is nearly the optimum length for a burst of duration 10 msec. The wavelength $\lambda = 0.545\mu$m is that of green argon ion lasers; the power $P_{in} = 1$ watt is a conservative estimate for the power available from such a laser. Laser and mirror technology do not hold out much hope at present of substantial reduction of λ. So our hopes for advancing beyond this entry-level sensitivity will rest on maximizing the input optical power P_{in}.

7   Radiation Pressure Noise

If shot noise were the only limit to precision determined by the optical power P_{in}, then we could in principle achieve arbitrary precision simply by using a sufficiently powerful laser. But in addition to the substantial technical problems of high power lasers, this line of reasoning neglects one of the deep truths of quantum mechanics. We've treated so far the limit that a quantized world sets on the precision of a measurement. We need to *complement* (to use a term of Niels Bohr's) that discussion with a description of how the measurement process disturbs the system under measurement. It is convenient to make a mental division of a gravitational wave interferometer into two parts. Call the nearly freely-falling mirrored test masses (and the space-time between them) the "system to be measured", and the laser, light beams, and photodetector the "measuring apparatus". There is a deep analogy between such an interferometer and the archetypal quantum mechanical measurement problem called the "Heisenberg microscope". Bohr gave a particularly clear description of it, using a semi-classical treatment. In his 1928 essay "The Quantum Postulate and the Recent Development of the Quantum Theory", Bohr wrote: [12]

> In using an optical instrument for determinations of position, it is necessary to remember that the formation of the image always requires a convergent beam of light. Denoting by λ the wave-length of the radiation used, and by ϵ the so-called numerical aperture, that is, the sine of half the angle of convergence, the resolving power of a microscope is given by the well-known expression $\lambda/2\epsilon$. Even if the object is illuminated by parallel light, so that the momentum h/λ of the incident light quantum is known both as regards magnitude and direction, the finite value of the aperture will prevent an exact knowledge of the recoil accompanying the scattering. Also, even if the momentum of the particle were accurately known before the scattering process, our knowledge of the component of momentum parallel to the focal plane after the observation would be affected by an uncertainty amounting to $2\epsilon h/\lambda$. The product of the least inaccuracies with which the positional co-ordinate and the component of

momentum in a definite direction can be ascertained is just given by [the uncertainty relation].

We are hardly dealing with a microscopic system here: the test masses are likely to range in size from 10 kg to 1 tonne or so. Yet because we aspire to such extreme precision of measurement, it is crucial to consider the sort of quantum effects usually relevant only for processes on the atomic scale. Note that we are not satisfied to know our test masses' precise positions at one moment only; we want to know the history of the path length difference of the interferometer. Perturbations of the momenta of the masses cannot be ignored, therefore, since the value of the momentum at one time affects the position later.

In the Heisenberg microscope, the phenomenon conjugate to the registration of the arrival of a photon bouncing off an atom is the recoil of the atom caused by the change in the photon's momentum upon reflection. In a gravitational wave interferometer we register an arrival rate of photons that we interpret as determined by the difference in phase between electromagnetic fields returning from the two arms. We can recognize the conjugate phenomenon by looking for a fluctuating recoil that can affect the same degree of freedom that we measure. Fluctuating radiation pressure on the test masses causes them to move in a noisy way. The resulting fluctuation in the length difference between the two arms shows how this effect can alter the measurement, identifying it as the conjugate phenomenon.

To estimate the size of this effect, first recall that the force exerted by an electromagnetic wave of power P reflecting normally from a lossless mirror is

$$F_{rad} = \frac{P}{c}. \tag{35}$$

The fluctuation in this force is due to shot noise fluctuation in P, discussed in the previous section. That is

$$\sigma_F = \frac{1}{c}\sigma_P, \tag{36}$$

or, in terms of an amplitude spectral density

$$F(f) = \sqrt{\frac{2\pi\hbar P_{in}}{c\lambda}} \tag{37}$$

independent of frequency.

This noisy force is applied to each mass in an arm. For now, let us consider a simple "one-bounce" interferometer like the first version of the Michelson interferometer; we allow the mirrors at the ends of the arms to be free masses, but assume that the beam splitter is much more massive than the other mirrors. The fluctuating radiation pressure from the power $P_{in}/2$ causes each mass to move with a spectrum

$$\begin{aligned} x(f) &= \frac{1}{m(2\pi f)^2}F(f) \\ &= \frac{1}{mf^2}\sqrt{\hbar P_{in}/8\pi^3 c\lambda}. \end{aligned} \tag{38}$$

The power fluctuations in the two arms will be anti-correlated. (In a semi-classical picture, one additional photon into one arm means one less into the other arm.) This doubles the effect on the output of an interferometer, since the phase shift is proportional to the difference in length of the two arms. Thus we have the *radiation pressure noise*

$$h_{rp}(f) = \frac{2}{L}x(f)$$
$$= \frac{1}{mf^2L}\sqrt{\hbar P_{in}/2\pi^3c\lambda}. \tag{39}$$

Thus we have two different sources of noise associated with the quantum nature of light. Note that they have opposite scaling with the light power — shot noise declines as the power grows, but radiation pressure noise grows with power.

If we choose to, we can consider these two noise sources to be two faces of a single noise; call it *optical readout noise*, given by the quadrature sum

$$h_{o.r.o.}(f) = \sqrt{h_{shot}^2(f) + h_{rp}^2(f)}. \tag{40}$$

At low frequencies, the radiation pressure term (proportional to $1/f^2$) will dominate, while at high frequencies the shot noise (which is independent of frequency, or "white") is more important. We could improve the high frequency sensitivity by increasing P_{in}, at the expense of increased noise at low frequency. At any given frequency f_0, there is a minimum noise spectral density; clearly, this occurs when the power P_{in} is chosen to have the value P_{opt} that yields $h_{shot}(f_0) = h_{rp}(f_0)$. The power that gives this relation is

$$P_{opt} = \pi c\lambda mf^2. \tag{41}$$

P_{opt} is typically quite large. Within this (too simple) interferometer model, let's make an indicative number. Take $m = 10$ kg, $f = 100$ Hz, and $\lambda = 0.545$ μm. Then, P_{opt} is about a half MegaWatt. Even though this number is reduced by the optical path folding schemes we'll talk about in the next section, it will still be true that early generation interferometers will generally run at lower than the optimum power, and so will have shot noise larger than the optimum.

When we plug this expression for P_{opt} into our formula for $h_{o.r.o.}$ we find

$$h_{QL}(f) = \frac{1}{\pi fL}\sqrt{\frac{\hbar}{m}}. \tag{42}$$

I have renamed this locus of lowest possible noise $h_{QL}(f)$, for "quantum limit", to emphasize its fundamental relationship to quantum mechanical limits to the precision of measurements. Note that the expression does not depend on P_{in} or λ, or any other feature of the read-out scheme, even though such details were useful for our derivation. This result shows that the end result of this examination of the workings of our Heisenberg microscope has been to provide an instrument-specific derivation of Heisenberg uncertainty principle. And it reminds us of the truth Bohr's remarks expressed, that in any measurement the uncertainty principle emerges from the specific mechanism of the measurement.

(A word of warning about notation: the left hand side of Eq. 42 is written as if this "quantum limit" noise had a particular spectral density. But please note that the object we call $h_{QL}(f)$ is instead the locus of the lowest possible points of the family of spectra $h_{o.r.o.}(f)$, with that family parametrized by, say, f_0. For this reason, $h_{QL}(f)$ is sometimes called a "pseudo-spectral density",[13] as we often want to plot it on a graph of noise spectra.)

8   Shot Noise in Classical and Quantum Mechanics

We actually determine the output power by letting the light fall on a photodetector, typically a silicon photodiode. An individual photon promotes, with probability η, an electron to the conduction band in the semiconductor. (It is not hard to achieve $\eta \approx 0.8$.)The resulting current of photoelectrons is the physical quantity we can amplify and measure. So the shot noise of interest is actually that of the photocurrent

$$I_p = \eta \bar{n} e = \eta e P_{in} \frac{\lambda}{2\pi\hbar c}. \tag{43}$$

All currents are made up of individual electrons, but not all currents exhibit shot noise of the form we have derived. We made the fundamental assumption that the individual events (photon arrivals, or photoelectron creations) are independent of one another. In the semi-classical view of light we are taking here, independence of photon arrivals is a consequence of the superposition principle for electromagnetic fields. For the photoelectrons, it depends on absence of saturation effects.

Shot noise is a fundamental consequence of a quantum, rather than continuum, picture of the world. Yet the only feature of quantum mechanics we used was the existence of photons. At some level, it is surprising that we can get a correct answer using a semi-classical view of the world, instead of the full apparatus of quantum mechanics. It is even more surprising when we consider that a standard quantum mechanical treatment tempts one to get the wrong answer for the radiation pressure noise. Consider the following passage from Dirac's *The Principles of Quantum Mechanics*:[14]

> Suppose we have a beam of light which is passed through some kind of interferometer, so that it gets split up into two components and the two components are subsequently made to interfere. We may, as in the preceding section, take an incident beam consisting of only a single photon and inquire what will happen as it goes through the apparatus. This will present to us the difficulty of the conflict between the wave and corpuscular theories of light in an acute form.
>
> Corresponding to the description that we had in the case of the polarization, we must now describe the photon as going partly into each of the two components into which the incident beam is split.
>
> ...Some time before the discovery of quantum mechanics people realized that the connexion between light waves and photons must be of

a statistical character. What they did not clearly realize, however, was that the wave function gives information about the probability of *one* photon being in a particular place and not the probable number of photons in that place. The importance of the distinction can be made clear in the following way. Suppose we have a beam of light consisting of a large number of photons split up into two components of equal intensity. On the assumption that the intensity of a beam is connected with the probable number of photons in it, we should have half the total number of photons going into each component. If the two components are now made to interfere, we should require a photon in one component to be able to interfere with one in the other. Sometimes these two photons would have to annihilate one another and other times they would have to produce four photons. This would contradict the conservation of energy. The new theory, which connects the wave function with probabilities for one photon, gets over the difficulty by making each photon go partly into each of the two components. Each photon then interferes only with itself. Interference between two different photons never occurs.

We used semi-classical reasoning to predict that there should be a radiation pressure contribution to the noise in an interferometer. But if we interpret Dirac's words naively, then it would seem that there could be only correlated fluctuations of radiation pressure in the two arms. Thus there would be no interferometer noise from radiation pressure fluctuations (since the interferometer output is sensitive only to differences between the arms). With only shot noise in the problem, we could always, in principle, increase P_{in} without limit. Thus there would be no quantum limit.

We seem to have a paradox, predicting the violation of quantum mechanics by using quantum mechanics. Some physicists, for a time, defended the interpretation that there was no quantum measurement limit in an interferometer. (Weiss refers to a "lively but unpublished controversy" in the community.[15]) On the other hand, the same section of Dirac's work includes the following statement:

Let us consider now what happens when we determine the energy in one of the components. The result of such a determination must be either the whole photon or nothing at all. Thus the photon must change suddenly from being partly in one beam and partly in the other to being entirely in one of the beams. This sudden change is due to the disturbance in the translational state of the photon which the observation necessarily makes. It is impossible to predict in which of the two beams the photon will be found. Only the probability of either result can be calculated from the previous distribution of the photon over the two beams.

One could carry out the energy measurement without destroying the component beam by, for example, reflecting the beam from a movable mirror and observing the recoil. Our description of the photon allows us to infer that, *after* such an energy measurement, it would not be possible to bring about any interference effects between the two components. So long as the photon is partly in one beam and partly in the other, interference can occur when the two beams are superposed, but this possibility disappears when the photon is forced entirely into one of the beams by an observation. The other beam then no longer enters into the description of the photon, so that it counts as being entirely in the one beam in the ordinary way for any experiment that may subsequently be performed on it.

So it seems that allowing for the possibility of measuring recoil may allow our semi-classical intuition to have some truth to it after all. We don't actually measure the recoil of the masses, so we don't cause the sort of "collapse of the wave function" to which Dirac's remark refers. But allowing the masses to recoil, with a subsequent observable effect on the net interferometer phase, does make our situation more subtle than the simple case in our first selection from Dirac's book.

A clear explanation of how to understand an interferometer's quantum limit was provided by Caves in 1980.[16] He showed that a proper quantum mechanical treatment of photon shot noise in an interferometer required consideration of the possibility of noise (in the form of *vacuum fluctuations* of the electromagnetic field) entering the interferometer from the output port. When this is treated properly, then we can construct a complete and consistent quantum mechanical description of the optical read-out noise. Such a picture justifies the semi-classical derivation given above.

Caves' remarkably fruitful idea not only clarified a longstanding confusion, but it also pointed the way toward manipulating the trade-off between intensity shot noise and radiation pressure noise that leads to the quantum limit. The key idea is to inject into the output port some so-called *squeezed light,* in which the noise in the electromagnetic field is not evenly and randomly distributed between the two conjugate quadratures of the oscillation. (Recall that quantum mechanical uncertainty relations only govern the product of the uncertainties in two *conjugate variables.*We are familiar with position and momentum of a mass as a pair of conjugate variables. The $\sin \omega t$ and $\cos \omega t$ quadratures of an oscillating electromagnetic field have the same sort of relationship.) Application of squeezed light holds out the prospect of shifting noise from intensity noise to radiation pressure noise, thus for example allowing one to achieve the quantum limit at high frequencies with less power than would otherwise be required. Some possibilities for actually surpassing the quantum limit have also been discussed. (For further information on this topic, the reader should consult the chapter by Brillet *et al.* in Blair's *The Detection of Gravitational Waves.*[17])

9   The Remarkable Precision of Interferometry

We can not leave this discussion of the fundamental measurement noise without remarking on how very finely we can distinguish motion of a macroscopic mass using interferometry. Far from being limited to a precision of order a tenth or so of a fringe, we have seen that the ultimate limit to interferometer read out precision may be pushed to 10^{-9} of a fringe, or perhaps even finer. Indeed, for signals in the vicinity of 1 kHz, measurements better than $10^{-6}\lambda$ are quite routine,[18] and $10^{-8}\lambda$ sensitivity has been demonstrated in special instruments.[19]

Note that this means that position noise can easily be smaller than a nuclear diameter in one second of integration time. How is this possible? Is the position of a mass even well-defined at this level of precision? In fact it is defined to a precision much finer than atomic dimensions, because the wavefront of the reflected light beam has its phase determined by the average position of all of the atoms across the beam's width. Small irregularities contribute to scattering of a small fraction of the light out of the beam, but otherwise the effects of atom-scale irregularities are negligible.

The mass does vibrate, though; at a minimum, it must shake with an energy of $k_B T$ in its internal normal modes. (The quantum zero point energy $\frac{1}{2}\hbar\omega$ is actually much smaller, by many orders of magnitude, than that associated with Brownian motion at room temperature.) It may take some care to keep Brownian motion noise smaller than the optical readout noise of the interferometer.

References

[1] C. W. Misner, K. S. Thorne, and J. A. Wheeler, *Gravitation* (W. H. Freeman, San Francisco, 1973).

[2] A. A. Michelson and E. W. Morley, *Am. J. Sci.* **34** (1887) 333.

[3] H. A. Haus, *Waves and Fields in Optoelectronics* (Prentice-Hall, Englewood Cliffs, New Jersey, 1984).

[4] M. E. Gertsenshtein and V. I. Pustovoit, *Sov. Phys. JETP* **16** (1962) 433.

[5] G. E. Moss, L. R. Miller, and R. L. Forward, *Appl. Opt.* **10** (1971) 2495; R. L. Forward, *Phys. Rev.* **D17** (1978) 379.

[6] R. Weiss, *Prog. Rep. MIT Research Lab of Electronics* **105** (1972) 54.

[7] B. F. Schutz and M. Tinto, *M.N.R.A.S.* **224** (1987) 131.

[8] S. P. Boughn, private communication.

[9] K. S. Thorne, in *300 Years of Gravitation*, eds. S. W. Hawking and W. Israel (Cambridge University Press, Cambridge, 1987).

[10] J. Weber, *Phys. Rev.* **117** (1960) 306.

[11] W. B. Davenport, Jr. and W. L. Root, *An Introduction to the Theory of Random Signals and Noise* (McGraw-Hill, New York, 1958).

[12] N. Bohr, *Nature* **121** (1928) 580; reprinted in J. A. Wheeler and W. H. Zurek, eds., *Quantum Theory and Measurement* (Princeton University Press, Princeton, New Jersey, 1983).

[13] C. M. Caves, in *Quantum Measurement and Chaos*, eds. E. R. Pike and S. Sarkar (Plenum, New York, 1987).

[14] P. A. M. Dirac, *The Principles of Quantum Mechanics* (Clarendon Press, Oxford, 1958).

[15] R. Weiss, in *Sources of Gravitational Radiation*, ed. L. Smarr (Cambridge University Press, Cambridge, 1979) p. 7.

[16] C. M. Caves, *Phys. Rev. Lett.* **45** (1980) 75.

[17] A. Brillet, J. Gea-Banacloche, G. Leuchs, C. N. Man and J. Y. Vinet, in *The Detection of Gravitational Waves*, ed. D. G. Blair (Cambridge University Press, Cambridge, 1991).

[18] A. Abramovici, W. E.Althouse, R. W. P. Drever, Y. Gürsel, S. Kawamura, F. J. Raab, D. Shoemaker, L. Sievers, R. E. Spero, K. S. Thorne, R. E. Vogt, R. Weiss, S. E. Whitcomb, and M. E. Zucker, *Science* **256** (1992) 325.

[19] D. Shoemaker, R. Schilling, L. Schnupp, W. Winkler, K. Maischberger, and A. Rüdiger, *Phys. Rev.* **D38** (1988) 423.

Is the Wavefunction Real?[*]

W. G. Unruh

CIAR Cosmology Program
Dept. of Physics
University of B. C.
Vancouver, Canada V6T 2A6

Using a simple version of the model for the quantum measurement of a two level system, the contention of Aharonov, Anandan, and Vaidman that one must in certain circumstances give the wavefunction an ontological as well as an epistemological significance is examined. I decide that their argument that the wave function of a system can be measured on a single system fails to establish the key point and that what they demonstrate is the ontological significance of certain operators in the theory, with the wave function playing its usual epistemological role.

Aharonov, Anandan, and Vaidman[1] have recently argued that in addition to its usual epistemological role, the wave function in quantum mechanics in certain situations also has an ontological role. In other words, in addition to acting as a device in the theory to encode the conditions (our knowledge of the world) it must also, in certain circumstances, be regarded as real, in the sense that one can completely determine an unknown wavefunction of a single system. Certainly if their claim were true, that one could take a single system with an unknown wavefunction, and completely determine that wave function on that single system, one would have to accord the wave function a reality on the grounds that that which is measurable is real. In the course of this paper I will argue that they have failed to establish the measurability of the wave function, and thus have failed in their attempt to demonstrate the reality of the wave function. The argument is however subtle. Thus the plan of this paper will be to first discuss the problem of reality in quantum mechanics, to set stage for the question that they are trying to answer.

I will then go through their argument in some detail for a simple system which will I hope clarify their argument. In particular I hope it will clarify the key term in their paper, namely "protection". I will finally argue that they have failed to

V. de Sabbata and H. Tso-Hsiu (eds.), Cosmology and Particle Physics, 257–269.
© *1994 Kluwer Academic Publishers. Printed in the Netherlands.*

establish a key requirement of their argument. To use their term, I will argue that "protection" is not an active attribute, in that one cannot protect a wave function. Rather protection is an attribute that a system can have, and that the wave function will play its usual epistemological role in stating the condition that the system has that property. It is in their failure to establish the active, rather than the passive sense of protection that I feel their argument fails.

I am presenting this paper at this school because I was asked to speak of issues of quantum non-demolition, the process by which one can make measurements of a quantity without in the process altering that quantity. In their process of protected measurements, AAV have presented another technique for carrying out such a measurement. In the ordinary approach to quantum non-demolition, the protection is carried out by essentially making measurements on variables which are constants of the motion for the system. In that way the effect of the mesurement on the non-measured dynamical degrees of freedom of the system do not subsequently alter the motion of the variables which one is measuring. AAV give another technique for such measurements, namely by coupling the measuring instrument adiabatiacally to the system under consideration. In such adiabatic couplings, not only do the measurements leave the system unaltered (if as I will argue they are in special states) but also allow one to make much more complete measurements on the system.

Quantum mechanics is a strange theory which even today, seventy years after its invention/discovery, causes immense difficulties to students and physicists alike. In particular, many find it difficult to establish a reliable quantum picture of reality. Physicists are wedded to the proposition that their field of study is one concerned with reality, that their mathematical formalism is an accurate model of something which exists in the outside world, and is not simply a free invention of the human mind.

In classical physics the relation between the mathematical theory and reality is relatively direct. The terms in the theory correspond in a relatively direct way with physical reality. The dynamical variables in the theory can be assigned values, and so can the attributes of objects in the world. The assigning of values to attributes of the real world in the process of measurement can be modeled directly in the assigning of values to the variables in the theory. The variables in the theory can therefore be taken as having a direct correspondence with the attributes in the world.

Quantum mechanics on the other hand does not behave in such a neat manner. There seems to be no direct correspondence between the structures in the theory and the attributes of the real world. It is this more than anything else which has lead to the widespread unease with quantum mechanics as an ultimate theory of nature. There are two separate structures in quantum mechanics, the operators on some Hilbert space, and the vectors which live in that Hilbert space. Both of these mathematical structures play a role in the theory, but how do they relate to the structure of the physical world?

The operators represent the dynamical variables, and thus correspond in the most direct way to the attributes of the physical world. However, one cannot simply ascribe a number to an operator. The correspondence between the values that we know physical attributes can have, and the assigning of values to the operators is far less straightforward than in classical physics. The values of the operators are its eigenvalues, and one can assign a value to an operator only when it is operating on special vectors in Hilbert space, namely the eigenvectors. Furthermore the various operators cannot all simultaneously have values. This feature of quantum theory is of course well known. The state vector, or wave function thus tells us which of the various operators and thus which of the physical attributes can be assigned a definite value. It furthermore plays the mysterious role of assigning probabilities to the various values that other dynamical variables can have. It thus plays an epistemological role in representing our knowledge about the system under consideration.

But what role does it play in physical reality? Is there some aspect of the physical world which corresponds to the wave function, some extra aspect of the world separate from the dynamic variables, or is it an ideal element of the theory, with no physical reality in its own right, and existing solely in the theory to represent our knowledge of the actual state of the physical world? Is the the wave function real or is it simply a device within the theory to incorporate our knowledge of the world, without it in itself corresponding to anything in the real world? In the former case it would have an ontological role, while in the latter a purely epistemological role.

That physicists have long wanted to regard the wave function as having an ontological role is clear. That desire underlies the unease surrounding the "projection postulate", and the oft heard lament that the wave function cannot simultaneously obey deterministic dynamical equations of evolution, and indeterministic collapse in the ill defined "measurement" process. As an object with an ontological role, as an object corresponding to some element of physical reality such schizophrenic behaviour is highly unsatisfactory. On the other hand if the wave function is simply a tool by which we encode our knowledge into the theory, the change of the wavefunction under a change in knowledge is perfectly rational. It leaves one, however with uncomfortable questions about how knowledge differs from other physical processes.

AAV try to answer the question about the reality of the wave function by asking whether or not the wave function is measurable. If not just the dynamic variables, but the wave function itself can be measured, then the wave function must correspond to an aspect of reality, must have an ontological as well as an epistomelogical aspect within the theory. Thus the question they ask is whether one can start with a single system and on that system determine what its wave function is? For example, if one can determine the expectation value of a sufficiently large number of operators, one can determine the wave function. Now, the expectation value can be determined if we start with a sufficiently large ensemble of systems, all in the same state. But this would at best give the wave function a role in ensembles, not in single systems.

To give it a true ontological meaning, one would like the wave function to have a physical significance, to be measurable, on a single system. Thus the question is "Can one take a system in some initial *unknown* state and completely determine that state by a sequence of measurements on that single system?" If one can, then one would be forced to the position that the the wave function of a single system is measurable and that the wave function therefor was a real attribute of that single system. This is what AAV claim. What they show however, is that, if the wavefunction is known beforehand to be the eigenstate of the unknown Hamiltonian of that system then it is possible to determine the properties of that eigenstate. What their analysis shows is that one can determine some of the properties of an unknown Hamiltonian on a single system, if one knows that the system is in an eigenstate of that operator.

Let us therefor analyze the AAV contention that such a measurement is possible under certain conditions. To do so, I will concentrate on a specific, simple, system, namely a two level (spin $\frac{1}{2}$) system. Can one determine what the wave function of such a system is on a single exemplar of that system? AAV claim the answer to be yes, if one can , to use their word, protect the wave function.

The difficulty with the usual measurement procedure is that the wave function in general changes as the result of measurements. In the view of measurement as a primitive of the theory, this change is just the projection postulate, the reduction of the wave function. In the view of measurement as a physical process, it is the result of the interaction of the measuring apparatus with the system being measured. Thus after a single measurement, the system is no longer in the state it began with, and there is no way in which one can recover that original state. Furthermore a single measurement of any time is not sufficient to determine the state of the system.

However, they argue that if one can "protect" the wave function, ie interact with the system in such a way that its wave function remains unchanged after the measurement, but still affects the measuring apparatus, that then one can make a succession of such measurements on the system, and finally completely determine its wave function.

The key to their "protection" was to realize that if a system is in an energy eigenstate, and if the interaction between the system and the rest of the world was adiabatic, that then the wave function after the measurement would still be that same energy eigenstate. The wave function would not have changed even though the system would have interacted with the rest of the world, and even though its interaction could have changed that outside world in a manner that depended on the specific state that the system was in. Thus one could measure the system (ie determine its effect on the outside world, and thus the state it was in) without changing the wave function of the system. This allows one to make repeated measurements on the system, and completely determine the state of the system.

To make these ideas clear, let us use a simple two level system as our system of interest. It is assumed to be in some pure state $|\psi_0\rangle$. Now, for a two level system any pure state can be represented by a three dimensional vector $\vec{\rho}$, such that $P = \frac{1}{2}(1+\vec{\rho}\cdot\vec{\sigma})$ is the projection operator onto that pure state. Thus given an unknown wave function $|\psi_0\rangle$, if we can determine its associated vector $\vec{\rho}$, we will have completely determined that wave function.

Let us now assume that we can arrange that this system has $H_0 = P$ as its bare Hamiltonian. This is the crucial assumption. It is exactly this condition which is the condition of "protection" introduced by AAV.

Our measuring apparatus will be a free particle coupled to the spin of the system. The coupling to to the measuring apparatus will be via an interaction of the form

$$H_{int} = \epsilon(t)\vec{L}(x) \cdot \vec{\sigma}$$

where $\vec{L}$ is a known position dependent vector, and x is the position operator of the particle. $\epsilon(t)$ is a time dependent coupling which is assumed to be slowly varying. The full Hamiltonian for the system plus measuring apparatus is then given by

$$H = \vec{p} \cdot \vec{\sigma} + \epsilon(t)\vec{L}(x) \cdot \vec{\sigma} + \frac{1}{2m}|\vec{p}|^2$$

To solve the Schroedinger equation for this Hamiltonian, we will begin by making the adiabatic assumption for the spin. Ie, we will assume that the particle begins in an instantaneous spin eigenstate of the Hamiltonian, and remains in the adiabatic eigenstate throughout. We will examine when this approximation is valid. Define

$$\vec{B}(x,t) = \vec{p} + \epsilon(t)\vec{L}(x), \tag{1}$$

the scalar

$$B(x,t) = |\vec{B}(x,t)| \tag{2}$$

and the unit vector

$$\hat{B} = \frac{1}{B}\vec{B} \tag{3}$$

Let us choose at each point in the space a basis $|+\rangle$ and $|-\rangle >$ which are unit norm spin eigenstates of the operator $\vec{B}(x) \cdot \vec{\sigma}$.

$$\vec{B}(x,t) \cdot \vec{\sigma}|+\rangle >= B(x,t)|+\rangle$$

and

$$\vec{B}(x,t) \cdot \vec{\sigma}|+\rangle >= -B(x,t)|+\rangle$$

In general we have that

$$\partial_i|+\rangle = i\alpha_{+i}|+\rangle + \beta_{+i}|-\rangle$$

$$\partial_i|-\rangle = i\alpha_{-i}|-\rangle + \beta_{-i}|+\rangle$$

Since the two states are orthonormal, we have that the α_is must be purely real, and that $\beta_{+i} = -\beta_{-i}^*$.

Using the definition of the states, we find that

$$\beta_i \equiv \beta_{+i} = -\beta^*_{-i} = \langle -|\hat{B} \cdot \vec{\sigma}|+\rangle \qquad (4)$$

We also define α_0 and β_0 by

$$\frac{\partial}{\partial t}|+\rangle = i\alpha_{+0}|+\rangle + \beta_{+0}|-\rangle \qquad (5)$$

and similarly for the $|-\rangle$ state. Again α_0s are purely real, and the beta obey

$$\beta_0 \equiv \beta_{+0} = -\beta^*_{-0} = \langle -|\hat{B}(x,t) \cdot \vec{\sigma}|+\rangle$$

We still have the freedom to choose the phases of the functions appropriately. It would be nice if one could choose the phases so that the α_i are zero, but this is in general impossible. Under a change in phase

$$|+\rangle \to e^{i\phi_+}|+\rangle \qquad (6)$$

we have

$$\alpha_{+i} \to \alpha_{+i} + \partial_i\phi_+ \qquad (7)$$

We could completely eliminate the α_{+i} if $\alpha_{+[i,j]} = 0$, by setting the phase angle $\phi_+ = -\int \alpha_{+i}dx^i$. However, from the definition of α we find that

$$\alpha_{+[i,j]} = -\alpha_{-[i,j]} = i\beta_{[i}\beta^*_{j]} \qquad (8)$$

which in general is not equal to zero. (Note that this fact is intimately related to the Berry's phase phenomenon.) One must therefor make some simple continuous smooth slowly varying choice of phase throughout the region. We can, however, choose the phases so that

$$\alpha_i \equiv \alpha_{+i} = -\alpha_{-i}. \qquad (9)$$

We then write the wave function in the form

$$\psi(t,x) = h_+(t,x)|+\rangle + h_-(t,x)|-\rangle \qquad (10)$$

and one obtains the equation of motion for h_+ and h_- of

$$i\dot{h}_\lambda = \frac{1}{2m}\left(\nabla^2 h_\lambda + 2i\vec{\nabla}h_\lambda \cdot \vec{\alpha} + h_\lambda(i\vec{\nabla} \cdot \vec{\alpha} - |\vec{\alpha}|^2 - |\vec{\beta}_\lambda|^2)\right) + \lambda(B + \alpha_0)h_\lambda$$
$$+ \frac{1}{2m}\lambda\vec{\nabla}h_{-\lambda} \cdot \vec{\beta} - i\beta_0 h_{-\lambda} \qquad (11)$$

If the terms corresponding to $h_{-\lambda}$ are negligable for all values of $\lambda = \pm$, the system is adiabatic with respect to the spin variables. Note that this will also depend on the state of the system. For example if teh state starts off initially in the upper spin state $(h_- - 0)$, then the validity of the adiabatic assumption will depend on the form of h_+.

Ie, for the adiabatic assumption to be valid, we need

$$0 \approx -i\beta_0 h_+ - \frac{1}{m}\vec{\nabla}h_+ \cdot \vec{\beta} \qquad (12)$$

This will be true as long as ϵ changes sufficiently slowly, and as long as the "field" direction ($\hat{B}$) is a sufficiently slowly varying function of position.

Thus in the case that the adiabatic assumption is valid, and we assume that the spin state is initially in the $|+\rangle$, we have

$$i\dot{h}_+ = -\frac{1}{2m}\left(\nabla^2 h_+ + 2i\vec{\nabla}h_+ \cdot \vec{a}\right) + \left((\alpha^2 + \beta^2)/2m + B(x) + \alpha_0(x)\right)h_+ \qquad (13)$$

Let me now simplify the above, and choose $\vec{L}(x) = \vec{x}$, with ϵ a sufficintly small and slowly varying function of time. Let me furthermore assume that m is sufficiently large that we can neglect terms which go as $1/m$ during the time that ϵ is non zero.

We thus need to calculate α_0 and for later use β_0. Defining

$$\cos(\theta) = \vec{\rho} \cdot \hat{B} \qquad (14)$$

and ϕ as an azimuthal angle angle about ρ, we can choose

$$\alpha_0 = \dot{\phi}\cos(\theta)/2$$

and

$$\beta_0 = -\frac{1}{2}(\dot{\theta} - i\dot{\phi}\sin(\theta))$$

Keeping terms to lowest order in ϵ, we get

$$\alpha_0 = 0$$

and

$$\beta_0 \approx -\frac{1}{2}\dot{\epsilon}\sqrt{|x|^2 - \vec{\rho}\cdot\vec{x}^2}$$

During that time therefor, h_+ will obey

$$i\dot{h}_+ = (|\vec{\rho} + 2\epsilon\vec{x}|)h_+$$

Now, assuming that ϵ is small, and that the $h_+(x)$ is confined to a region in which $\epsilon|\vec{x}|$ is much less than unity, we can expand the right hand side to give

$$i\dot{h}_+ = (1 + \epsilon\vec{\rho}\cdot\vec{x})h_+ \qquad (15)$$

and

$$h_+(t) = e^{-i(t+\int_0^t \epsilon(t)dt\vec{\rho}\cdot\vec{x})}h_+(0) \qquad (16)$$

The term $e^{-i\int_0^t \epsilon(t)dt\vec{\rho}\cdot\vec{x}}$ is just the momentum displacement operator on the function h_+, so that if we write h_+ in the momentum representation, we get

$$h_+(t,\vec{p}) = h_+(0,\vec{p} - \int_0^t \epsilon(t)dt\vec{\rho}).$$

Thus by measuring the momentum change of the apparatus, we can determine ρ, the unknown vector representing the initial state and the unknown Hamiltonian.

It is in fact possible in principle to measure ρ to arbitrary precision in a single measurement of on the spin system. Assume that we can place the apparatus into an arbitrary initial state. I will choose a Gaussian with width d. Ie,

$$h_+(0,x) = exp(-(|\vec{x}|^2/2d^2)/(\pi d^2)^{\frac{3}{4}}$$

or

$$h_+(0,p) = exp(-(d|\vec{p}|)^2/2)/(\pi/d^2)^{\frac{3}{4}}$$

To order ϵ we have

$$h_+(t,p) \approx \frac{1}{(\pi/d)^{\frac{3}{4}}}(exp(-|d(\vec{p} - \int_0^t \epsilon dt\vec{\rho})|^2/2) \tag{17}$$

The next quadratic order term in the Hamiltonian is of order $\epsilon^2 x^2$ which leads to a phase shift of order $\int \epsilon^2 dt\ d^2$, which we want to be small(much less than π. The momentum change is $\int \epsilon$, which we want to be much larger than $1/d$, the initial uncertainty in the momentum. In fact defining the accuracy δ by the ratio of the initial uncertainty in the momentum to the displacement in momentum space, we get

$$\delta \approx 1/\int \epsilon dt\ d.$$

Thus, in order that the second order contribution remain small, we require that $\epsilon d/\delta$ be much less that π. But, by making ϵ sufficiantly small, for a given d, this can be made as small as desired. Thus, one can determine the direction $\vec{\rho}$ to arbitrary accuracy with a single measurement via this adiabatic procedure.

Let us also check on the adiabatic assumption. The requirement is the h_- be negligable. Teh equation for h_- is

$$i\dot{h}_- = \beta_0 h_+ = -i\frac{1}{2}\epsilon\sqrt{|x|^2 - \vec{\rho}\cdot\vec{x}^2}h_+ \tag{18}$$

switching ϵ on and off slowly, and keeping it constant in the intermediate times, the only contribution to h_- thus comes from the initial and final switching times. Under our assumptions, the particle does not move appreciably during these times and so the change in x will be negligible. Thus the contribution to h_- from this term will also be negligible.

We thus see at least in principle that it is possible to measure the state of a spin system to arbitrary accuracy with a single (long time) interaction with a measuring apparatus, and furthermore, we see that the state at the end of the measurement is exactly the same as it was at the beginning of the measurement. Although the measuring apparatus changes state (the momentum has been shifted by the interaction), the system we are measuring, the spin $\frac{1}{2}$ system, has not changed it's state at all.

The AAV argument in a nutshell is, that by protecting the wave function, ie, choosing the free Hamiltonian of the system so that the initial state of the system is an eigenfunction of the Hamiltonian, and thus protecting it, the wave function can be determined and measured on a single system (and, as I have shown, with a single measurement).

The crucial step is the protection postulate. Can a wave function be protected or is it rather a feature that the wave function has, like being the eigenstate of some operator. Is one measuring the wave function in the above procedure, or is one measuring one of the dynamic variables of the system. for example, it is well known that one can measure various properties of a single system. In particular, one can measure the properties of the Hamiltonian of a single system. The detailed Hamiltonian of a single atom enclosed within a Penning trap for example can be measured with arbitrary accuracy on that single system. The crucial point about the measurement of the wave function is whether or not one can measure an arbitrary wave function of a given system, and this they fail to show. In the above example, the arbitrariness of the wavefunction was encoded in the arbitrariness of the vector $\vec{\rho}$. The assumption was made that the Hamiltonian of the system had the same vector ρ as did the wave function. If one could take an arbitrary wave function and without knowing what that wave function was, one could choose the Hamiltonian for the system so that it was given by that same ρ, then one could genuinely say that one had measured the wavefunction by the above technique. However, the situation we have is that the Hamiltonian for the system is given by some unknown ρ, and we can choose the state so that it is described by the same ρ, ie is an eigenfunction of that same Hamiltonian.

One may be able to ensure that the particle is in the eigenstate of the unknown Hamiltonian (although one's ability to do even this may be questionable, especially if the state in question is not the ground state of that unknown Hamiltonian), but one cannot ensure that the Hamiltonian has the given but unknown state as its eigenstate. To do the latter, to take the system in an unknown state, and to then choose a Hamiltonian for the system so that that unknown state is its eigenstate would mean that the equation of evolution of the wave function would be non-linear. The Hamiltonian driving the evolution would be a function of the initial state of the system. If we could carry out that procedure for our example, in the Schroedinger representation, one would have

$$i\frac{\partial \psi}{\partial t} = \vec{\rho}(\psi) \cdot \vec{\sigma}\psi \tag{19}$$

Because the Hamiltonian depends on the wave function itself, the evolution is no longer a linear evolution, but is highly non-linear. Instead one has

$$i\frac{\partial \psi}{\partial t} = (\vec{\rho}_0 \cdot \vec{\sigma})\psi \tag{20}$$

where ρ_0 is some constant independent of ψ.

This point lies at the heart of their paper. If one comes upon the system after the initial selections have been made, one could say that all one has is a state and a Hamiltonian. Both are unknown, but it is known that the state is an eigenstate of the Hamiltonian. What does it matter how they got that way? What does it matter whether the physicists began with an unknown Hamiltonian and forced the system to be in an eigenstate of the hamiltonian, or one started with an unknown state, and chose the Hamiltonian so be the projection operator onto that state say? The end result is identical. However, if one is to regard the wave function as real, as having an ontological significance for that single system, the two are crucially different. We can force the wave function to be an eigenstate of an operator, we cannot force the operator to have the unknown wavefunction as its eigenstate. It is precisely the impossibility of the latter that leads to the conclusion that the wave function is not real, does not have the necessary permanence and independence to be considered as real.

What happens if we present their system with the spin half system in a truly unknown state? The system has some (unknown) Hamiltonian, which we cannot force to have the unknown state as its eigenstate. Thus the bare Hamiltonian is given by

$$H_0 = \vec{\rho}_0 \cdot \vec{\sigma} \tag{21}$$

where $\vec{\rho}_0$ is unknown but is **not** equal to $\rho = <\psi|\vec{\sigma}|\psi>$. Given this as the Hamiltonian for the system, and given the same coupling to the apparatus, one finds that the initial wave function for the system will be

$$|\psi_0\rangle = \cos(\theta)|+\rangle + \sin(\theta)|-\rangle \tag{22}$$

where

$$\cos(\theta/2) = \vec{\rho} \cdot \vec{\rho}_0$$

and the phase of $|-\rangle$ has chosen appropriately (it will not enter into the following analysis). $|+\rangle, |-\rangle$ are the eigenstates of $\vec{\rho}_0 \cdot \vec{\sigma}$. After the interaction, the final wave function is given by

$$|\Psi\rangle = \cos(\theta)|+\rangle\phi(\vec{p} - \int \epsilon dt \vec{\rho}) + \sin(\theta)|-\rangle\phi(\vec{p} + \int \epsilon dt \vec{\rho}) \tag{23}$$

If $\int \epsilon dt \vec{L} \cdot \vec{\rho}_0$ is sufficiently large and the wave function $\phi(\vec{p})$ sufficiently peaked, then a measurement of $\vec{p}$ after the interaction will tell one that the system was in the $|+\rangle$ state with probability $\cos^2(\theta)$ and in the $|-\rangle$ state with probability $\sin^2(\theta)$. Ie, the measurement envisaged by AAV leads to the standard result that the measurement gives us one of the two possible outcomes for the polarisation of the spin half system along the direction of ρ_0 with the usual probabilities. Their protected measurements are simply another form of the standard quantum measurement theory. They measure the polarisation of the particle along the direction of the (possibly unknown) direction ρ_0.

Note that this measurement procedure is somewhat different from the standard von Neumann analysis of the measurement process. There the measurement is of the operator of the system which couples the measuring apparatus to system. In this problem that would be the vector operator $\vec{\sigma}$. Rather the measurement is one of the free Hamiltonian of the system, H_0. But such situations were already well known from the standard analysis of the Stern Gerlach experiment in which the particle is measured to be in the eigenstates of spin along the constant magnetic field (which does not couple to the translational degrees of freedom of the particle) rather than the spin along the direction of the inhomogeneous field, which does couple to the translational degrees of freedom by applying a force to the particle. In fact as the above analysis makes clear, our system and measuring apparatus of the standard Stern Gerlach experiment[2], with $\vec{L}$ just the matrix $\vec{B}$, and the measuring apparatus being the translational degrees of freedom of the particle.

I commented that the deflection of the measuring apparatus was proportional to $\vec{\rho}$, which could be written as the expectation value of the operator $\vec{\sigma}$ in the special "protected" state. This is one of the key points which led AAV to conclude that this process can be considered to be a measurement of the state of the system. However, the change in momentum of the apparatus can also be written as being proportional to $Tr H_0 \vec{\sigma}$. In fact when we choose the state to be different from the eigenstate of the Hamiltonian, we realize that this is a better expression for the deflection, because it applies for an arbitrary initial state. The deflection remains the same regardless of the state of the system, as long as H_0 remains the same. Thus, one would be more accurate to regard the the deflection is a property not of the state of the system, but of the dynamic operators which define the system and the measuring apparatus. One is not measuring a property of the state of the system, namely the expectation value of some operators. Rather one is measuring properties of the dynamical operators. It is they, not the wave function which have ontological property. The wave function, in their analysis, plays its usual epistemological role in telling us that we are in an eigenstate, that the system has a certain definite energy. Just because we do not happen to know exactly what the Hamiltonian is that the system is in the eigenstate of, does not give the wave function any ontological property that it would not have if we knew what the Hamiltonian was.

Thus, I have shown that the analysis of AAV is crucially flawed in demonstrating

268

the reality of the wavefunction. If a system has an unknown Hamiltonian, and if one knows that the state of the system is in an eigenstate of that unknown Hamiltonian (ie, the wave function plays its usual epistemological role in designating knowledge we have about the system), then one can determine that eigenstate— ie that property of the unknown Hamiltonian– on that single system. However, if the system is in a completely unknown state, then they have not demonstrated that they can protect that state, and there is no measurement or sequence of measurements which will tell us what that state is. I.e., an unknown state is not measurable, and thus is not real or ontological.

ACKNOWLEDGMENTS

I would like to thank J. Anandan for sending me a copy of the paper before publication, and Leslie Ballentine for his helpful comments on previous versions of this manuscript. I would also like to thanks the Canadian Institute for Advanced Research and the Natural Science and Engineering Research Council for support while this work was being done.

REFERENCES

* This paper is a slightly expanded version of one submitted to Phys. Rev. A (June 30,1993)

[1] Y.Aharonov, J. Anandan, L. Vaidman Phys Rev A **47** 4616(1993)

[2] See for example the treatment of the Stern Gerlach experiment in D. Bohm *Quantum Mechanics* p593*ff* (Prentice Hall, Englewood Cliffs, N.J.)(1951) although he unnecessarily restricts the gradient in the field to lie in the same direction as the field itself. (For a discussion of the relation between the direction of polarisation actually measured in the Stern Gerlach experiment see M. Bloom, K. Erdman,Can J Phys **40**, 179(1962))

SUPERCOLLIDER GRAVITATIONAL EXPERIMENTS

J. WEBER
University of Maryland
College Park, Maryland 20742
and
University of California
Irvine, California 92717

ABSTRACT. The elementary particle interactions at supercolliders produce gravitational fields which may be observable.

1. Introduction

Gravitational radiation antennas have been operating for about thirty years. Over 100 coincident pulses were observed on the University of Rome and University of Maryland gravitational antennas during the Supernova 1987A rapid evolutionary period. Supernovae occur rarely and parameters such as the antenna orientation and source detector distance cannot be varied. Experiments at supercolliders can be carried out in a controlled way for relatively long periods.

Very considerable controversy has been associated with predictions and observations, for gravitational radiation. In this paper we will employ the widely accepted theory to compute observable forces. If such forces are observed, these will provide tests of the theory in new regimes of energy and time scales, and the observations needed to construct theories which give a unified description of gravitation and the other forces.

2. Expected Forces and Accelerations

A particle with mass m which is at rest at the point $\bar{r}'$ will have a gravitational potential for an observer at point $\bar{r}$ with

$$V(\bar{r}) = \frac{Gm}{|\bar{r} - \bar{r}'|} \tag{1}$$

according to the Newtonian gravitational theory. G is the gravitation constant 6.6×10^{-8} erg cms gm^{-1} in the C.G.S. system.

In modern elementary particle theory the gravitational interaction is described as associated with exchange or propagation of zero rest mass spin 2 gravitons. The

271

V. de Sabbata and H. Tso-Hsiu (eds.), Cosmology and Particle Physics, 271–278.
© 1994 *Kluwer Academic Publishers. Printed in the Netherlands.*

relativistic quantum theory requires a ten component field. The generalization of (1) is the second rank symmetric tensor[1] object

$$\varphi_{\mu\nu}(\bar{r}, t) = \frac{KGmU'_\mu U'_\nu}{c^2 U'_\alpha (r^\alpha - r'^\alpha)} \tag{2}$$

U'_μ is the four velocity of the particle. The repeated Greek index implies a sum over the three space coordinates and one time coordinate. Repeated Latin indices are summed only over the three space coordinates.

$$r^\alpha - r'^\alpha = r^i - r'^i, \quad c(t - t'); \quad t' = t - \frac{|\bar{r} - \bar{r}'|}{c} \tag{3}$$

K is a constant. In the appendix it is shown that if K is chosen to be 4, the theory given here is the widely accepted weak field approximation to Einstein's General Theory of Relativity. Time and space components of (2) are written in terms of the particle velocity v' as

$$\varphi_{ij}(\bar{r}, t) = \frac{4Gmv'_i v'_j}{c^4 |\bar{r} - \bar{r}'| \sqrt{1 - \frac{v'^2}{c^2}} \left(1 - \frac{v'_i |r^i - r'^i|}{c |\bar{r} - \bar{r}'|}\right)} \tag{4}$$

$$\varphi_{0j}(\bar{r}, t) = -\frac{4Gmv'_j}{c^3 |\bar{r} - \bar{r}'| \sqrt{1 - \frac{v'^2}{c^2}} \left(1 - \frac{v'_i |r^i - r'^i|}{c |\bar{r} - \bar{r}'|}\right)} \tag{5}$$

$$\varphi_{00}(\bar{r}, t) = \frac{4Gm}{c^2 |\bar{r} - \bar{r}'| \sqrt{1 - \frac{v'^2}{c^2}} \left(1 - \frac{v'_i |r^i - r'^i|}{c |\bar{r} - \bar{r}'|}\right)} \tag{6}$$

To compute the applied forces we will employ the formalism of General Relativity Theory.

The General Relativity field variables are the components of the metric tensor $g_{\mu\nu}$. The four dimensional squared interval is

$$-dS^2 = g_{\mu\nu} dx^\mu dx^\nu \tag{7}$$

For weak gravitational fields

$$g_{\mu\nu} = \delta_{\mu\nu} + h_{\mu\nu} \tag{8}$$

In (8) $\delta_{\mu\nu}$ is the Lorentz metric and $h_{\mu\nu}$ is a first order quantity. The $\varphi_{\mu\nu}$ of equations (2) − (6) are related to $h_{\mu\nu}$ by

$$\varphi^\nu_\mu = h^\nu_\mu - \frac{1}{2} \delta^\nu_\mu h \tag{9}$$

$h_{\mu\nu}$ is related to h_μ^ν by $h_\mu^\nu = g^{\nu\alpha} h_{\alpha\mu}$, and h is the trace given by the sum

$$h = h_\alpha^\alpha \tag{10}$$

$\delta_\alpha^\alpha = 4$. The trace $\varphi_\alpha^\alpha = \varphi$, is related to h, according to (9) and (10) as

$$\varphi = -h \tag{11}$$

Suppose now that we have particles moving in a circular orbit shown in Figure 1.

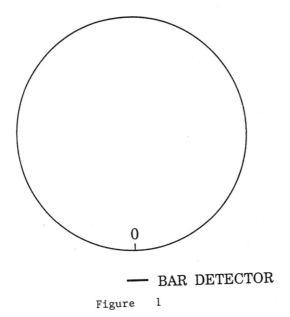

0

— BAR DETECTOR

Figure   1

We will calculate the transverse fields from the particles at 0, exerted on the particles in an appropriate detector.

For particles moving with velocities close to the velocity of light, (4) and (6) imply

$$\varphi_{xx} \approx \varphi_{00} \tag{12}$$

$$\varphi_0^0 = -\varphi_x^x, \qquad \varphi = 0 \tag{13}$$

$$\varphi_{00} \approx h_{00} \qquad \varphi_{11} \approx h_{11} \tag{14}$$

The higher derivatives of the potentials induce forces similar to the tidal forces of the sun and moon on the earth.

The covariance of General Relativity gives equations valid in arbitrary coordinates. Interpretation of the equations, and calculation of the order of magnitude of possible observables are facilitated by use of coordinates which differ only slightly from the coordinates of the Special Theory of Relativity. We choose therefore

$$x^\mu = 0 \tag{15}$$

$$g_{\mu\nu} = \delta_{\mu\nu} \tag{16}$$

at the center of mass of our detector (Figure 1). Also at the center of mass of our detector all first derivatives of $g_{\mu\nu}$ are taken to be zero.

Consider now the internal motion of a mass quadrupole oscillator consisting of two masses with separation vector ℓ^i. Let ξ^x be the relative displacement of one mass with respect to the second mass for the orientation implied by Figure 1. The equation of motion for ξ^x is then

$$\frac{d^2\xi^x}{dt^2} + \frac{Dd\xi^x}{dt} + k\xi^x = -c^2 R^x_{oio}\ell^i = -\frac{c^2}{2}\left(\frac{2\partial^2\varphi_{ox}}{c\partial x\partial t} - \frac{\partial^2\varphi_{00}}{\partial x^2} - \frac{\partial^2\varphi_{xx}}{c^2\partial t^2}\right)|\ell^i| \tag{17}$$

R^x_{oio} in (17) is the Riemann tensor defined in the appendix. Annihilation of an electron positron pair to give a Z^0 leads to a value for $\frac{\partial^2\varphi_{xx}}{\partial t^2}$ much larger than other terms on the right of (17). (4) gives

$$\frac{\partial^2\varphi_{xx}}{c^2\partial t^2} = \frac{3Gm}{c^2\,|\bar{r} - \bar{r}'|}\left(1 - \frac{v'^2}{c^2}\right)^{-5/2}\left(\frac{2v'dv'}{c^3dt}\right)^2 \tag{18}$$

In (18), the scalar product $v'_i(r^i - r'^i)$ has been set equal to zero for the orientation of Figure 1.

It is known that the entire annihilation and Z^0 creation occurs in a time dt_ω with

$$dt_\omega \approx \frac{\hbar}{\Delta E_{Z^0 \to e^+e^-}} \tag{19}$$

In (19) $\Delta E_{Z^0 \to e^+e^-}$ is the partial width of Z^0 for its production by e^+e^- annihilation. Since the Z^0 is at rest, $dv' \sim c$. With these assumptions (17) becomes

$$\frac{d^2\xi^2}{dt^2} + \frac{Dd\xi^x}{dt} + k\xi^x = \frac{6Gm}{c^2\,|\bar{r} - \bar{r}'|}\left(1 - \frac{v'^2}{c^2}\right)^{-5/2}\frac{\ell_i}{dt_\omega^2} \tag{20}$$

The right side of (20) differs from zero only during the very short time dt_ω, and no significant displacements or velocity change can occur over that short time. Therefore a good approximation is

$$\frac{d^2\xi^x}{dt^2} \approx \frac{6Gm}{c^2 \mid \bar{r} - \bar{r}' \mid} \left(1 - \frac{v'^2}{c^2}\right)^{-5/2} \frac{\ell_i}{dt_\omega^2} \tag{21}$$

There are no good reasons to doubt the approximations on which (20) is based. For LEP

$$\left(1 - \frac{v'^2}{c^2}\right) = 2 \times 10^5 \tag{22}$$

$$\Delta E_{Z^\circ \to e^+ e^-} \approx .085 GeV \tag{23}$$

This values give for (21), with $\mid \bar{r} - \bar{r}' \mid = 2000 \; cms \; \ell^i = 100 \; cms$,

$$\frac{d^2\xi^x}{dt^2} \approx 2.4 \times 10^{17} \; cms \; sec^{-2} \tag{24}$$

Considerably greater effects over very small volumes are predicted for the experimental arrangement shown in Figure 2. Now the quantity

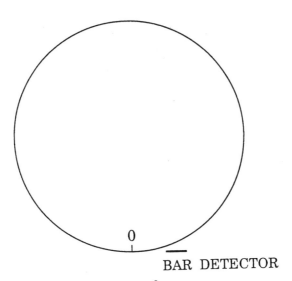

BAR DETECTOR

Figure   2

$$\frac{v_i'\left(r^i - r'^i\right)}{\left(\bar{r} - \bar{r}'\right)} \approx c$$

and

$$\frac{d^2\xi^x}{dt^2}\bigg|_{\text{max longitudinal}} = \frac{60Gm}{c^2\left|r - r'\right|}\left(1 - \frac{v'^2}{c^2}\right)^{-7/2}\left(\frac{dv'}{cdt_w}\right)^2 |\ell_i| \tag{25}$$

For LEP (25) is evaluated, again for an electron positron pair, $|r - r'| \approx 2000$

$$\frac{d^2\xi^x}{dt^2} \approx 10^{29} cms\ sec^{-2} \tag{26}$$

the condition

$$1 - \frac{v_i\left|r^i - r'^i\right|}{c\left|\bar{r} - \bar{r}'\right|} \approx 1 - \frac{v}{c}$$

is approximately valid over a solid angle $\Omega \approx 10^{-10}$ radians.

Therefore only a very small fraction of a finite mass detector might have acceleration (26) per electron positron pair.

The unusually large values (24) and (26) for a single electron positron annihilation 20 meters from the detector give strong encouragement for an experimental program.

(19) implies a time duration $\approx 7 \times 10^{-24}$ seconds. However the very large value (24) has a much smaller time, for which (22) is approximately valid. The primary objective of this paper is to invite attention to the very large values (24) and (26). A number of possible detectors are being studied for operation at such unusually small excitation times.

3. Conclusion

The LEP supercollider electron positron annihilations produce intense gravitational fields over very short times, which may be observable.

References

1. J. Weber, General Relativity and Gravitational Waves, Interscience Publishers Ltd., New York, London 1961.

4. Mathematical Appendix

4.1 General Relativity Theory

As noted earlier the four dimensional squared interval for two events separated by the four vector dx^μ is

$$-ds^2 = g_{\mu\nu} dx^\mu dx^\nu \tag{A1}$$

with repeated indices summed over the one time and three space coordinates. In Special Relativity

$$g_{\mu\nu} = \begin{vmatrix} -1 & 0 & 0 & 0 \\ 0 & 1 & 0 & 0 \\ 0 & 0 & 1 & 0 \\ 0 & 0 & 0 & 1 \end{vmatrix} \tag{A2}$$

For the General Theory of Relativity the $g_{\mu\nu}$ are functions of the coordinates. The Christoffel symbols $\Gamma^\mu_{\alpha\beta}$ are defined by

$$\Gamma^\mu_{\alpha\beta} = \frac{1}{2} g^{\mu\kappa} \left(\frac{\partial g_{\kappa\alpha}}{\partial x^\beta} + \frac{\partial g_{\kappa\beta}}{\partial x^\alpha} - \frac{\partial g_{\alpha\beta}}{\partial x^\kappa} \right) \tag{A3}$$

In (3) the $g^{\mu\nu}$ is defined by

$$g^{\mu k} g_{k\alpha} = \delta^\mu_\alpha \tag{A4}$$

The Riemann tensor $R^\mu_{\alpha\beta\gamma}$ is defined by

$$R^\mu_{\alpha\beta\gamma} = \frac{\partial \Gamma^\mu_{\alpha\gamma}}{\partial x^\beta} - \frac{\partial \Gamma^\mu_{\alpha\beta}}{\partial x^\gamma} + \Gamma^\sigma_{\alpha\gamma} \Gamma^\mu_{\sigma\beta} - \Gamma^\sigma_{\alpha\beta} \Gamma^\mu_{\sigma\gamma} \tag{A5}$$

The Ricci tensor $R_{\alpha\gamma}$ is defined by

$$R_{\alpha\gamma} = R^\mu_{\alpha\mu\gamma} \tag{A6}$$

The curvature scalar R is defined by

$$R = g^{\alpha\gamma} R_{\alpha\gamma} \tag{A7}$$

We may write Einstein's equations if some additional quantities are introduced. Let G be Newton's constant of gravitation and let $T_{\mu\nu}$ be the stress energy tensor of all matter fields except the gravitational field. Einstein's equations are

$$R_{\mu\nu} - \frac{1}{2} g_{\mu\nu} R = \frac{8\pi G}{c^4} T_{\mu\nu} \tag{A8}$$

The weak field solutions are obtained, writing

$$g_{\mu\nu} = \delta_{\mu\nu} + h_{\mu\nu} \qquad (A9)$$

$\delta_{\mu\nu}$ is the Lorentz metric and $h_{\mu\nu}$ is a first order quantity. New variables $\varphi_{\mu\nu}$ are defined by

$$\varphi_{\mu\nu} = h_{\mu\nu} - \frac{1}{2}\delta_{\mu\nu}h \qquad (A\,10)$$

In (A10) h is the trace of h_μ^ν, with

$$h_\mu^\nu = \delta^{\alpha\nu}h_{\alpha\mu} \qquad (A\,11)$$

Coordinates may be chosen such that

$$\phi_{\mu,\nu}^\nu = 0 \qquad (A\,12)$$

With these definitions and approximations Einstein's equations become

$$\Box\varphi_{\mu\nu} = -\frac{-16\pi G}{c^4}T_{\mu\nu} \qquad (A\,13)$$

If only gravitational forces are present, particles with spin zero move along geodesics. The position vector x^μ is described by

$$\frac{d^2x^\mu}{ds^2} + \Gamma_{\alpha\beta}^\mu\frac{dx^\alpha}{ds}\frac{dx^\beta}{ds} = 0 \qquad (A\,14)$$

For an extended body the center of mass may move along geodesics. The higher derivatives of $g_{\mu\nu}$ induce tidal forces - which cause relative displacements of particles of the body from each other and the center of mass. If we have a simple harmonic oscillator composed of two masses with equilibrium spacing defined by the vector r^β, coordinates may be chosen to give a relatively simple equation of motion for the relative displacement. For a Lorentz frame at the center of mass point, the relative displacement vector ξ^μ satisfies the equation

$$c^2\frac{d^2\xi^\mu}{ds^2} + D_\alpha^\mu\frac{d\xi^\alpha}{ds} + \kappa_\alpha^\mu\xi^\alpha = -c^2R_{o\beta o}^\mu r^\beta \qquad (A\,15)$$

New theoretical results of $2\nu\beta\beta$ decay with the Operator Expansion Method

X.R. Wu, M. Hirsch, A. Staudt and H.V. Klapdor-Kleingrothaus

Max-Planck-Institut für Kernphysik, D-6900 Heidelberg, Germany

and

Ching Cheng-rui * and Ho Tso-hsiu *

Centre for Theoretical Physics, CCAST (World Laboratory) and
Institute of Theoretical Physics, Academia Sinica, Beijing, China

Abstract   The half-lives for $2\nu\beta\beta$ decay for all potential $\beta\beta$-emitters with A$\geq$ 70 are calculated by the operator expansion method. Compared with the directly measured half-lives of ^{76}Ge, ^{82}Se, ^{100}Mo and ^{238}U the theoretical values are in excellent agreement with experimental ones.

Recently, more than half a century after Goeppert-Mayer first discussed $\beta\beta$ decay under Wigner's suggestion in 1935[1], direct counter experiments reported the observation of $2\nu\beta\beta$ decay for the nuclei ^{76}Ge [2,3], ^{82}Se [4], ^{100}Mo [5,6] and ^{238}U [7]. These measurements now supplement the earlier geochemical experiments for the $\beta\beta$ candidates ^{82}Se, ^{128}Te and ^{130}Te [8-11].

For the more interesting $0\nu\beta\beta$ decay mode, however, no evidence has been found so far, and only lower limits for the half-lives have been quoted in the literature [2-7]. At present new experiments like that of the Heidelberg-Moscow collaboration have been started[12,13], in a new attempt to look for $0\nu\beta\beta$ decay with increased sensitivity.

The amount of information which can be extracted from such experiments depends in a decisive way on the reliability of the theoretical estimates for the $0\nu\beta\beta$ decay matrix elements. The calculation of $2\nu\beta\beta$ decay is a valuable test for the accuracy which can be expected for the $0\nu\beta\beta$ transition matrix elements, since both calculations involve essentially the same nuclear physics. Here, we would like to report the calculation of $2\nu\beta\beta$ decay half-lives for all potential $\beta\beta$-emitters with $A \geq 70$, within a new model, the operator expansion method.[14-15]

The calculation of the $\beta\beta$ nuclear transition matrix elements between the initial even-even parent nucleus (A, Z) and the final state in the daughter nucleus $(A, Z + 2)$ includes

---

* Project supported partly by the Natural Science Foundation of China

V. de Sabbata and H. Tso-Hsiu (eds.), Cosmology and Particle Physics, 279–287.
© 1994 Kluwer Academic Publishers. Printed in the Netherlands.

a sum of - in principle - infinite number of intermediate states in the adjacent odd-odd nucleus $(A, Z + 1)$. The determination of these intermediate states is a difficult task and their treatment in shell model calculations[16,17] as well as calculations within the usual quasiparticle random-phase approximation (QRPA)[18-20] shows some weakness. As is well known, in shell model calculations the energies of the intermediate states are replaced by an average value (closure approximation) which has been shown to be doubtable for $2\nu\beta\beta$-decay. In QRPA the matrix elements for $2\nu\beta\beta$ decay are very sensitive to the particle-particle interaction parameter g_{pp} when g_{pp} is near its physical value of $g_{pp} = 1$. Therefore, recently an alternative method – the Operator Expansion Method (OEM), which can treat the sum over the infinite intermediate states in a more elegant way, has been proposed by Ching and Ho and applied to the calculation of ^{48}Ca, ^{76}Ge, ^{82}Se, ^{100}Mo and ^{130}Te with rather promising results[14-15]. Encouraged by these successes and as a further application of OEM, we have calculated the $2\nu\beta\beta$-decays half-lives for all possible $\beta\beta$-emitters with A$\geq$ 70.

If one assumes that the sum of the energies for each pair of emitted electron and neutrino can be replaced by the average value of $\Delta = (E_I - E_F)/2 = \frac{1}{2}Q_{\beta\beta} + m_e$, (for a discussion see for example ref.[21]) then, as is well-known, the half-life for $0^+ \rightarrow 0^+$ $2\nu\beta\beta$ decay can be expressed in a factorized form as

$$[T_{1/2}^{2\nu}]^{-1} = F^{2\nu}|M_{GT}|^2 \tag{1}$$

where $F^{2\nu}$ is a lepton phase-space integral. The matrix element is given by

$$M_{GT}^{2\nu} = \sum_N \frac{\langle 0_F^+|A^\alpha|1_N^+\rangle\langle 1_N^+|A^\alpha|0_I^+\rangle}{\Delta + (E_N - E_I)}. \tag{2}$$

$M_{GT}^{2\nu}$ of eq.(2) is a special case for $0^+ \rightarrow 0^+$ transition. The basic idea of OEM now is to transfer the energy denominator to the numerator by the following mathematical procedure. We take instead of $M_{GT}^{2\nu}$ the following matrix element

$$M_{GT}(Z) = \frac{1}{2}\sum_N \left\{ \frac{\langle F|A^\alpha|N\rangle\langle N|A^\beta|I\rangle}{\Delta + (E_N - E_I)Z} - \frac{\langle F|A^\beta|N\rangle\langle N|A^\alpha|I\rangle}{\Delta + (E_F - E_N)Z} \right\}, \tag{3}$$

where $|I\rangle$, $|N\rangle$ and $|F\rangle$ denote the wave functions of the initial, intermediate and final nuclear states, respectively and E_I, E_N, and E_F are the corresponding energies and Z is a complex variable. Mathematically $M_{GT}(Z)$ is a single-valued, regular function of Z in an open, connected region containing the origin and Z = 1, except some possible poles along the real axis. For small Z $M_{GT}(Z)$ can be expanded using the binomial theorem:

$$\frac{1}{\Delta + (E_N - E_I)Z} = \frac{1}{\Delta}\left\{ 1 - \frac{Z(E_N - E_I)}{\Delta} + \frac{Z^2(E_N - E_I)^2}{\Delta^2} - \cdots \right\}. \tag{4}$$

Then, after introducing the nuclear Hamiltonian H_S the following $M_{GT}(Z)$ is obtained

$$
M_{GT} = \frac{1}{2\Delta} \Big\langle F \Big| \Big\{ [A^\alpha, A^\beta] - \frac{Z}{\Delta}[A^\alpha, [H_S, A^\beta]]
$$
$$
+ \frac{Z^2}{\Delta^2} \Big[A^\alpha, [H_S, [H_S, A^\beta]]\Big] - \cdots \Big\} \Big| I \Big\rangle, \tag{5}
$$

where the summation over the intermediate states has been carried out. It is important to note that eq.(5) is an exactly equivalent formulation of the matrix element of eq.(3). On the other hand, mathematically eq.(5) is a divergent series for $2\nu\beta\beta$ with $Z = 1$ and thus one has to sum up all terms up to infinity for the calculation of the matrix element (see for example G.M. Hardy, in: Divergent Series [22]).

For the summation of the infinite series, however, we have to assume a specific form of the nuclear Hamiltonian, and we take

$$
H_S = \langle H_0 \rangle + V_S \tag{6}
$$

where

$$
V_S = \frac{1}{2} \sum_{i \neq j} \Big\{ v_0(r_{ij}) + v_\tau(r_{ij})\tau_i \cdot \tau_j
$$
$$
+ v_\sigma(r_{ij})\sigma_i \cdot \sigma_j + v_{\sigma\tau}(r_{ij})\sigma_i \cdot \sigma_j \tau_i \cdot \tau_j \Big\} \tag{7}
$$

which is the most general central static force. $v_0(r_{ij})$, $v_\sigma(r_{ij})$, $v_\tau(r_{ij})$ and $v_{\sigma\tau}(r_{ij})$ denote the radial parts of the corresponding interactions, respectively, and $r_{ij} = |\mathbf{r}_i - \mathbf{r}_j|$ stands for the relative coordinates of the nucleons.

The assumption that the unperturbed nuclear Hamiltonian can be replaced by its average value $\langle H_0 \rangle$, which is constant and thus does not contribute to the commutator series (5), is the main approximation of the present OEM. The accuracy of this approximation has still to be checked. However, one can argue [23] that for the case of a small model space H_0 is nearly degenerate, i.e. $\langle H_0 \rangle - H_0$ is small compared with V_S and thus the approximation should be reasonable.

Adopting this approximation, and after a fair amount of algebraic operations, one can sum up the infinite series (5) and simplify the $0^+ \to 0^+$ transition matrix element (2) into

$$
M_{GT} \equiv \Big\langle 0_F^+ \Big| \sum_{i \neq j} \mathcal{M}_{ij} + \sum_{i \neq j \neq k} \mathcal{M}_{ijk} + \cdots \Big| 0_I^+ \Big\rangle, \tag{8}
$$

where there are totally $(A - 1)$ operators from 2-body operator $\mathcal{M}_{ij}$ to A-body operator.

Finally for the two-body operator OEM gives the matrix element M_{GT} as the analytical continuation $M_{GT} = M_{GT}(Z)|_{Z=1}$,

$$\mathcal{M}_{ij}(Z) = \frac{12\big(v_\sigma(r) - v_\tau(r)\big)Z}{\Delta^2 - 16Z^2\big(v_\sigma(r) - v_\tau(r)\big)^2}\,\Omega_0(ij) + \frac{4Z\big(2v_{\sigma\tau}(r) - v_\sigma(r) - v_\tau(r)\big)}{\Delta^2 - 16Z^2\big(2v_{\sigma\tau}(r) - v_\sigma(r) - v_\tau(r)\big)^2}\,\Omega_1(ij)$$

(9)

where we have introduced the spin singlet and spin triplet operators

$$\Omega_1(ij) = \frac{3 + \sigma_i \cdot \sigma_j}{4}\,, \qquad \Omega_0(ij) = \frac{1 - \sigma_i \cdot \sigma_j}{4}.$$

(10)

The mathematical procedure just outlined is the so-called Euler's method of summation. We note that Simkovic and Gmitro [24] starting from the time-ordered product of two operators rederived the OEM result for the two-body operator (9). They obtained exactly the same expression for $\mathcal{M}_{ij}$.

In principle, all $A - 1$ terms can be given a compact form, but due to the technical difficulty to calculate the matrix elements of the many-body operators, only the two-body terms are taken into account. This is our second approximation. We expect this approximation to be reasonable, since for $\beta\beta$ decay the transition operator is of two-body character. However, we plan to investigate the influence of the three-body terms in a future publication.

We note that if $v_\sigma = v_\tau = v_{\sigma\tau}$ in the effective interaction (7), the Hamiltonian (7) would be of [0] irreducible tensor form of Wigner's SU(4) multiplet, and the two-body operator $\mathcal{M}_{ij}$ in eq (9) is exactly zero. The nonzero contributions come from differences of the interaction strengths v_σ, v_τ and $v_{\sigma\tau}$. We note that in ref.[25], a $H' = V(\tau_i \cdot \tau_j - \sigma_i \cdot \sigma_j)$ SU(4) symmetry-breaking Hamiltonian was taken into account and the M_{GT} given there by use of the SU(4) algebra perhaps corresponds to the contribution of the $\Omega_0(ij)$ term (the first term) in equation (9). But the Hamiltonian used in the present work is more realistic than those used in the SU(4) algebra techniques and thus more effects of SU(4) symmetry-breaking are picked up here.

To calculate the $2\nu\beta\beta$ matrix element (8) we only need to know the effective interactions for evaluating $\mathcal{M}_{ij}$ and the ground state wave functions of the initial and final states. This is an delighting advantage of the present approach.

The operator $\mathcal{M}_{ij}$ can be determined by the effective interaction, for which we take the Paris-potential [26]. In principle, one should take the eigenfunctions of H_S with the correct eigenvalues E_I and E_F, respectively. However, since most of the potential $\beta\beta$-emitters are located far away from closed shells, it is impossible at present to use shell model wave functions for the initial and final states. Thus, for the determination of $|0_I^+\rangle$ and $|0_F^+\rangle$ QRPA wave functions are used in the present work. For a description of the QRPA we refer to the original literature for brevity[18-20], for the combination of OEM and QRPA wavefunctions to [15].

We describe the initial $|0_I^+\rangle$ and final state $|0_F^+\rangle$ by the QRPA vacuum based on BCS states. Then we insert two sets of complete and orthogonal mathematical functions $|a\rangle$ and $|b\rangle$ into eq.(9). We construct these two sets of mathematical functions by the phonon operators acting on the QRPA vacuum, these phonons consisting of the quasiparticle proton-neutron-pair. Then we can directly use the QRPA techniques to calculate the transition matrix element M_{GT}.

The main advantage of the present approach lies in the fact that the calculated matrix element M_{GT} is only weakly sensitive to the choice of the particle-particle interaction parameter g_{pp}. This was demonstrated , for the example, in ref.[15] for ^{100}Mo. The similar behaviour of the matrix element holds also for ^{238}U. The OEM leads to matrix elements which are always smaller than the QRPA calculation for $g_{pp} = 0$, but do not exhibit a strong dependence on g_{pp}.

It is a natural feature of the present approach that M_{GT} is insensitive to g_{pp}. Recall that the two-body operator $\mathcal{M}_{ij}$ in eq.(9) is not explicitly dependent on the intermediate energy spectrum and that the wave functions of the intermediate states are only used as a complete and orthogonal set of mathematical functions. The numerical results of M_{GT} should only care of whether these two sets of functions are complete and orthogonal, and thus $M_{GT}^{2\nu}$ should be constant no matter how g_{pp} changes. On the other hand, our present calculation uses the QRPA vacuum for the description of the initial $|0_I^+\rangle$ and final state $|0_F^+\rangle$ and these depend on g_{pp}. However, the QRPA ground state is

$$|QRPA\rangle = |BCS\rangle + YX^{-1}|BCS\rangle + \cdots. \tag{11}$$

Clearly, the $|BCS\rangle$ state does not depend on g_{pp} and since in QRPA the coefficient necessarily fulfills the condition $YX^{-1} << 1$, in the first order, the QRPA vacuum and thus our calculated M_{GT} should only weakly depend on g_{pp}. On the other hand, in usual QRPA calculations the wave functions of the intermediate states should be the real wave functions of the odd-odd nucleus. The energy eigenvalues of these intermediate states depend on g_{pp}, and consequently in usual QRPA the results are more sensitive to g_{pp}.

We think that the strong dependence of the $2\nu\beta\beta$ decay half-lives on g_{pp} is a major disadvantage of QRPA calculations, as we would like to discuss for the special case of ^{238}U. As the same as in the case of ^{100}Mo, the $2\nu\beta\beta$ matrix element crosses zero for a certain value of g_{pp}, translating into an infinite half-life. For the parameter choice of [20] one obtains $T_{1/2}^{2\nu} = 1.5 \cdot 10^{23} y$, which is a factor of 100 larger than the experimental result of [7] but could easily be adapted to the experimental value by a small change of g_{pp}. On the other hand, the OEM result of $T_{1/2}^{2\nu} = 0.9 \cdot 10^{21} y$ is relatively stable against variations on g_{pp}, and also in good agreement with the experimental data.

Since the results of the present approach are not sensitive to a specific choice of the parameters g_{ph} and g_{pp} we can use the physical values $g_{pp} = g_{ph} = 1$ directly. However, in some cases (as indicated in table 1) the QRPA equation collapses for values below $g_{pp} = 1$. For these isotopes the g_{pp} value before the collapse points are used. The calculated matrix elements M_{GT} and corresponding half-lives for all potential $\beta\beta$-emitters are given in table 1. The half-lives given are for $g_A/g_V = -1.254$ and the phase-space factors of ref [21]. Note, however, that for the heaviest isotopes the phase space factor of [21] differs from the one used in the QRPA calculation of [20] by a factor of ~ 3. On the other hand, for medium heavy isotopes, such as ^{76}Ge, the phase factors of [21] and [20] are essentially equal.

For a comparison we listed also the experimental data available to us. For all nuclides which have been measured by direct counter experiments (^{76}Ge, ^{82}Se, ^{100}Mo and ^{238}U), the theoretical half-lives are in quite well agreement with the experimental ones. However for the Te isotopes our calculation gives a larger decay rate than the geochemical experiments.

To summarize, we have calculated $2\nu\beta\beta$ decay half-lives for all potential $\beta\beta$ emitters with $A \geq 70$. Our results are not sensitive to a specific choice for the particle-particle interaction parameter g_{pp}. The present approach thus overcomes problems of earlier QRPA calculations.

Acknowledgement

We thank Dr. K. Muto for many helpful discussions and for his QRPA code.

REFERENCES

[1] M. Goeppert-Mayer, Phys. Rev. 48, 512 (1935)

[2] H.S. Miley, F.T. Avignone, III, R.L. Brodzinski, J.I. Collar and J.H. Reeves, *Phys. Rev. Lett.* **65** (1990) 3092;

[3] F.T. Avignone III et al., *Phys. Lett.* **B256** (1991)559

[4] S.R. Elliott, A.A. Hahn, and M.K. Moe, *Phys. Rev. Lett.* **59**, (1987)2020.

[5] H. Ejiri et al., *Phys. Lett.***B258** (1991)17.

[6] K. Okada et al, Nucl Phys B (Proc. Suppl.) 19(1991)

[7] A.L. Turkevich, T.E. Economou and G.A. Cowan, *Phys. Rev. Lett.* **67** (1991)3211

[8] T. Kirsten, W. Gentner and O.A. Schaeffer, Z. Phys. 202 (1967) 273.

[9] T. Kirsten, Proc. Int. Symp on Nuclear Beta Decays and Neutrino, June 1986, Osaka, eds. T. Kotani, H.Ejiri and E. Takasugi (World Scientific, Singapore, 1986) p.81

[10] O.K. Manuel, as ref.[9] but p.103

[11] W.J. Lin et al., Nucl. Phys. A481 (1988) 477

[12] H.V. Klapdor-Kleingrothaus, in: Proc. Second Int. Workshop on Theoretical and Phenomenological Aspects of Underground Physics (TAUP 91), Sep. 9-13 1991, Tolédo Spain, in press.

[13] M. Beck, J. Bockholt, J. Echternach, G. Heusser, H.V. Klapdor-Kleingrothaus, B. Maier, F.Petry, A. Piepke, U. Schmidt-Rohr, H. Strecker, R. Zuber, A. Balysh, S.T. Belyaev, A. Demehin, A. Gurov, I. Kondratenko, V.I. Lebedev, submitted to Phys. Lett. B.

[14] C.R. Ching and T.H. Ho, Commun. Theor. Phys. 11, (1989)433
C.R. Ching, T.H. Ho and X.R. Wu, Phys. Rev. C40, (1989)304; Commun. Theor. Phys. 12, (1989)167; in Proceeding of the BIMP Symposium on Heavy Flavor Physics, Eds. Chao Kuangta et al, World Scientific, Singapore, 1989, p359

[15] X.R. Wu, S. Staudt, H.V. Klapdor-Kleingrothaus, C.R. Ching and T.H. Ho, Phys. Lett. **B272** (1991) 169

[16] W.C. Haxton and G.J. Stephenson, Jr., Progress in Particle and Nuclear Physics 12, (1984)409.

[17] J.D. Vergados, Phys. Rep. 133 (1986)1

[18] P. Vogel and M.R. Zirnbauer, Phys. Rev. Lett. 57, (1986)3148.
O. Civitarese, A. Faessler and T. Tomoda, Phys. Lett. B194, (1987)11.
K. Muto and H.V. Klapdor, Phys. Lett. B201, (1988)420
T. Tomoda, Rep. Prog. Phys. 54, (1991)53.

[19] H.V. Klapdor and K. Grotz, Phys. Lett. B142 (1984)323 ; K. Grotz and H.V. Klapdor, Phys. Lett. B153 (1985)1, Phys. Lett. B157 (1985)242 and Nucl. Phys. A460 (1986)395

[20] A. Staudt, K. Muto and H.V. Klapdor-Kleingrothaus, Europhys. Lett. 13 (1990)31
A. Staudt, T.T.S. Kuo and H.V. Klapdor-Kleingrothaus, Phys. Lett. B242 (1990)17

K. Muto, E. Bender and H.V. Klapdor, Z. Phys. A334 (1989)177 and 187

[21] M. Doi, T. Kotani and E. Takasugi, Prog. Theor. Phys. Suppl. 83, (1985)1

[22] G.M. Hardy, Divergent Series, Clarendon, Oxford, 1949

[23] X.R. Wu, S. Staudt, T.T.S. Kuo and H.V. Klapdor-Kleingrothaus, Phys. Lett. (in press)

[24] F. Simkovic, JINR Rapid Commun. 39 (1989)21 and M. Gmitro and F. Simkovic, Izv. AN SSSR 54 (1990)1780.

[25] J. Bernabeu et al., Z. Phys. C46 (1990) 792.

K.T. Hecht, S.C. Pang, J. Math. Phys. 8 (1969) 1571

[26] M. Lacombe et al. Phys. Rev. C21, (1980) 861.

[27] A.S. Barabash, V.V. Kuzminov et al, Sov. J. Nucl. Phys. 51 (1990)1

[28] E. Bellotti et al, Phys. Lett 266B (1989)209

[29] A.A. Klimenko et al, Nucl. Instrum. and Meth. B16 (1986) 446

Table 1

Nuclei	M_{GT}	Half life (yrs)	Experimental half life (years)
^{70}Zn	0.0918†	1.44×10^{24}	
^{76}Ge	0.3351	2.61×10^{20}	$(1.1^{+0.6}_{-0.3}) \times 10^{21}$ [a], $(9.2^{+0.7}_{-0.4}) \times 10^{20}$ [b]
^{80}Se	0.3424	2.68×10^{29}	
^{82}Se	0.1019	0.848×10^{20}	$(1.1^{+0.8}_{-0.3}) \times 10^{20}$ [c], $(1.30 \pm 0.05) \times 10^{20}$ [d]
^{86}Kr	0.0580	3.42×10^{23}	
^{94}Zr	0.0117†	1.68×10^{24}	
^{96}Zr	0.0314†	2.02×10^{20}	
^{98}Mo	0.0800	6.16×10^{30}	
^{100}Mo	0.1065†	3.58×10^{19}	$(1.15^{+0.3}_{-0.2}) \times 10^{19}$ [e], $(1.16^{+0.34}_{-0.22}) \times 10^{19}$ [f]
^{104}Ru	0.1162†	3.09×10^{22}	
^{110}Pd	0.0879†	1.24×10^{21}	
^{114}Cd	0.1642†	9.84×10^{24}	
^{116}Cd	0.0171†	1.57×10^{22}	
^{122}Sn	0.1712	1.25×10^{26}	
^{124}Sn	0.0391	1.49×10^{21}	
^{128}Te	0.1462	2.11×10^{23}	$> 5 \times 10^{24}$ [g], $(1.8 \pm 0.7) \times 10^{24}$ [h]
^{130}Te	0.1006	0.787×10^{20}	$(1.5 - 2.8) \times 10^{21}$ [d], $(7.5 \pm 0.3) \times 10^{20}$ [h]
^{134}Xe	0.1286	2.69×10^{23}	
^{136}Xe	0.0280	1.01×10^{21}	$> 8.4 \times 10^{19}$ [i], $> 1.6 \times 10^{20}$ [j]
^{142}Ce	0.0400	3.30×10^{22}	
^{146}Nd	0.3285†	7.31×10^{30}	
^{148}Nd	0.0548†	1.19×10^{21}	
^{150}Nd	0.0441	1.66×10^{19}	$\geq 1.8 \times 10^{19}$ [k]
^{154}Sm	0.0793	1.49×10^{22}	
^{160}Gd	0.0454	2.81×10^{21}	
^{170}Er	0.1685	2.46×10^{23}	
^{176}Yb	0.1593†	4.92×10^{21}	
^{186}W	0.1506†	1.30×10^{24}	
^{192}Os	0.1777	2.40×10^{24}	
^{198}Pt	0.0741	1.14×10^{22}	
^{204}Hg	0.0510	1.81×10^{25}	
^{232}Th	0.1263	4.03×10^{21}	
^{238}U	0.0785	0.914×10^{21}	$(2.0 \pm 0.6) \times 10^{21}$ [l]

R-W COSMOLOGICAL SOLUTION FOR POLYTROPES

ZHANG, ZHENJIU**, HUANG, HUANRAN*,
HUANG, XINTANG*, GUO, KEXIN* and YU, LIJUN*
**Centre For Relativity Studies
*Department of Physics
Central China (Huazhong) Normal University
Wuhan, 430070, P.R. China

ABSTRACT. Einstein's theory of general relativity has given modern physics a consistent and fruitful framwork in which to study cosmology.

Upon the works of 3-6, in the case of $K = \Lambda = 0$, we solve R-W cosmological evolution equations with polytropes field $P = \gamma \rho^m$ (where γ and m are constants), and finds the general expression of exact solution with ρ being a parameter. Finally, we disccusses the existing conditions of the conformal solution of the equations.

1. The Cosmological Principle

A large portion of modern cosmological theory is built on the cosmological principle, the hypothesis that our universe is spatially homogeneous and isotropic (averaged over cells of diameter 10^8 to 10^9 light years and over all of its history from observation). The cosmic microwave radiation appears to be highly isotropic and it has been shown that the universe cannot be isotropic about every point without also being homogeneous.

V. de Sabbata and H. Tso-Hsiu (eds.), Cosmology and Particle Physics, 289–299.

1.1. HOMOGENEITY

We define x^μ ($\mu = 0, 1, 2, 3$) as the cosmic standard coordinate system. A different set of coordinates x'^μ may be considered equivalent, if the coordinate transformation $x \to x'$ is an isometry, every cosmic field $g_{\mu\nu}, T_{\mu\nu}$, and so on, is form-invariant under this transformation and cosmic standard time $t = t$, is necessarily used.

We split spacetime (as a 4-manifold M) up into a family of 3-space -like submanifolds-hypersurfaces of constant time t. The metric tensor g of spacetime is positive-definite on all vectors tangent to a hypersurface. Let G be Lie group of isometries of the manifold M with metric tensor g. The Lie algebra of G is that of Killing vector fields of g. Elements of G are mappings of M onto itself (diffeomorphisms) . The manifold M is said to be homogeneous if its isometry group acts transitively on it. That is, the geometry is the same everywhere in M. In other words, every point x^μ in spacetime is on some fundamental trajectory $x^i = X^i(t)$, which fill up all space at any time t and are determined by three independent parameters $a^i \equiv X^i(t)$. Thus homogneity means that there is a three-parameter set of coordinates for origin on X^i (t; a)

$$y'^i(X(t, a), t; a) = 0. \qquad (1.1)$$

1.2. ISOTROPY

Suppose there are elements of G which leave some point P of M fixed. Then the product of any two also leaves P fixed, and since the identity e is one of them, they form a subgroup I of G called the isotropy group of P. These are the familiar rotations about an axis through P. It is just SO(4).

Then the assumption of spatial isotropy can be formulated as for the origin

$$x'^i_{\mu\nu}(0, t; \theta) = 0, \qquad (1.2)$$

where we can think of the three parameters θ as, for example, Euler angles that specify the orientation of the x'^i relative to the x^i.

1. 3. MAXIMAL SYMMETRY

It can be shown[1] that a maximally symmetric connected manifold is homogeneous and a maximally symmetric space- like manifold is isotropic. The isotropic group for space-like three -manifolds is SO(3), and we say the manifold is spherically symmetric about any point. The Killing vectors of SO(3) define spheres, with the usual θ and Φ on each sphere and a third 'radial' coordinate labelling spheres. The metric of M induces a metric tensor, which defines a volume two-form and a total area S (integral of the volume two-form). We define the radial coordinate r of a sphere by

$$r = (S/4\pi)^{1/2}. \tag{1.3}$$

For each r, we have a sphere S^2. At every point q, there is a vector n orthogonal to the sphere at q, $g(n, v) = 0$ for any tangent vector v in the tagent space $T_q(S^2)$, normalized to unity, $g(n, n) = 1$, and point away from q. This vector field is called the unit normal vector field. Choose the pole of any particular S^2 arbitrarily and then fix the poles of all the others by demanding they lie on the integral curve of n through the original pole. Therefor any integral curve of n is a coordinate line of radial coordinate with constant θ and ϕ.

Then for the spherecally symmetric three-space, we have

$$\text{diag}(g) = [f(r), r^2, r^2 \sin^2 \theta]. \tag{1.4}$$

For a spherecally symmetric manifold, the Lie algebra of its Killing vector fields has a subalgebra, which is the Lie algebra of SO(3). We define the space of functions $L^2(S^2)$ to be the Hilbert space of all complex-valued functions on S^2, which are squre-integrable over the usual area element of the sphere. Every irreducible subspace of $L^2(S^2)$ is characterized by an integer $1 > 0$ and has dimention $2l+1$. The spherical harmonics, $\{Y_{1m}, m = -1, \ldots, 1\}$, are basis for this subspace. The two sets of vector spherically harmonics Y^+_{1m} and Y^-_{1m} form a complete set for representing vectors on the two-sphere. We use them to construct all the Killing vector fields of S.

There are maximally six Killing vectors on S. Those vectors can be expressed by the spherical harmonics. The reqirement of satisfying Killing equations gives the solution for f(r). We have

$$\text{diag}(g) = [(1-Kr^2)^{-1}, r^2, r^2 \sin^2 0].$$ (1. 5)

K = (-1, 0, +1) corresponds to a closed (finite), flat and an open (infinite) universe respectively. This shows that the geometry depends only on the sigh of K.

2. Dynamics of a Homogeneous, Isotropic Universe

When Einstein's equatons (we include cosmological constant term for later use)

$$G_{\mu\nu} = R_{\mu\nu} - (1/2) R g_{\mu\nu} + \lambda g_{\mu\nu} = 8\pi T_{\mu\nu}$$ (2. 1)

are supplied with initial data which are homogeneous and isotropic (both for the geometry and matter variables), then the subsequent evolution of the universe maintains the symmetry. It follows that the only aspect of the geometry which can change with time is the scale factor K. A change in K produces a change in the distance between points, and this is meant by an expanding universe. All three kinds of universe begin with zero 'volume' (K = ∞) and expands away from this 'big bang'.

2. 1. STANDARD MODEL OF THE UNIVERSE AND ROBERTSON-WALKER METRIC

The metric can be written as[8a]

$$ds^2 = -d\tau^2 + R^2(\tau) \{dr^2/(1-Kr^2) + r^2(d\theta^2 + \sin^2\theta d\varphi^2)\},$$ (2. 2)

which is known in cosmology as Robertson-Walker metric, and R(t) is called the cosmic scale factor. The affine connection is

$$\Gamma^{\lambda}_{\mu\nu} = Kx^{\lambda} g_{\mu\nu}$$ (2. 3)

and the differential equation for a geodesic is

$$d^2 x^{\mu} / d\tau^2 + K x^{\mu} = 0. \tag{2.4}$$

2.2 THE EVOLUTION OF OUR UNIVERSE

We can obtain a deeper insight into the behavior of matter in a Robertson-Walker universe by applying the Cosmological Principle to the tensor that describes the average state of cosmic matter. It can be proven[2] that the energy-momentum tensor of the universe necessarily takes the same form as for a perfect fluid:

$$T_{\mu \nu} = (\rho + p) u_{\mu} u_{\nu} + p g_{\mu \nu}, \tag{2.5}$$

where ρ is the (average) mass density of matter and radiation; while p is the pressure.

We can compute $R_{\mu \nu}$, R and therefore $G_{\mu \nu}$ from the metric, equation (2.2), and equate it with $8\pi T_{\mu \nu}$, equation (2.5). Then we obtain the solutions of Einstein's equations (2.1) for $\lambda = 0$.

The first striking result is that the universe cannot be static, provided only that $\rho > 0$ and $p \geqslant 0$[3].

3. Robertson-Walker Cosmological Solutions With Different State Equations

3.1. INTRODUCTION

The solutions of Einstein's equations have been studied under Robertson-Walker metric with perfect fluid source, for the state equation $\rho = 3P$, and K = 0. The conformally flat solution of Einstein's equations has been gotten by Z. Zhang[4] and by D. Song, D. Li and Z. Zhang[5] under the assumptions that the source of the gravitational field is perfect fluid obeying the equation of state $\rho = nP$ ($n \geqslant 3$), and that Weyl-tensor vanishes, i.e. $g_{\mu \nu} = e^{2B} \eta_{\mu \nu}$, where B = $B(b_{\mu} X^{\mu})$, b_{μ} - constant, and also K = 0. This is the case of cosmology, that the energy density of the universe is contributed by high relativistic particles. Of course, this implies that the fluid is rotation-free, shear-free and moving along geodesic. From the viewpoints of large scale, the stress-energy tensor for the universe can be that of the perfect fluid, the same as equation (2.5):

$$T_{\mu\nu} = (P + \rho)u_{\mu}u_{\nu} + P g_{\mu\nu}$$

where u_{μ} -- 4-velocity at the given event in spacetime, i. e. the average 4-velocity of galaxy near the event, also the velocity that the observer should move in order to measure the microwave background radiation to be isotropic; where ρ -density, including the energy density of matter(rest mass of galaxies plus negligible kinetic energy, rest and kinetic energy of cosmic ray, rest mass and thermal energy of galaxies gas) and radiation energy density (radiation of photons, neutrinors and gravitation); where P-pressure of both matter and radiation as above ρ. In the universe, for the radiation, always $\rho = 3P$, but for the matter of our today universe, $\rho >> P$.

Collins and Wainwright have considered solution of the gravitational field equations for an uncharged perfect fluid satisfying the equation of state $P = P(\rho)$, $P + \rho \neq 0$, which undergoes irrotational sheer-free expantion or contraction. They found that the only possible solutions are the Friedmann-Lemaitre-Robertson-Walker models, the Wyman solution, and new class of plane-symmetric solution[6]

J. Shi and Z. Zhang have generalized the conformal solution of Einstein's equations with perfect fluid source from $\rho = 3P$ into $\rho = nP$ when $K = \pm 1$. They have also presented the solution of the Einstein's equations with $\Lambda \neq 0$ and $K = 0$, when $\rho = nP$. Using the results, they discuss the radial null geodesic propagating in a R-W cosmology[7].

Here, upon the works before, in the case of $K = \Lambda = 0$, we solve the R-W cosmological evolution equations with polytropes field $P = \gamma \rho^{m}$ and finds the general expression of exact solution with ρ being a parameter. Finally, we disccusse the existing conditions of the conformal solution of the equations.

3. 2. SOLUTIONS UNDER R-W METRIC OF K = 0

Robertson-Walker metric, equation (2. 2) is

$$ds^2 = -d\tau^2 + R^2(\tau) \ \{dr^2/(1-Kr^2) + r^2(d\theta^2 + \sin^2\theta \, d\varphi^2)\}$$

where $R(\tau)$ satisfies the equations[8b]

$$3(dR/d\tau)^2 R^{-2} = 8\pi\rho - 3KR^{-2} + \Lambda \qquad (3.1)$$

$$3(d^2R/d\tau^2)R^{-1} = -4\pi(\rho + 3P) + \Lambda \qquad (3.2)$$

$$d\rho/d\tau + 3(\rho + P)(dR/d\tau)R^{-1} = 0 \qquad (3.3)$$

These are general evolution equations of R-W cosmology. We consider the polytropes fields

$$P = \gamma\rho^m. \qquad (3.4)$$

We revise (3. 1) and get

$$3(dR/d\tau)R^{-1} = (8\pi\rho - 3KR^{-2} + \Lambda) \qquad (3.5)$$

Substituting (3.4) and (3.5) into (3.3), we have

$$d\tau = \{-\rho(1 + \gamma\rho^{m-1})(8\pi\rho - 3KR^{-2} + \Lambda) \ \}^{-1} d\rho. \qquad (3.6)$$

In the case of $K = \Lambda = 0$, $m = 1$, we obtain

$$\rho = [(1+\gamma)\tau]^{-2}/(6\pi), \qquad (3.7)$$

$$\rho R^{3(1+\gamma)} = C_1 (\text{integral constant}). \qquad (3.8)$$

This is coincident with (23) of the work[5].
In the case of $K = \Lambda = 0$, $m \neq 1$ but an integer, we obtain

$$\rho R^3 (1 + \gamma\rho^{m-1})^{-1/(m-1)} = C_2 (\text{integral constant}), \qquad (3.9)$$

$$\tau = -(6\pi)^{-1/2} (A + B + C) \qquad (3.10)$$

296

where

$$A = -\rho^{-1/2} \tag{3.10a}$$

$$B = -\gamma^{1/2(m-1)} / (2m-2) \sum_{n=1}^{m-1} \{\cos \frac{\pi (2n-1)}{2 (m-1)}$$

$$\ln[1 - 2\gamma^{-1/2(m-1)} \rho^{-1/2} \cos \frac{\pi (2n-1)}{2 (m-1)}$$

$$+ \gamma^{-1/(m-1)} \rho^{-1}]\}$$

$$+ \gamma^{1/2(m-1)} / (m-1) \sum_{n=1}^{m-1} \{\sin \frac{\pi (2n-1)}{2 (m-1)}$$

$$\text{arctg}[(\gamma^{-1/2(m-1)} \rho^{-1/2} - \cos \frac{\pi (2n-1)}{2 (m-1)})$$

$$(\sin\frac{\pi (2n-1)}{2 (m-1)})^{-1}]\}, \tag{3.10b}$$

$$C = [\gamma^{1/2(m-1)}] / (m-1) \sum_{n=1}^{m-1} \{\sin \frac{\pi (2n-1)}{2 (m-1)}$$

$$\text{arctg} [(\cos\frac{\pi (2n-1)}{2 (m-1)})$$

$$(\sin\frac{\pi (2n-1)}{2 (m-1)})^{-1}] \}. \tag{3.10c}$$

3.3. CONFORMAL SOLUTION

For conformal solution, let

$$d\tau = Rd\eta \tag{3.11}$$

where η is a conformal time. When $K = \Lambda = 0$, and m is an integer, equation (3.6) does not always exist exact conformal solution. But we know from discussion below, when m takes some special values, it exists exact conformal solutions. Now let us seperate equations into two parts, $m = 1$ and $m \neq 1$ (here m is not always restricted as integers).

(1). When $K = \Lambda = 0$ and $m = 1$, we have

$$d\eta = -[3(1+\gamma)\rho(8\pi\rho/3)^{1/2}(D\rho)^{-1/3(1+\gamma)}]^{-1}d\rho, \qquad (3.12)$$

from which the existing of exact conformal solution is apparent. We solve it and obtain the solution as

$$D^{-1} = \rho R^{3(1+\gamma)} \qquad (3.13)$$

$$\eta = [2/(3\gamma+1)](3/8\pi)^{1/2}D^{1/(3+3\gamma)}\rho^{-(3\gamma+1)/(6+6\gamma)}+E, \quad (3.14)$$

where E is an integral constant depending upon the γ.

(2). When $K = \Lambda = 0$, and $m \neq 1$ (here m is not always restricted as integer), from (3.6), (3.9) and (3.11), we have

$$d\eta = (-1/3)(3/8\pi)^{1/2}\rho^{-7/6}(1+\gamma\rho^{m-1})^{(2-3m)/3(m-1)}d\rho. \quad (3.15)$$

Generally, this equation does not always exist exact solution. But it does exist exact solutions under the following special conditions below:

    a. $m = (5 + 6n)/6(n + 1)$, (n-integer, but $n \neq -1$);

    b. $m = (2 + 3p)/3(p + 1)$, (p-integer, but $p \neq -1$);

    c. $m = (3 + 2q)/2(q + 2)$, (q-integer, but $q \neq -2$).

4. Conclusion

The different types of the conformal solutions of Einstein's equations under Robertson - Walker metric for different equation of state and for different cases of K and Λ can be classified as:

a. $\rho = 3P$, $g_{\mu\nu} = e^{2B}\eta_{\mu\nu}$, $K = \Lambda = 0$. (Z. Zhang, 1980)

b. $\rho = nP$ $(n \geqslant 3)$, $g_{\mu\nu} = e^{2B}\eta_{\mu\nu}$, $K = \Lambda = 0$. (D. Song, D. Li and Z. Zhang, 1983)

c. $P = P(\rho)$, $\rho + P \neq 0$, $(n \geqslant 3)$, $g_{\mu\nu} = e^{2B}\eta_{\mu\nu}$, $K = \Lambda = 0$. (B. Mashhoon, M. H. Partovi, 1984)

d. $\rho = nP$ $(n \geqslant 3)$, $g_{\mu\nu} = e^{2B}\eta_{\mu\nu}$, $K = 0$, $\Lambda \neq 0$. (J. Shi, Z. Zhang, 1986)

e. $\rho = nP$ $(n \geqslant 3)$, $g_{\mu\nu} = e^{2B}\eta_{\mu\nu}$, $K = \pm 1$, $\Lambda = 0$. (J. Shi, Z. Zhang, 1986)

f. $P = \gamma\rho^{m}$, $g_{\mu\nu} = e^{2B}\eta_{\mu\nu}$, $K = \Lambda = 0$, $m \neq 1$. (Z. Zhang, H. Huang, X. Huang, K. Guo, L. Yu, 1993)

For the case,

$P = \gamma\rho^{m}$, $g_{\mu\nu} = e^{2B}\eta_{\mu\nu}$, $K \neq 0$, or $\Lambda \neq 0$, and $m \neq 1$,

up to now, we have not found the solutions.

References

1. B. Schutz, Geometrical Methods of Mathematical Physics,
   Cambridge University Press, 1980
2. S. Weinberg, Gravitation and Cosmology, John Wiley & Sons,
   New York, 1972
3. C. M. Misner, K. S. Thorne and J. A. Wheeler, Gravitation, Freeman,
   San Francisco, (1973) 722.
4. Z. Zhang, Journal of Central China Normal University
   (Natural Science Edition), Vol. 3, (1980) 1
5. D. Song, D. Li and Z. Zhang, Contributed papers of 10th GRG.
   eds. B. Bertoti et al., Rome, Vol. 1, (1984) 353
6. B. Mashhoon and M. H. Partovi, Phys. Rev. D30, (1984) 2695; (BR) 15
7. J. Shi and Z. Zhang, Proceedings of the 26th Liege
   International Astro-physical Colloquim, Belgium, (1986) 383
8. R. M. Wald, General Relativity, The University of Chicago Press,
   (1984), a. 97; b. 116

Unification of Standard Model Gauge Couplings: Meaning, Status and Perspectives

A. ZICHICHI

CERN, 1211 Geneva 23, Switzerland

ABSTRACT

One of the most interesting conceptual developments of particle physics is the understanding that not only gauge couplings but also masses and other dynamical variables run with energy. The running of these quantities in itself has limited interest. The main goal is to find out how the unification of these quantities at high energy is linked with the solution of the many problems left open by the Standard Model in the energy range where our facilities operate. And this means if the new physics beyond the Standard Model can be seen now. This review lecture has the purpose of clarifying and indicating the correct way out. In fact what is needed are not qualitative arguments but a set of detailed calculations with definite prediction. The "geometrical" convergence of the three gauge couplings, $\alpha_1, \alpha_2, \alpha_3$, the origin of M_{SUSY}, and the reasons why this meaningless quantity, together with its "geometrical" consequences, should be abandoned, are discussed. The first step is to describe the convergence of the three gauge couplings with a set of seventeen differential non-linear equations, coupled by the radiative effects of the underlying physics: the inputs being the high precision LEP data, together with other experimental limits. The unification of the Standard Model gauge forces, $SU(3)_c \times SU(2)_L \times U(1)_Y$, is described by a set of Figures which allow a simple understanding of the various interconnections which exist between measured quantities and new Physics. These interconnections are often hidden in the complex mathematical formalism, but their knowledge is badly needed. Finally, the motivations why Supergravity is the right way out to overcome the limitations and problems of the Standard Model, are summarized. Understanding the Physics at the Planck scale tells us that the key for new Physics beyond the Standard Model could be at the Fermi scale and therefore the search for new Physics with existing facilities (Fermi-lab, LEP (I,II), HERA, Gran Sasso, SKK) should continue and be encouraged, as shown with detailed calculations.

V. de Sabbata and H. Tso-Hsiu (eds.), Cosmology and Particle Physics, 301–326.

Contents

1 Introduction

To unify all fundamental forces of Nature is the dream of cosmologists and of particle physicists. Important developments in this direction have taken place during the last two decades: they are named supersymmetry and string theory. The first puts bosons and fermions on equal footing. The second abandons the point-like structure of our world in favour of a, at least, one-dimensional element, called string. A nice feature of string theory is that it needs to be supersymmetric, thus the hope that we are on the right track. The advent of LEP has allowed to measure, at M_Z, the three gauge couplings of the Standard Model, $SU(3)_c$, $SU(2)_L$, $U(1)_Y$, with unprecedented accuracy:

$$
\begin{aligned}
\alpha_1^{-1} &= 58.83 \pm 0.11 \\
\alpha_2^{-1} &= 29.85 \pm 0.11 \\
\alpha_3^{-1} &= 8.47 \pm 0.57
\end{aligned}
$$

Furthermore, the number of families has been determined to be

$$N_f = 3.00 \pm 0.03$$

From these experimental data the Renormalization Group Equations (RGEs) allow to span as many orders of magnitude as wanted, provided we know the virtual phenomena to be accounted for in these large ranges of energy, starting at M_Z.

The key features in this field of physics are therefore the quantities above and the RGEs. From these ingredients we would like to know if it is possible to predict the threshold for the lightest detectable supersymmetric particle. It is a field which I have studied in the late 70's with my friend André Peterman. We realized the importance of the new degree of freedom (the threshold for supersymmetric particle production) introduced in the running of the gauge couplings $(\alpha_1^{-1}, \alpha_2^{-1}, \alpha_3^{-1})$ for the convergence of these couplings towards a unique value. However, we also noticed the many problems to be overcome for a correct approach to the key issue of understanding where the energy level for the superparticles could be. This review paper is organized as follows. In section 2 the problem of the convergence of the gauge couplings $\alpha_1, \alpha_2, \alpha_3$ is discussed. In section 3 the constraints from the unification conditions are presented in order to allow the reader an understanding of the physical consequences hidden in the mathematical formalism. Section 4 discusses the popular quantity M_{SUSY} and the reasons why it should be abandoned. Section 5 presents the need to promote global supersymmetry to be a local one thus giving rise to the supergravity description of low-energy phenomena and in section 6 we present the conclusions.

2 High precision LEP data and convergence of couplings: Physics is not Euclidean geometry

Experimentally we know that (see *e.g.*, [1]),

$$\alpha_e^{-1}(M_Z) = 127.9 \pm 0.2, \tag{1}$$

$$\alpha_3(M_Z) = 0.118 \pm 0.008, \tag{2}$$

$$\sin^2 \theta_W(M_Z) = 0.2334 \pm 0.0008. \tag{3}$$

The $U(1)_Y$ and $SU(2)_L$ gauge couplings are related to these by $\alpha_1 = \frac{5}{3}(\alpha_e / \cos^2 \theta_W)$ and $\alpha_2 = (\alpha_e / \sin^2 \theta_W)$. As mentioned above, the values of the three gauge couplings of the Standard Model derived from (1),(2), and (3), are:

$$at \quad Q = M_Z \quad \left\{ \begin{array}{rcl} \alpha_1^{-1} & = & 58.83 \pm 0.11 \\ \alpha_2^{-1} & = & 29.85 \pm 0.11 \\ \alpha_3^{-1} & = & 8.47 \pm 0.57 \end{array} \right. \tag{4}$$

These three gauge couplings evolve with increasing values of the scale Q in a logarithmic fashion, and may become equal at some higher scale, signaling the possible presence of a larger gauge group. However, this need not be the case: the three gauge couplings may meet and then depart again. Conceptually, the presence of a unified group is essential in the discussion of unification of couplings. In this case, the newly excited degrees of freedom will be such that all three couplings will evolve together for scales $Q > E_{GUT}$, and one can then speak of a unified coupling [2].

The running of the gauge couplings is prescribed by a set of first-order non-linear differential equations: the renormalization group equations (RGEs) [3] for the gauge couplings. In general, there is one such equation for each dynamical variable in the theory (*i.e.*, for each gauge coupling, Yukawa coupling, and sparticle mass). These equations give the rate of change of each dynamical variable as the scale Q is varied. For the case of a gauge coupling, the rate of change is proportional to (some power of) the gauge coupling itself, and the coefficient of proportionality is called the *beta function*. The beta functions encode the spectrum of the theory, and how the various gauge couplings influence the running of each other (a higher-order effect). Assuming that all supersymmetric particles have a common mass M_{SUSY}, the RGEs (to two-loop order) are:

$$\frac{d\alpha_i^{-1}}{dt} = -\frac{b_i}{2\pi} - \sum_{j=1}^3 \frac{b_{ij}\alpha_j}{8\pi^2}. \tag{5}$$

where $t = \ln(Q/E_{GUT})$, with Q the running scale and E_{GUT} the unification mass. The one-loop (b_i) and two-loop (b_{ij}) beta functions are given by

$$b_i = \left(\frac{33}{5}, 1, -3 \right), \tag{6}$$

$$b_{ij} = \begin{pmatrix} \frac{199}{25} & \frac{27}{5} & \frac{88}{5} \\ \frac{9}{5} & 25 & 24 \\ \frac{11}{5} & 9 & 14 \end{pmatrix}. \tag{7}$$

These equations are valid from $Q = M_{SUSY}$ up to $Q = E_{GUT}$. For $M_Z < Q < M_{SUSY}$ an analogous set of equations holds, but with beta functions which reflect the non-supersymmetric nature of the theory (*i.e.*, with all the sparticles decoupled),

$$b'_i = \left(\tfrac{41}{10}, -\tfrac{19}{6}, -7 \right), \tag{8}$$

$$b'_{ij} = \begin{pmatrix} \tfrac{199}{50} & \tfrac{27}{10} & \tfrac{44}{5} \\ \tfrac{9}{10} & \tfrac{35}{6} & 12 \\ \tfrac{11}{10} & \tfrac{9}{2} & -26 \end{pmatrix}. \tag{9}$$

The non-supersymmetric equations are supplemented with the initial conditions given in Eq. (4).

If the above is all the physics which is incorporated in the study of the convergence of the gauge couplings, then it is easy to see that the couplings will always meet at some scale E_{GUT}, provided that M_{SUSY} is tuned appropriately. This is a simple consequence of euclidean geometry, as can be seen from Eq. (5). Neglecting the higher-order terms, we see that as a function of t, α_i^{-1} are just straight lines. In fact, the slope of these lines changes at $Q = M_{SUSY}$, where the beta functions change. The convergence of three straight lines with a change in slope is then guaranteed by euclidean geometry, as long as the point where the slope changes is tuned appropriately. (This fact was pointed out by A. Peterman and myself in 1979 [4].) What is non-trivial about the convergence of the couplings is that with the initial conditions given in Eq. (4), the change in slope needs to be $M_{SUSY} \sim 1\,TeV$ [5].

This prediction [5] for the likely scale of the supersymmetric spectrum (*i.e.*, $M_{SUSY} \sim 1\,TeV$ [5]) is in fact *incorrect* [2, 6, 7, 8]. The reason is simple: the physics at the unification scale, which is used to predict the value of M_{SUSY}, has been ignored completely. In fact, such a geometrical picture of convergence of the gauge couplings is physically inconsistent, since for scales $Q > E_{GUT}$ the gauge couplings will depart again, as can be seen in Fig. 1 of Amaldi et al. [5]. One must consider a unified theory to be assured that the couplings will remain unified, as shown in Fig. 2. This entails the study of a new kind of effect, namely the influence on the running of the gauge couplings of the degrees of freedom which are excited near the unification scale (*i.e.*, the *heavy threshold effects*) [2, 8, 9]. In fact, the whole concept of a single unification point needs to be abandoned. The upshot of all this is that the theoretical uncertainties on the values of the parameters describing the heavy GUT particles are such that the above prediction for M_{SUSY} [5] is washed out completely [2, 6, 7, 8]. Furthermore, the insertion of a realistic spectrum of sparticles at low energies (as opposed to an unrealistic common M_{SUSY} mass) blurs the issue even more [8, 10, 11]. *Thus, it is perfectly possible to obtain the unification of the gauge couplings, with supersymmetric particle masses as low as experimentally allowed.* The most complete analysis of a unified theory is shown in Fig. 3. Note the unification of the gauge couplings which continues above E_{GUT}. Notice also that "light" and "heavy" thresholds have been duly accounted for, plus other detailed effects like the

306

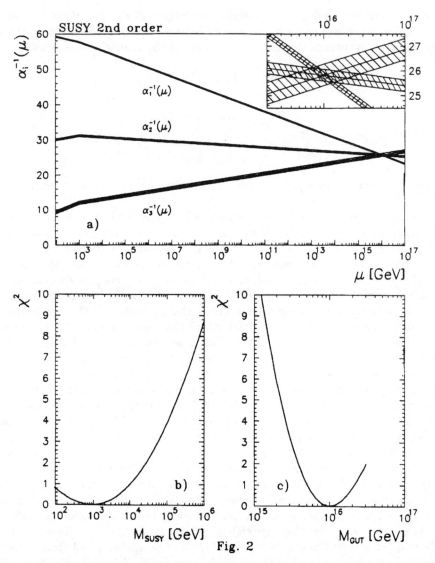

Fig. 2

Figure 1: This is the most publicized example of unification of couplings with divergence above E_{GUT} (μ and M_{GUT} in the figure). It is Fig. 2 of ref. [5]. Neither "light" nor "heavy" thresholds are considered. Nor are other effects needed to address the difficult task of "predicting" where the Superworld could start showing evidence for its existence. Notice that the best fit for the "geometrical" convergence of the couplings "predicts" M_{SUSY} at 10^3 GeV. The value for $\alpha_3(M_Z)$ is 0.108, not the world average. Notice the purely geometrical definition of the χ^2 of ref. [5]: $\chi^2 = \sum_{i=i}^{3} \frac{[\alpha_i(\mu) - \alpha_{GUT}(\mu)]^2}{\sigma_i^2}$, where $\alpha_{GUT} = \sum_{i=i}^{3} \frac{\alpha_i(\mu)}{3}$. The aim of the minimization of such a χ^2 is to make the three curves $\alpha_i(\mu)$ intersect in a point, without any care for their divergence after.

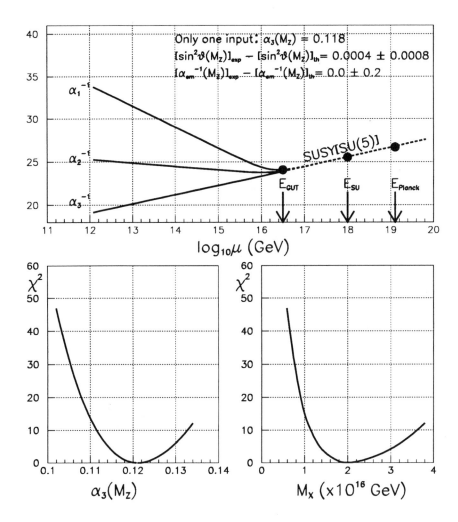

Figure 2: The convergence of the gauge couplings $(\alpha_1, \alpha_2, \alpha_3)$ at E_{GUT} is followed by the unification in a unique α_{GUT} above E_{GUT}. The RGEs include the heavy and light thresholds plus the evolution of gaugino masses. These results are obtained using as input the world-average value of $\alpha_3(M_Z)$ and comparing the predictions for $\sin^2\theta(M_Z)$ and $\alpha_{em}^{-1}(M_Z)$ with the experimental results. The χ^2 constructed using these two physically measured quantities allows to get the best E_{GUT}, $\alpha_{GUT}(E_{GUT})$, and $\alpha_3(M_Z)$ (shown). The RGEs can go down to M_Z without any need of introducing a change of slope at $E \approx 10^3$ GeV as would be required if the various effects mentioned above are neglected. The χ^2 definition is based on physical quantities and is the following: $\chi^2 = \dfrac{\{[\sin^2\theta(M_Z)]_{exp} - [\sin^2\theta(M_Z)]_{th}\}^2}{[\sigma_s]^2} + \dfrac{\{[\alpha_{em}(M_Z)]_{exp} - [\alpha_{em}(M_Z)]_{th}\}^2}{[\sigma_e]^2}$

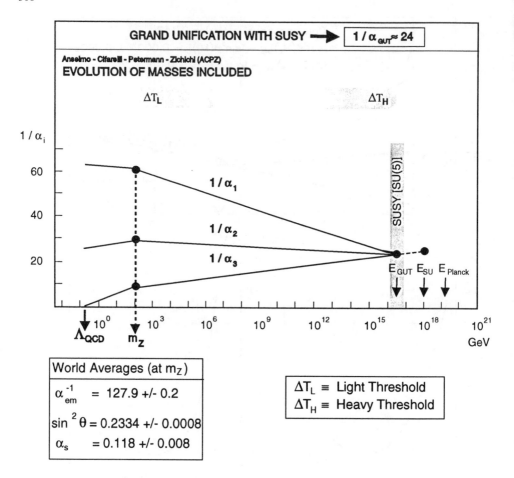

Figure 3: This is the best proof that the convergence of the gauge couplings can be obtained with M_{SUSY} at an energy level as low as M_Z. Notice that the effects of "light" and "heavy" thresholds have been accounted for, as well as the Evolution of Gaugino Masses [10,12]. This figure is Fig. 2 of ref. [13]. E_{SU} is the string unification scale.

evolution of gaugino masses (EGM) [10]. This effect has in fact been calculated at two loops [12].

A related point is that LEP data do not uniquely demonstrate that the gauge couplings must unify at a scale $E_{GUT} \sim 10^{16}\, GeV$ [9]. This is probably the simplest conclusion one could draw. However, this conclusion is easily altered by for example considering all experimental and theoretical uncertainties. In fact, once this is done, the value of E_{GUT} can reach the string unification scale, i.e., $E_{GUT} \sim 10^{18}\, GeV$, as shown in Fig. 4.

3 Interconnections between the measured quantities due to Unification

The convergence of the gauge couplings implies that given α_e and α_3, one is able to compute the values of $\sin^2 \theta_W$, the unification scale E_{GUT}, and the unified coupling α_U. In lowest-order approximation (i.e., neglecting all GUT thresholds, two-loop effects, and taking $M_{SUSY} = M_Z$) one obtains

$$\ln \frac{E_{GUT}}{M_Z} = \frac{\pi}{10}\left(\frac{1}{\alpha_e} - \frac{8}{3\alpha_3}\right), \tag{10}$$

$$\frac{\alpha_e}{\alpha_U} = \frac{3}{20}\left(1 + \frac{4\alpha_e}{\alpha_3}\right), \tag{11}$$

$$\sin^2 \theta_W = 0.2 + \frac{7\alpha_e}{15\alpha_3} . \tag{12}$$

These equations provide a rough approximation to the actual values obtained when all effects are included. Nonetheless, they embody the most important dependences on the input parameters. In Fig. 5 we show the relation between E_{GUT} and α_3 for various values of $\sin^2 \theta_W$. One can observe that:

$$\alpha_3 \uparrow \quad \Rightarrow \quad E_{GUT} \uparrow \quad for\ fixed\ \sin^2 \theta_W \tag{13}$$

$$\sin^2 \theta_W \uparrow \quad \Rightarrow \quad E_{GUT} \uparrow \quad for\ fixed\ \alpha_3 \tag{14}$$

$$\alpha_3 \uparrow \quad \Rightarrow \quad \sin^2 \theta_W \downarrow \quad for\ fixed\ E_{GUT} \tag{15}$$

These are the most important systematic correlations, which are not really affected by the neglected effects. These correlations are evident in Eqs. (10–12) and in Fig. 5. In this figure we also show the lower bound on E_{GUT} which follows from the proton decay constraint. Clearly a lower bound on $\alpha_3(M_Z)$ results, which allows the world-average value. Another interesting result is the anticorrelation between E_{GUT} and M_{SUSY}. This is shown in Fig. 6, where for fixed $\sin^2 \theta_W(M_Z)$ we see that increasing $\alpha_3(M_Z)$ increases E_{GUT} (as already noted in Eq. (13)) and decreases M_{SUSY}. Taking for granted this approach (i.e., all supersymmetric particle masses degenerate at M_{SUSY}) for comparison with the large amount of papers published following this logic, in Fig. 7 we see the narrow band left open once the experimental limits on τ_p and M_{SUSY} are

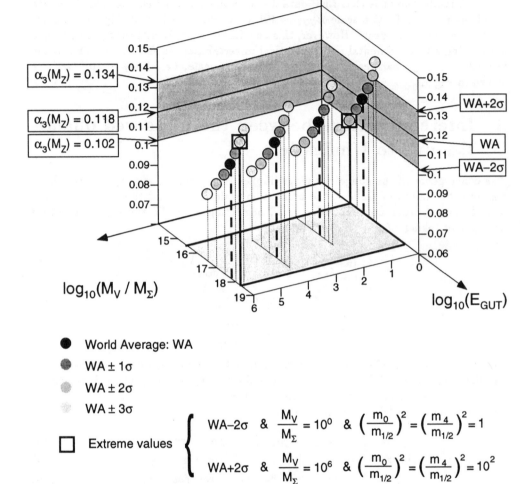

Figure 4: The dependence of E_{GUT} on $\alpha_3(M_Z)$ and on the ratio of the two crucial heavy-threshold masses, M_V/M_Σ. Note that the extreme value for E_{GUT} is above 10^{18} GeV.

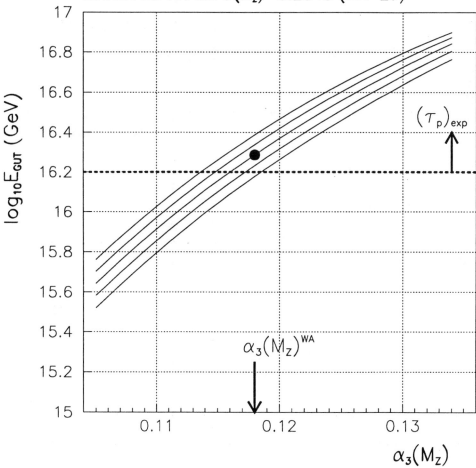

Figure 5: The unification scale E_{GUT} versus $\alpha_3(M_Z)$ for various values of $\sin^2\theta_W(M_Z)$ within $\pm 2\sigma$ of the world-average value. Also indicated is the lower bound on E_{GUT} from the lower limit on the proton lifetime.

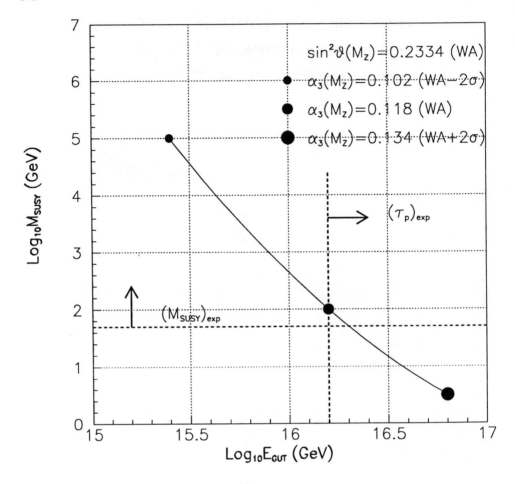

Figure 6: The unification scale E_{GUT} versus M_{SUSY} for different values of $\alpha_3(M_Z)$ and fixed $\sin^2 \theta_W(M_Z)$. Note the anticorrelation between M_{SUSY} and E_{GUT}. The experimental lower bound on M_{SUSY} is shown. The lower bound on E_{GUT} from Fig. 5 is also indicated.

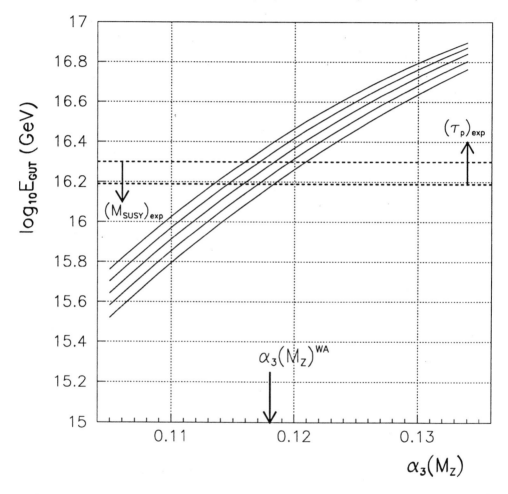

Uppermost curve: $\sin^2\vartheta(M_z)=0.2350$ (WA+2σ)
Central curve: $\sin^2\vartheta(M_z)=0.2334$ (WA)
Lowest curve: $\sin^2\vartheta(M_z)=0.2318$ (WA−2σ)

Figure 7: The correlation between all measured quantities, $\alpha_3(M_Z)$, $\sin^2\theta_W(M_Z)$, τ_p, the limits on the lightest detectable supersymmetric particle (here represented by M_{SUSY}) and the unification energy scale E_{GUT}.

314

imposed. Figure 7 is a guide to understand the qualitative interconnection between the basic experimentally measured quantities, $\alpha_3(M_Z)$, $\sin^2\theta_W(M_Z)$, τ_p, M_{SUSY} and the theoretically wanted E_{GUT}. The experimental lower bounds on the proton lifetime $(\tau_p)_{exp}$ and on M_{SUSY} produce two opposite bounds (lower and upper, respectively) on the unification energy scale E_{GUT}. Note that, in order to make definite predictions on the lightest detectable supersymmetric particle, a detailed model is needed. In particular, it is necessary to incorporate the evolution of all masses. This has been done in ref. [11] and an example of spectra is shown in Fig. 8. Let me emphasize that the study of the correlations between the basic quantities, as exemplified in Fig. 7, is interesting but should not be mistaken as example of prediction for the superworld. In particular, the introduction of the quantity M_{SUSY} is really misleading.

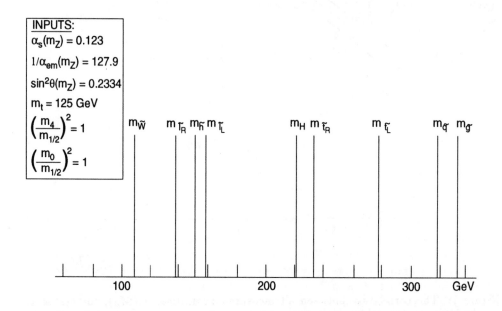

Figure 8: A detailed spectrum of SUSY particles showing how misleading is to think of a unique quantity M_{SUSY} in order to describe a SUSY particle spectrum.

4 The origin of M_{SUSY} and why it should be abandoned: masses and spectra

The quantity M_{SUSY} has been around for too long. It was introduced at the time of one-loop RGEs for the gauge couplings. The running of the gauge couplings $\alpha_1^{-1}, \alpha_2^{-1}, \alpha_3^{-1}$ was described by straight lines (see equations 5) whose slopes had to change due to the change in the beta functions from the SUSY-regime (6),(7) to the no-SUSY regime (8),(9). The change of slope was, for simplicity, described by a unique parameter M_{SUSY}, but in no case was M_{SUSY} supposed to represent a physical quantity. It is in fact out of the question that the real spectrum of the supersymmetric particles (an example is in Fig. 8) can be degenerate and therefore be represented by a single mass value.

The calculations which we have done in ref. [11], attempted to determine the supersymmetric particle spectrum, by fitting the various sparticle masses in order to obtain the "best possible" unification picture. These calculations are the most extensive analysis of the problem with the simultaneous evolution of seventeen dynamical variables with the radiative effects consistently computed. It is interesting to see how the masses of the superpartners of the various gauge particles change as a function of the experimentally measured quantities $\alpha_3(M_Z), \sin^2\theta_W(M_Z), \alpha_e(M_Z)$. Examples for the $\tilde{W}$ mass, $m_{\tilde{W}}$, and the gluino mass, $m_{\tilde{g}}$, are shown in Figs. 9 and 10, respectively. We also show in Figs. 11 and 12 how the squark and slepton masses change as a function of the basic quantities quoted above $(\alpha_3(M_Z), \sin^2\theta_W(M_Z), \alpha_e(M_Z))$. Let me emphasize that these are the only experimentally known input parameters in our system of seventeen coupled equations [11]. A quantity which will, hopefully, be soon determined is the top-quark mass m_t. Our system of seventeen coupled equations allows to determine the $\tilde{W}$ mass versus m_t. This is shown in Fig. 13. Notice that increasing m_t corresponds to lowering the value of $m_{\tilde{W}}$.

It is interesting to quantitatively see the effect of the three basic quantities $(\alpha_3(M_Z), \sin^2\theta_W(M_Z), \alpha_e(M_Z))$ on the variation of $m_{\tilde{W}}$. This is summarized in Table 1: the dominant effect is $\alpha_3(M_Z)$. It should be noted that these results are given for illustrative purposes: i.e., in order to understand how radiative effects influence the masses of the superparticles when the input measured quantities $(\alpha_3(M_Z), \sin^2\theta_W(M_Z), \alpha_e(M_Z))$ change within their limits of errors.

This program has large inherent uncertainties. As mentioned above, and first of all, because of the fact that we cannot ignore the great uncertainties in the physics at the GUT scale. Furthermore, even if we take the simplest approach and assume that the physics at the GUT scale is represented by a unique threshold, then the introduction of experimental and theoretical uncertainties in the seventeen evolution equations relating gauge couplings and masses, corresponds to a variation (within 2σ) of the sparticle spectra from GeV to PeV, as shown in Fig.14. Nevertheless, given a GUT model, it is possible to compute the GUT threshold effects (notice the spread of the spectra in Fig. 14) and the "best fit" to deduce the corresponding light supersymmetric spectrum in order to find out what is the lightest detectable

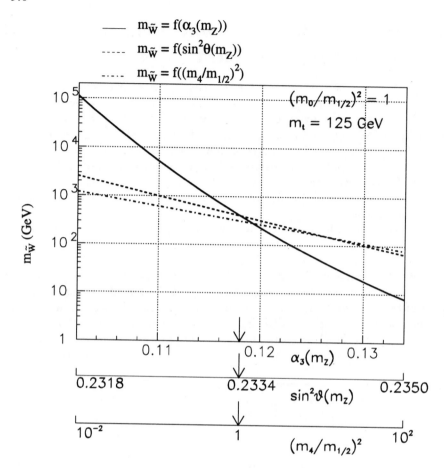

Figure 9: The value of the W-ino mass, $m_{\tilde{W}}$, $vs.\alpha_3(M_Z)$, $sin^2\theta(M_Z)$, and $(m_4/m_{1/2})^2$. Each curve is the result of our iterative system of seventeen evolution equations. The curve showing the variation of $m_{\tilde{W}}$, $vs.\alpha_3(M_Z)$, or $sin^2\theta(M_Z)$, or $(m_4/m_{1/2})^2$ is computed keeping the other parameters fixed at the values indicated by the arrows. For the experimentally measured values, $\alpha_3(M_Z)$, $sin^2\theta(M_Z)$, the world averages have been chosen. For the ratio of primordial masses the value $(m_4/m_{1/2})^2 = 1$ has been taken. The top mass is kept at $m_t = 125$ GeV. Please note that our analysis is valid only above M_Z. The $m_{\tilde{W}}$ values shown below this limit are for illustrative purposes.

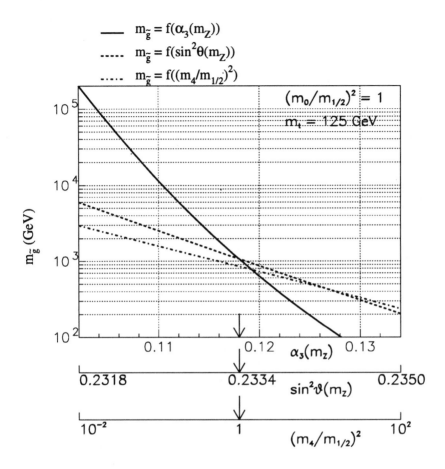

Figure 10: Same as figure 9 when the gluino mass, $m_{\tilde{g}}$, is computed.

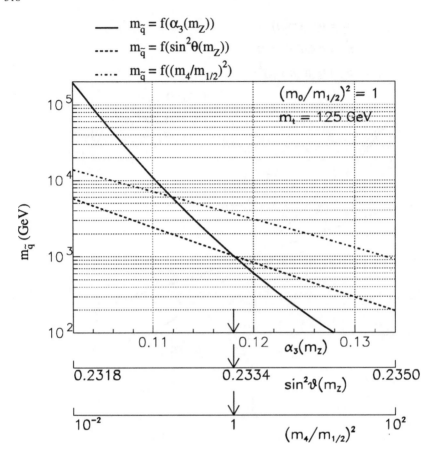

Figure 11: Same as figure 9 when the squark mass, $m_{\tilde{q}}$, is computed.

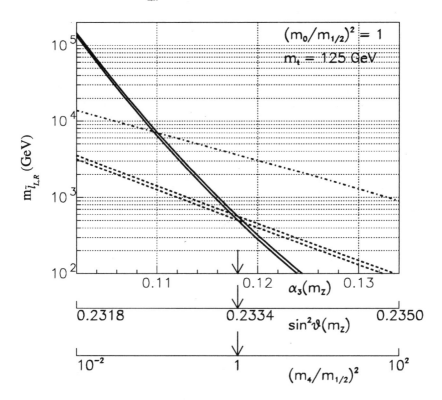

Figure 12: Same as figure 9 when the slepton masses are computed. The left (upper curve) and right (lower curve) slepton masses almost coincide.

$\alpha_3(m_z) = 0.118$ (WA)

$\sin^2\vartheta(m_z) = 0.2334$ (WA)

$m_0/m_{1/2} = 1$

$m_t/m_{1/2} = 1$

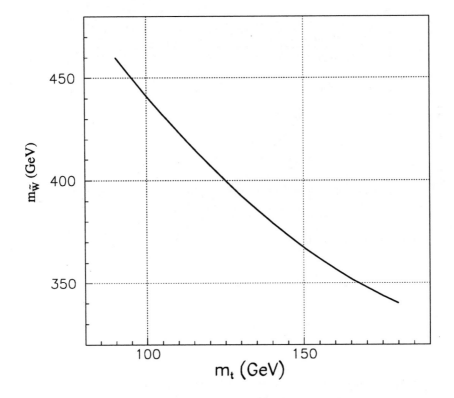

Figure 13: The value of the W-ino mass $vs. m_t$ for given values of the other inputs, as indicated. This is the only case $(m_{\tilde{W}})$ where the dependence on m_t is shown. We do not show the result for $m_{\tilde{g}}, m_{\tilde{q}}, m_{\tilde{l}}$.

$$\frac{m_{\tilde{W}}(\alpha_3(M_Z)^{WA}-2\sigma)}{m_{\tilde{W}}(\alpha_3(M_Z)^{WA}+2\sigma)} \simeq 2 \times 10^4$$

[WA values for the experimental inputs: $\sin^2\theta(M_Z)$, $\alpha_{em}(M_Z)$]

$$\frac{m_{\tilde{W}}(\sin^2\theta(M_Z)^{WA}-2\sigma)}{m_{\tilde{W}}(\sin^2\theta(M_Z)^{WA}+2\sigma)} = 20 \div 50$$

[WA values for the experimental inputs: $\alpha_3(M_Z)$, $\alpha_{em}(M_Z)$]

$$\frac{m_{\tilde{W}}(1/\alpha_{em}(M_Z)^{WA}-2\sigma)}{m_{\tilde{W}}(1/\alpha_{em}(M_Z)^{WA}+2\sigma)} = 1.2 \div 1.3$$

[WA values for the experimental inputs: $\sin^2\theta(M_Z)$, $\alpha_3(M_Z)$]

Allowed range of variation for primordial SUSY breaking mass ratios:

$$(m_0/m_{1/2})^2 = 10^{-2} \div 10^2$$

$$(m_4/m_{1/2})^2 = 10^{-2} \div 10^2$$

Table 1: The variation of the W-ino mass corresponding to $\pm 2\sigma$ variation of the experimental inputs, $\alpha_3(M_Z), \sin^2\theta(M_Z), \alpha_{em}(M_Z)$, with respect to their world average values, and to the variation of $(m_0/m_{1/2})^2$ and $(m_4/m_{1/2})^2$ in the indicated range. The data are derived from our system of seventeen coupled evolution equations. The dominant effect is clearly due to $\alpha_3(M_Z)$. Please note that our equations are valid only above the Z^0 mass. Nevertheless the results in terms of $m_{\tilde{W}}$ ratios are given, for illustrative purposes, even when the $+2\sigma$ limit of the experimental inputs pushes $m_{\tilde{W}}$ below M_Z.

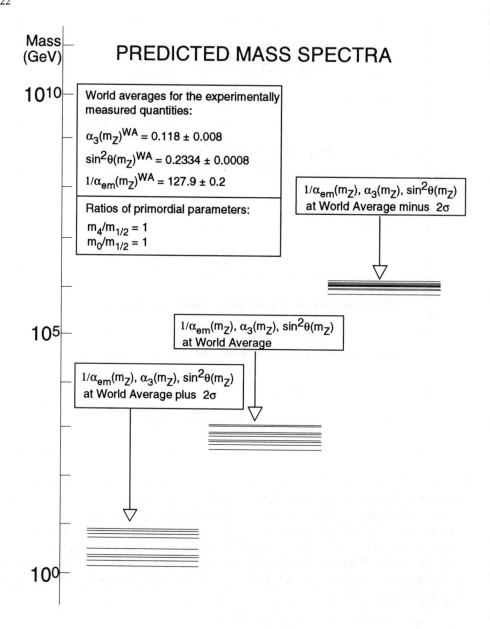

Figure 14: The predicted SUSY mass spectrum for three cases, when the measured quantities $(\alpha_3(M_Z), \sin^2\theta(M_Z), \alpha_{em}(M_Z))$ are taken at their world average values and at $\pm 2\sigma$. The ratios of the primordial parameters are kept equal to one. Note the large range where the SUSY spectra could be on the basis of our best experimental and theoretical knowledge.

supersymmetric particle. And especially if the threshold for its detection is reachable with existing facilities. However this is not very satisfying since one would like to know why the supersymmetric spectrum should be the way the fit would require it to be. In other words, the real question is: what determines the values of the sparticle masses? And why should these be below $\sim 1\,TeV$, so that the gauge hierarchy problem is not re-introduced?

5 The new step forward: Supergravity

In order to answer this question we must abandon "global" SUSY and promote this symmetry of nature to be "local". It is local supersymmetry, *i.e.*, supergravity which provides the means to compute the masses of the sparticles in a non arbitrary *ad hoc* fashion. In fact, the crucial point is the breaking of local supersymmetry [14]. In supergravity the breaking occurs in a "hidden sector" of the theory, where "gravitational particles" (those introduced when the supersymmetry was made local) may grow vacuum expectation values (vevs) [15] which break supersymmetry spontaneously in the hidden sector. These vevs are best understood as induced dynamically by the condensation of the supersymmetric partners of the hidden sector particles [16] when the gauge group which describes them becomes strongly interacting at some large scale. The splitting of the particles and their partners would then be generated, and would be of the order of the condensation scale ($\sim 10^{12-16}\,GeV$). However, such huge mass splittings will be transmitted to the "observable" (the normal) sector of the theory through gravitational interactions, since it is only through these interactions that the two sectors communicate. The dampening in the transmission mechanism is such that the splittings in the observable sector are usually much more suppressed than those in the hidden sector, and suitable choices of hidden sectors may yield realistic low-energy supersymmetric spectra. This picture of hidden and observable sectors becomes completely natural in the context of superstrings [17], where models typically contain both sectors and one can study explicitly the predicted spectrum of supersymmetric particles at low energies.

In a large number of models, the supersymmetric particle masses at the unification scale are also "unified". This situation is called *universal soft-supersymmetry-breaking*, and the masses of all scalar partners (*e.g.*, squarks and sleptons) take the common value of m_0, the gaugino (the partners of the gauge bosons) masses are given by $m_{1/2}$, and there is a third parameter (A) which basically parametrizes the mixing of stop-squark mass eigenstates at low energies. The breaking of the electroweak symmetry is obtained dynamically in the context of these models, through the so-called *radiative electroweak symmetry breaking mechanism*, which involves the top-quark mass in a fundamental way [17]. After all these well motivated theoretical ingredients have been incorporated, the models depend on only four parameters: $m_{1/2}, m_0, A$, and the ratio of the two Higgs vacuum expectations values ($\tan\beta$) [18], plus the top-quark mass (m_t).

In generic supergravity models the five-dimensional parameter space is constrained by phenomenological requirements, such as sparticle and Higgs-boson masses not in conflict with present experimental lower bounds, a sufficiently long proton lifetime, a sufficiently old Universe (a cosmological constraint on the amount of dark matter in the Universe today), various indirect constraints from well measured rare processes, etc. In addition, further theoretical constraints can be imposed which give m_0 and A as functions of $m_{1/2}$, and thus reduce the dimension of the parameter space down to just two (plus the top quark mass).

It should be pointed out that a rather interesting framework occurs in the so-called *no-scale* scenario [19, 20], where all the scales in the theory are obtained from just one basic scale (*i.e.*, the unification scale or the Planck scale) through radiative corrections. These models have the unparalleled virtue of a vanishing cosmological constant at the tree-level *even after supersymmetry breaking*, and in their unified versions predict that, at E_{SU}, the universal scalar masses and trilinear couplings vanish (*i.e.*, $m_0(E_{SU}) = A(E_{SU}) = 0$) and the universal gaugino mass $(m_{1/2})$ is the only seed of supersymmetry breaking. Moreover, this unique mass can be determined in principle by minimizing the vacuum energy at the electroweak scale. The generic result is $m_{1/2} \sim M_Z$ [19, 20], in agreement with theoretical prejudices (*i.e.*, "naturalness"). Finally, it should not be forgotten that no-scale supergravity is infrared solution of superstring theory [21].

6 Conclusions

To predict the energy level where the superworld should show up is the key issue of High Energy Physics. The search for this energy level cannot be based on geometrical properties of three straight lines. Moreover, it is not true that the "best" convergence of the gauge couplings requires this energy level to be in the TeV range. The crucial problem is to know what are the effects to be detected with the existing facilities. This needs detailed calculations. For this reason we have studied in details what could be observed at Fermi-lab [22], at LEP (I,II) [23], at HERA [24], and in the underground facilities [25] (Gran Sasso, Superkamionkande, DUMAND, AMANDA) [26]. The general conclusion of our works is that the lightest detectable sparticle could be at reach with existing facilities: Fermilab, LEP (I,II), HERA. Furthermore, the underground facilities can search - in addition to nuclear stability - for neutralinos [26] and other cosmic phenomena. Consequently the search for supersymmetric particles should not be postponed, but encouraged right now as much as possible.

Acknowledgements

The results reported in this review paper would not have been possible without the dedicated work of all my collaborators: F. Anselmo, L. Cifarelli, J.L. Lopez, D.V. Nanopoulos, A. Peterman, H. Pois, and Xu Wang. To all of them I would like to express my sincere appreciation and gratitude.

References

[1] See *e.g.*, P. Langacker and M. Luo, Phys. Rev. D **44** (1991) 817; S. Bethke, J. Phys. **G17** (1991) 1455; G. Burgers and F. Jegerlehner, Z Physics at LEP I, editors G. Altarelli, R. Kleiss and C. Verzegnassi, CERN Report 89-08, Vol. **1** (1989), p. 55.

[2] F. Anselmo, L. Cifarelli, A. Peterman, and A. Zichichi, Nuovo Cim. **105A** (1992) 1025.

[3] E.C.G. Stueckelberg and A. Peterman, Helv. Phys. Acta **26** (1953) 499.

[4] A. Zichichi, EPS Conference, Geneva 1979, Proceedings; and Nuovo Cimento Rivista **2** (1979) 1. This is the first instance in which it was pointed out that supersymmetry may play an important role in the convergence of the gauge couplings.

[5] U. Amaldi, W. de Boer, and H. Fürstenau, Phys. Lett. B **260** (1991) 447.

[6] F. Anselmo, L. Cifarelli, A. Peterman, and A. Zichichi, Nuovo Cim. **104A** (1991) 1817.

[7] F. Anselmo, L. Cifarelli, and A. Zichichi, Nuovo Cim. **105A** (1992) 1357.

[8] J. Ellis, S. Kelley, and D. V. Nanopoulos, Phys. Lett. B **249** (1990) 441; Phys. Lett. B **260** (1991) 131; Nucl. Phys. B **373** (1992) 55.; Phys. Lett. B **287** (1992) 95. R. Barbieri and L. Hall, Phys. Rev. Lett. **68** (1992) 752.

[9] F. Anselmo, L. Cifarelli, and A. Zichichi, Nuovo Cim. **105A** (1992) 1335.

[10] F. Anselmo, L. Cifarelli, A. Peterman, and A. Zichichi, Nuovo Cim. **105A** (1992) 581.

[11] F. Anselmo, L. Cifarelli, A. Peterman, and A. Zichichi, Nuovo Cim. **105A** (1992) 1179.

[12] F. Anselmo, L. Cifarelli, A. Peterman, and A. Zichichi, Nuovo Cim. **105A** (1992) 1201.

[13] A. Zichichi, *"Understanding where the supersymmetry threshold should be"*, CERN-PPE/92-149, 7 September 1992.

[14] See *e.g.*, chapter 4 of *"The Superworlds of SU(5) and SU(5) × U(1): A Critical Assessment and Overview"* by J. L. Lopez, D. V. Nanopoulos, and A. Zichichi, in Proceedings of the XXX Course of the "E. Majorana" International School of Subnuclear Physics: "From Superstrings to the Real Superworld", Erice, Italy, 14-22 July 1992 (to be published by World Scientific Pub. Co., Singapore); CERN-TH.6934/93, June 1993.

[15] I. Antoniadis, J. Rizos, and K. Tamvakis, Phys. Lett. B **278** (1992) 257.

[16] J. L. Lopez, and D. V. Nanopoulos, Phys. Lett. B **251** (1990) 73; Phys. Lett. B **256** (1991) 150; Phys. Lett. B **268** (1991) 359; S. Kalava, J. L. Lopez, and D. V. Nanopoulos, Phys. Lett. B **278** (1992) 257.

[17] For a review see A. B. Lahanas and D. V. Nanopoulos, Phys. Rep. **145** (1987) 1.

[18] See *e.g.*, S. Kelley, J. L. Lopez, D. V. Nanopoulos, H. Pois, and K. Yuan, Nucl. Phys. B **398** (1993) 3.

[19] J. Ellis, A. Lahanas, D. V. Nanopoulos, and K. Tamvakis, Phys. Lett. B **134** (1984) 429.

[20] J. Ellis, C. Kounnas, and D. V. Nanopoulos, Nucl. Phys. B **241** (1984) 406, Nucl. Phys. B **247** (1984) 373.

[21] E. Witten, Phys. Lett. B **155** (1985) 151.

[22] J. L. Lopez, D. V. Nanopoulos, X. Wang, and A. Zichichi, *"Supersymmetry tests at Fermilab: A proposal"*, Phys. Rev. D **48** (1993) 2062.

[23] J. L. Lopez, D. V. Nanopoulos, H. Pois, X. Wang, and A. Zichichi, *"Sparticle and Higgs Production and Detection at LEPII in two Supergravity Models"*, CERN-PPE/93-16, 8 February 1993.

[24] J. L. Lopez, D. V. Nanopoulos, X. Wang, and A. Zichichi, *"SUSY Signals at HERA in the No-scale Flipped SU(5) Supergravity Model"*, Phys. Rev. D **48** (1993) 4029.

[25] J. L. Lopez, D. V. Nanopoulos, H. Pois, and A. Zichichi, *"Proposed Tests for Minimal SU(5) Supergravity at Fermilab, Gran Sasso, SuperKamiokande, and LEP"*, Phys. Lett. B **299** (1993) 262.

[26] R. Gandhi, J. L. Lopez, D. V. Nanopoulos, K. Yuan, and A. Zichichi, *"Scrutinizing Supergravity Models through Neutrino Telescopes"*, CERN-TH.6999/93, September 1993.

INDEX

Acceleration
    maximal, 117, 118, 121
    minimal, 113, 117, 118, 121
Annihilation e^+e^-, 274
Antenna
    gravitational wave, 207, 271
Axionic instanton, 132

Background
    cosmic microwave, 216
    quantization, 14
    space-time, 10
    structure, 14
Baryonic dark matter, 162, 174
Bianchi identities, 22
Big bang, 51
Black holes
    critical phenomena, 183
    gravitational thermodynamic, 189
Bose-Einstein correlation, 59
Breaking of local supersymmetry, 323

Canonical
    energy-momentum complex, 19
    formulation of Einstein theory, 13
    formulation of the non-Riemannian part of gravity, 13
    Gibbs distribution, 11
    momentum, 11
    quantization, 13, 14
    variables, 15
Cauchy data, 4